Jochen Kieninger
Electrochemical Methods

Also of Interest

Carbon for Micro and Nano Devices
Sharma, 2024
ISBN 978-3-11-062062-7, e-ISBN 978-3-11-062063-4

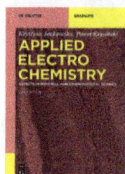

Applied Electrochemistry
Aspects in Material and Environmental Science
Jackowska, Krysiński, 2024
ISBN 978-3-11-116034-4, e-ISBN 978-3-11-116098-6

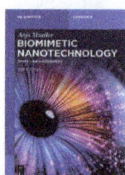

Biomimetic Nanotechnology.
Senses and Movement
Mueller, 2023
ISBN 978-3-11-077918-9, e-ISBN 978-3-11-077919-6

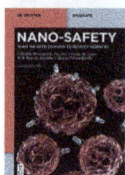

Nano-Safety.
What We Need to Know to Protect Workers
Fazarro, Sayes, Trybula, Tate, Hanks (Eds.), 2024
ISBN 978-3-11-078182-3, e-ISBN 978-3-11-078183-0

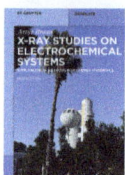

X-Ray Studies on Electrochemical Systems.
Synchrotron Methods for Energy Materials
Braun, 2024
ISBN 978-3-11-079400-7, e-ISBN 978-3-11-079403-8

Jochen Kieninger

Electrochemical Methods

For Biosensors, MEMS, Nanotechnology, Neuroscience,
Renewable Energy, Batteries

2nd, Completely Revised and Extended Edition

DE GRUYTER
OLDENBOURG

Author
Dr. Jochen Kieninger
University of Freiburg – IMTEK
Department of Microsystems Engineering
Georges-Köhler-Allee 106
79110 Freiburg
Germany
book@electrochemical-methods.org

Dr.-Ing. Jochen Kieninger studied Microsystems Engineering and wrote his Ph. D. thesis on "Electrochemical Microsensor System for Cell Culture Monitoring". He works as a Senior Scientist and Lecturer in the Department of Microsystems Engineering (IMTEK) at the University of Freiburg, where he teaches on electrochemical methods, BioMEMS, and sensor technology.
The author's current research interests comprise electrochemical sensors, biosensors, microsensors for neurotechnology, environmental monitoring, corrosion monitoring, and electrochemical methods for MEMS applications.

ISBN 978-3-11-148761-8
e-ISBN (PDF) 978-3-11-148884-4
e-ISBN (EPUB) 978-3-11-148953-7

Library of Congress Control Number: 2024950858

Bibliographic information published by the Deutsche Nationalbibliothek
The Deutsche Nationalbibliothek lists this publication in the Deutsche Nationalbibliografie; detailed bibliographic data are available on the Internet at http://dnb.dnb.de.

© 2025 Walter de Gruyter GmbH, Berlin/Boston
Cover image: ©Andreas Weltin
Typesetting: VTeX UAB, Lithuania

www.degruyter.com
Questions about General Product Safety Regulation:
productsafety@degruyterbrill.com

To my father
Emil Kieninger
(1937–2020)

Preface

Traveling is highly essential to broaden our horizons, discover new landscapes, and understand the perspective of the locals. Sometimes the journeys are adventurous or challenging; in most cases, they are pure pleasure and fun—if not in the situation itself at least in retrospective. Academic careers in interdisciplinary fields are similar to traveling to different cultures, striving for the best of all by understanding the different languages, thinking about the concepts, and listening to the different points of view. My own experience being a microsystems engineer with a background in physics and electronics, exposed to electrochemistry while developing sensors for biomedical applications is a typical course in this sense.

The outcome of my journey between the disciplines resulted in a class on electrochemical methods for master students in Microsystems Engineering. After teaching the topic a few semesters, the outline improved while students from different subjects such as Functional Materials, Embedded Systems Engineering, and Sustainable Systems Engineering attended the lecture and provided their feedback. The selection of topics proved to fit together perfectly. I realized that such a combination of a consistent electrochemical theory, coherent description of methods, insight into the instrumentation, and implications for the different applications in microelectromechanical systems (MEMS) and nanotechnology is unique. The positive feedback encouraged me to write a full textbook based on this class.

A vast and further growing field of application for electrochemical methods is the energy sector. Two primary drivers stick out: For systems smaller in size, it is the need for portable energy storage for mobile computers, smartphones, smartwatches, headphones, and any kind of wearables. Secondly, the increasing awareness of climate change and new strategies for "greener power" asks for efficient energy storage and energy conversion for both the power grid and automotive application. At the same time, elements like lithium commonly used for energy storage are not an optimal solution when thinking of widespread usage in large batteries, e. g., for electromobility. Here, other materials are needed to replace lithium and cobalt because of their limited resources, environmental concerns, and to avoid further exploitation of poorer nations. Beyond different materials for batteries also other storage forms of energy, e. g., as hydrogen gas, appears a promising alternative.

Nanotechnology allows us to fabricate materials for energy conversion and storage, electrochemical sensors, and to some extent, electrochemical actuators. Additionally, electrochemical methods are employed to make the nanosystem itself or for the characterization of the obtained materials. Studies with engineered nanostructures lead to an awareness of effects on the nanoscale, even when dealing with microscopic or macroscopic structures.

Electrochemistry seems far from natural processes when looking from the perspective of material science or energy systems. However, considering how information transfer happens while you read this preface, it is mainly about electrochemistry. There are

https://doi.org/10.1515/9783111488844-201

some optical effects to bring the information onto the retina of your eye (slightly different if you are reading the printed or electronic book), or maybe an acoustic pathway if you use a screenreader. But as soon as the information is within you, it is all about the interplay between (bio-)chemistry and electricity—the core domain of electrochemistry. With these thoughts in the back, it is clear why electrochemical methods are so powerful when considering applications in the biomedical field, in neurotechnology, and generally in the life sciences.

Many current MEMS devices incorporate an electrochemical principle with either sensor or actuator functionality. That is especially true for bioMEMS applied in life science. While most MEMS are designed using state-of-art microfabrication technologies, electrochemical features are often rudimentary developed, and hence lower overall system performance. Electrochemical analytical methods are widely used for sophisticated purposes in laboratories, often with ultramicroelectrodes exploiting microspecific features within macroscopic setups. However, there is little transfer of advanced electrochemical methods to lab-on-chip systems, although the performance of devices would increase by orders of magnitude with the same fabrication effort.

I hope you enjoy the journey described in this book as I did during writing— at least in retrospective—and be able to apply the knowledge to future micro- and nanosystems on which you are working. Besides, take your time to attend scientific conferences— they can be a good starting point for travel in its literal sense as well as to communicate the powerfulness of electrochemical methods in MEMS and nanotechnology.

Jochen Kieninger

Preface to the second edition

The first edition's title, "Electrochemical Methods for the Micro- and Nanoscale," highlighted where the magic happens: at the interface between the electrode and the electrolyte with characteristic dimensions in the micrometer or nanometer range. But at the same time, the title was misleading when looking at devices in some applications being far from tiny, even many meters, thinking of stationary energy storage, for example. So, I skipped the micro- and nanoscale from the title, but I assure you that it is primarily this domain you should consider to understand what happens at your square meter-sized electrodes. Even without the dimensions mentioned in the title, the book is especially useful when working on microsystems or nanotechnology.

Besides, the focus and compilation of the topics received much positive feedback, motivating me to keep the general outline. Readers who used the book in their jobs told me I hit the sweet spot between being too theoretical and too superficial to apply the method to their applications successfully. Many inputs, especially from students, helped to improve the didactics in some parts. A few sections were revised for clarity. Several aspects were deepened, and some were updated to reflect recent progress.

I hope you equally enjoy the book's second edition and start or continue your journey in research or development with passion and success.

Jochen Kieninger

https://doi.org/10.1515/9783111488844-202

About the book

The book has three parts. The first part contains electrochemical theory, discussion of the instrumentation, and information about the lab tools. Details are provided only to the level, which is needed to understand and apply the electrochemical methods. The second and major part explains classical and advanced techniques. All equations are adjusted to be consistent with the fundamentals. The last part discusses several fields of application and explains the therein used electrochemical methods with the specific terms of the area.

Undergraduate or master students can read the book linearly as a comprehensive textbook. They might discover in the third part of the book their future topic. For Ph. D. students, postdoctoral researchers as well as for researchers in the industry, the book will help by its clear structure to get fast answers from a specific section.

Throughout the chapters, you will find several tasks on different levels in a box with a pen in its pictogram. The given problems should help you to familiarize yourself with the topic, both by understanding the derivation of essential equations and by learning how calculations can be applied. On page 383, there is an overview of all tasks.

The difficulty level refers to students from a Master's program in Engineering:

■□□ Simple task, brings you in touch with the topic.
■■□ Medium level task, require more thinking or previous knowledge.
■■■ Difficult task, needs substantial time and effort to find the solution.

Depending on your background, you might feel the difficulty level differently and benefit from following the solutions. Nevertheless, I strongly recommend trying to solve the tasks on your own before checking the answers.

Please check www.electrochemical-methods.org. Here, you can find solutions for all tasks. Additionally, the webpage provides updates to the book and further information.

https://doi.org/10.1515/9783111488844-203

Contents

Part I: **Fundamentals**

Part II: **Methods**

Part III: **Applications**

Part I: **Fundamentals**

"Nothing is too wonderful to be true, if it be consistent with the laws of nature."

Laboratory journal entry by Michael Faraday (1849)

1 Introduction

Electrochemistry can be understood as the part of chemistry in which electricity is involved. Classical textbooks often define electrochemistry from this perspective: "Electrochemistry is the branch of chemistry concerned with the interrelation of electrical and chemical effects" [1]. Consequently, in many faculties, the electrochemistry is accounted for physical chemistry. In contrast, it is mainly scientists with a background in microsystems engineering, nanoscience, material science, electrical engineering, or physics who are developing microelectromechanical systems (MEMS) and nanotechnology applications. Often the perspective is from the electrical effect, which extends from the electron-conducting phase (the electrical wire) to the ion-conducting phase (the electrolyte). In this context, we can understand electrochemistry as the interfacing discipline, which describes the transition between ion and electron conduction.

Looking back in history, researchers made the first observations of electrochemical phenomena during biological studies. However, the development of the discipline separated from biology until recent applications in electrophysiology, biosensors, or neuroscience. In the following section, we start a short and highly selective journey through history to follow the path from the biological interpretation to the modern understanding of electrochemical effects. A brief discussion of the current fields of applications follows illustrating the completely different environments in which electrochemical methods play a role. This chapter closes with an introduction to the electrochemical cell describing in more detail the arena in which the theory, methods, and applications discussed throughout this book play a role.

1.1 Short history of electrochemistry

It is surprising how late electrochemical effects were described as such – especially considering that many processes in nature, most prominently physiological mechanisms rely on electrochemical principles. With this in mind, it is quite ironic that Galvani attributed his observation of electrochemical phenomena as a biological effect, which later proved to be independent of biology.

1.1.1 Luigi Galvani – animal electricity

It was the 18th century when many scientists became enthusiastic about electrostatic effects. Luigi Galvani (1737–1798), graduated in medicine and philosophy, worked at the University of Bologna on "medical electricity." He was rubbing frog skin to provoke electrostatic effects. The legend says that while doing those experiments in 1780, his assistant touched the sciatic nerve of a frog leg lying around on the table with a metal scalpel,

https://doi.org/10.1515/9783111488844-001

which might have picked up charges from previous experiments. Suddenly, the frog leg kicked as it still would be alive. Galvani attributed this action to what he called *animal electricity* and assumed the existence of an "animal electric fluid," which is flowing inside the nerves. This view was opposing the well-established balloonist theory dating back to the Greek physician Galen described in the second century. Galen believed that fluid flows in hollow nerves, causing muscle contraction. Replacing this model by animal electricity was a big step forward toward nowadays understanding of *bioelectricity*.

! To remember the meaning of oxidation, imagine a lion called LEO who says GER. LEO reminds you of "Loss of Electrons is Oxidation," and GER stands for "Gain of Electrons is Reduction."

Galvani optimized his experiments as shown in Figure 1.1. He connected a copper wire to the nerve and a zinc wire to the lower part of the leg. The frog leg twitched whenever he brought the wires in contact. Today we can easily understand what happened in Galvani's experiment. Copper itself does not take part in the reaction but catalyzes the reduction of hydrogen ions to form molecular hydrogen while zinc gets oxidized and dissolves:

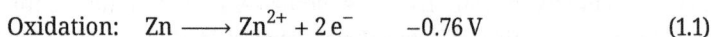

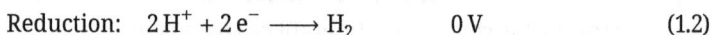

$$\text{Oxidation:} \quad \text{Zn} \longrightarrow \text{Zn}^{2+} + 2\,\text{e}^- \quad -0.76\,\text{V} \tag{1.1}$$

$$\text{Reduction:} \quad 2\,\text{H}^+ + 2\,\text{e}^- \longrightarrow \text{H}_2 \quad\quad 0\,\text{V} \tag{1.2}$$

The frog leg acts as both the electrolyte and the recipient for the electrical stimulation. The cell voltage can be up to 0.76 V, which is sufficient to cause twitching of the frog leg.

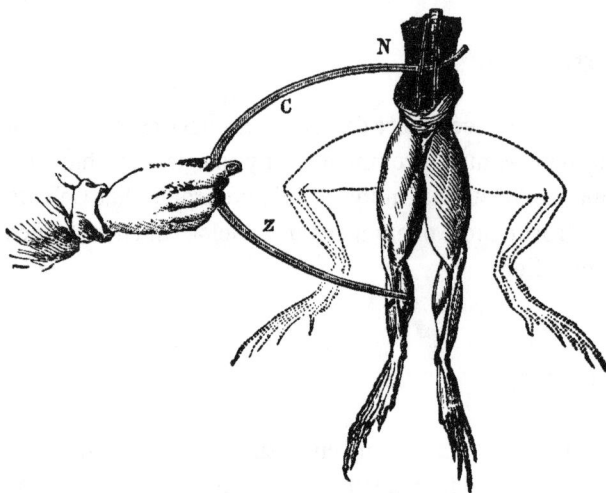

Figure 1.1: Galvani's famous experiment to demonstrate what he called animal electricity: A copper wire (C) connects to the sciatic nerve (N) and a zinc wire (Z) to the lower part of a frog leg. Upon contact of the two wires, the leg twitches [2].

Galvani also did other experiments with frog legs. Inspired by Benjamin Franklin's work (1706–1790), he connected frog legs hanging outside his house with one wire to the roof and another cable to a deep well. When a thunderstorm approached, he observed twitching of the frog legs even before the actual lightning. Imagine how people looked at those activities at the end of the 18th century – it is not surprising that Galvani's work was later one of Mary Shelley's (1797–1851) inspirations writing her novel "Frankenstein."

1.1.2 Allesandro Volta – pile to disprove animal electricity

At the same time, Allesandro Volta (1745–1827) did research in experimental physics at the University of Pavia. Inspired by Galvani's findings on animal electricity, he repeated the frog leg experiments with the copper and zinc wires. However, he concluded that the observed effect is not an inherent property of the animal but can be generated chemically with two metals in a salt solution.

To proof his position, he constructed at the turn of the century a demonstrator, which we know today as the *Voltaic pile* (Figure 1.2).

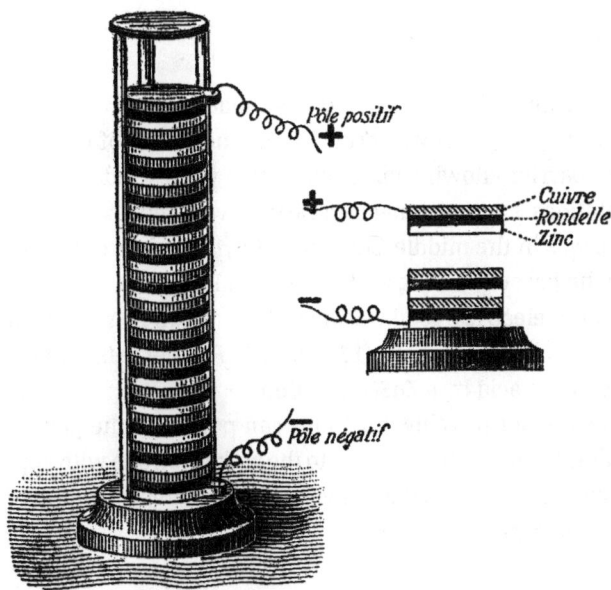

Figure 1.2: Voltaic pile with 14 stacked cells. Each cell consists of a copper electrode ("Cuivre"), a disk ("Rondelle") of cardboard soaked with salt water and a zinc electrode ("Zinc"). Image: Leçons de Physique (1904), courtesy of Éditions Vuibert.

During the experiments to find the optimal electrode materials, Volta realized that a pair of the same metals do not show any electricity, while copper and zinc showed the highest effect. He thereby discovered the electrochemical series. Initially, Volta used vessels in which he kept brine (a high concentration aqueous NaCl solution) with the metal pairs forming an electrochemical cell. Several connected elements in series results in a higher effect, measured in what we today call *electromotive force* (emf). In the Voltaic pile, cardboard soaked with brine was the electrolyte, and only stacking connected the individual cells. An inherent disadvantage of this early realization of a battery was the electrode blocking with hydrogen gas bubbles at the copper electrode.

1.1.3 John Daniell – early battery

In the following decades, many scientists attempted to improve the Voltaic pile with the goal of a constant power supply for more extended periods. In 1836, the British scientist, John Frederic Daniell (1790–1845), presented a two-fluid battery that was later called the *Daniell cell.*

Initially, Daniell used a copper vessel filled with $CuSO_4$ solution. This setup replaced the reduction reaction causing hydrogen gas formation by copper plating:

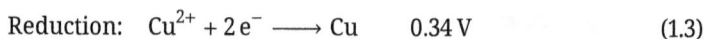

$$\text{Reduction:} \quad Cu^{2+} + 2\,e^- \longrightarrow Cu \quad 0.34\,\text{V} \tag{1.3}$$

To shield the zink electrode from the copper ions, Daniell mounted it into an ox-gullet filled with H_2SO_4 solution. The ox-gullet was brought into the center of the copper vessel and acted as a porous barrier allowing ionic conduction while hindering the mixing of the two electrolyte solutions. Later on, unglazed earthenware replaced the ox-gullet, and the copper electrode was in the middle, surrounded by an excess of $CuSO_4$. The zink electrode was outside the porous barrier, as illustrated in Figure 1.3.

The overall cell voltage without electrical load was up to 1.1 V, as can be seen from the difference in equilibrium potentials of equations (1.1) and (1.3). Further refinements comprised the replacement of sulfuric acid by a $ZnSO_4$ solution.

Recharging of the Daniell cell is not possible. Cu^{2+} ions can penetrate the porous barrier resulting in contamination of the electrolyte next to the zink electrode with copper ions. Upon application of a charging voltage, the copper ions would deposit onto the zink electrode destroying the battery's performance.

1.1.4 William Grove – early fuel cell

Sir William Robert Grove (1811–1896) was a fellow of the Royal Society, as was Daniell. The United Kingdom's national academy of sciences was a breeding ground for numerous scientific advances in Europe since the 17th century, and became after Italy, the next hotspot for electrochemical developments. Grove improved the Daniell cell to obtain a

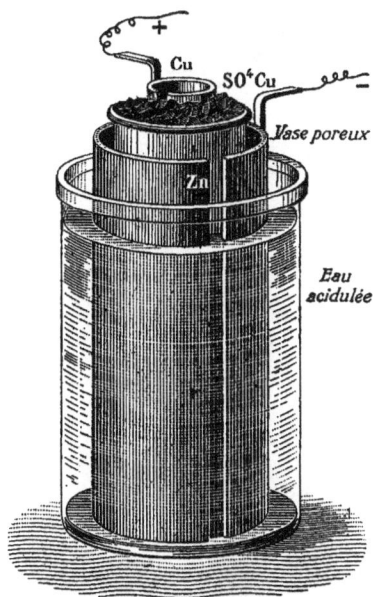

Figure 1.3: The Daniell cell was a significant improvement to the Voltaic pile. The copper electrode (Cu) surrounded by electrolyte with copper sulfate ("SO^{4}Cu") was in the center. A porous vessel ("Vase poreux") separated this electrolyte from the zink electrode in an acidic electrolyte ("Eau acidulée"). Image: Leçons de Physique (1904), courtesy of Éditions Vuibert.

higher cell voltage. He used platinum in aqueous nitric acid solution instead of the copper in a copper ion containing solution. However, the reduction reaction resulted in the release of the toxic gas nitrogen dioxide, which was one reason a slightly modified Daniell cell without platinum was later the preferred choice for the American telegraph system, by that time an important field of application for batteries.

In 1839, Grove came up with his gas-voltaic battery, what we nowadays call a *fuel cell*.[1] The gas voltaic battery contained two half-cells partially filled with gas as the fuel and an electrolyte. At the negative terminal, hydrogen gas was around the electrode. On the other side, oxygen was at the inert metal. The redox scheme providing electrical energy results in no waste products but water:

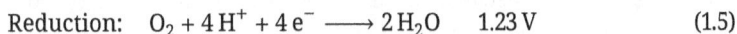

$$\text{Oxidation:} \quad H_2 \longrightarrow 2\,H^+ + 2\,e^- \qquad\qquad 0\,V \qquad\qquad (1.4)$$

$$\text{Reduction:} \quad O_2 + 4\,H^+ + 4\,e^- \longrightarrow 2\,H_2O \quad 1.23\,V \qquad\qquad (1.5)$$

---

1 Sir William Robert Grove is often named as the inventor of the fuel cell. However, already one year earlier, the German chemist, Christian Friedrich Schönbein (1799–1868), by then a professor in Switzerland, demonstrated the working principle of a fuel cell paving the way for the technical realization by Grove.

Practically, Grove stacked several of those cells together to achieve a higher cell voltage and even split again water in an electrolysis cell (Figure 1.4). He also stacked tens of such batteries to operate a carbon arc lamp.

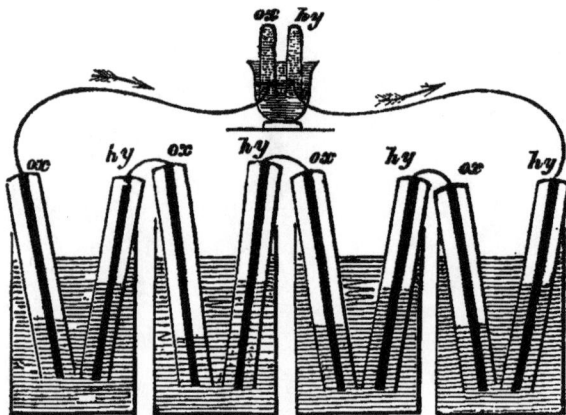

Figure 1.4: A series connection of four Grove's gas-voltaic batteries, which we would call today fuel cells. Each half-cell contains a platinum wire with acidic electrolyte in a partially gas-filled vessel. Each battery included a half-cell with oxygen ("ox") and hydrogen ("hy"). The smaller setup on top is an electrolysis cell proving the batteries' function [3]. Reprinted from Grove, On a Gaseous Voltaic Battery. Philosophical Mag Ser 3, 1842, 21 (140), 417–420, by permission of Taylor & Francis Ltd, http://www.tandfonline.com.

Today's fuel cells rely on the same principle. The reduction reaction uses oxygen either as a pure gas or from the air. Besides pure hydrogen gas, several hydrocarbons feed the oxidation reaction. A more detailed treatment on fuel cells can be found in Section 11.1.1.

1.1.5 Robert Bunsen – economic electrode material

Only two years later, Robert Wilhelm Eberhard Bunsen (1811–1899) came up with an improved setup nowadays referred to as Bunsen battery replacing the costly platinum with carbon, and thus paving the way for large scale applications. Many students today may recognize his name because of the Bunsen burner he invented. However, from an economic perspective, Bunsen's focus on cheaper electrode materials influences our daily lives much more.

The issue of high metal cost is a hot topic until today. In contrast to Bunsen's original approach, today's catalysts are not made of pure carbon but often are sophisticated functional materials with micro or nanosized features and, in many cases, decorated with metallic catalysts. Additionally, those materials may also offer more extended stability in harsh conditions or provide a high catalytic active surface area because of arrangement as a hierarchical micro-nanostructures (Section 9.2.2).

1.1.6 Michael Faraday – quantitative experiments

One of the most excellent experimentalists was Michael Faraday (1791–1867). Many significant technical and scientific breakthroughs go back to him, such as the invention of the dynamo, visualization of magnetic field lines, or the description of a magneto-optical effect, which we today call the Faraday effect.

Faraday was the first who made electrochemical experiments quantitative. In particular, he figured out that he needed a charge of 96 500 A s to deposit 108 g (what is 1 mol) of silver in a one-electron process. This number was an exact result compared to the modern formulation of what we today call *Faraday's law* in the more general writing

$$m = \frac{M}{zF} \cdot Q \tag{1.6}$$

with the constant $F = 96\,485\,\mathrm{A\,s\,mol^{-1}}$ attributed as *Faraday constant*, which is nowadays linked to fundamental constants rather than experiments. m is the mass, M the molar mass, z the valency, and Q the charge (Section 2.2.1).

Faraday was also a great science communicator. In his Christmas lecture series "On the Chemical History of a Candle," which he gave 1848/49 and repeated 1860/61, he explained physics, chemistry, and electrochemistry by considerations starting around a burning candle. According to reports, the lectures attracted not only juveniles, students, and scientists but also celebrities (Figure 1.5). So another thing to learn from history is that great scientists should be able to explain their topics simple to reach out and make their ideas persistent. In contrast, refusing honors like the offering of knighthood or becoming president of the Royal Society did not affect his legacy. Faraday was very religious and followed strict ethical standards. Being asked by the British government to support the production of chemical weapons, he denied it.

1.1.7 Walther Nernst – thermodynamics

Walther Hermann Nernst (1864–1941) was a German physicist and chemist, who worked most of his career at the University of Göttingen. During his habilitation in Leipzig, he linked the by then mainly experimentally driven electrochemistry with thermodynamics. He modeled the concentration by pressure assuming the electrolyte as an ideal gas, which allowed him to derive what we know as the Nernst equation (Section 2.2.3).

While Nernst today is mainly recognized for the Nernst equation and maybe the Nernst diffusion layer, it was the Nernst heat theorem, now better known as the third

Figure 1.5: Michael Faraday in one of his Christmas lectures at the Royal Institution of Great Britain; also Prince Albert and his sons were in the audience. Wood engraving, 1856, after Alexander Blaikley. Image: Wellcome Collection (public domain).

law of thermodynamics for which he received the Nobel Price in 1920. Nernst was very serious about everything he did. While he made an outstanding contribution to thermodynamics and electrochemistry, Nernst worked with a similar passion for chemical weapons development throughout World War 1. During the Third Reich, he distanced himself from the Nazi regime and spoke for his Jewish colleagues what finally required him to retire in 1933.

For the decades after World War II, the shorter distance to the present makes it less easy to clearly identify the hallmarks. Significant contributions were possible because of advances in the instrumentation. More recently, improved computational capabilities led to novel approaches to model electrode reaction, such as density-functional theory (DFT) calculations, ab-initio molecular dynamics (MD), and cross-scale simulations. While the exact field of applications changed over the centuries, analysis of biological processes and energy storage respective conversion are the common threads. If you became interested, you can find a more detailed storyline of historical scientists shaping our nowadays understanding of electrochemistry from Shukla et al. [4].

1.2 Fields of applications

Electrochemical methods for the micro and nanoscale may seem to be a very narrow topic. However, looking at the many fields of application in which the methods play a crucial role, one might wonder why not all engineering curricula include a compulsory lecture on this topic. This section gives you an overview of a wide range of different areas. The intention here is to emphasize the broad range of applications rather than to provide a comprehensive list.

1.2.1 Biomedical sensors and point-of-care systems

Glucose monitoring is a big success story for widespread applications of biosensors. Since the 1970s, diabetes patients benefit from blood glucose measurements with finger pricking and later on complemented by continuous glucose monitoring (CGM) devices leading to better life quality and higher life expectancy. Section 10.1.5 discusses the various electrochemical methods playing together in a commercial glucose meter to detect glucose in a blood droplet.

Figure 1.6 shows a state-of-art CGM system. Diabetes patients can wear such systems for up to a week on the skin, e. g., on the belly or the arm. A tiny needle allows for continuous measurement of the tissue glucose concentration helping the patient control their glucose levels. In contrast to finger pricking alone, CGM provides continuous data, also during sleep. Additionally, the system can warn the patient when values exceed thresholds before the situation gets critical.

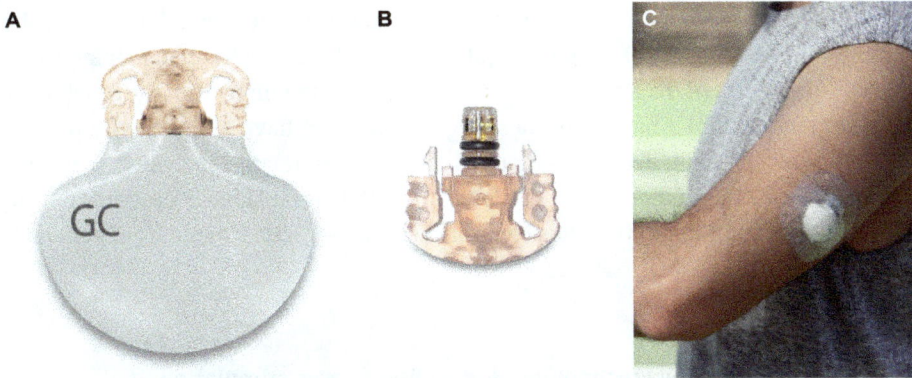

Figure 1.6: The Guardian Sensor 4 system allows continuous glucose monitoring (CGM) from interstitial fluid. The transmitter (A) incorporates a disposable sensor unit (B) with a tiny cannula, which the patient can place on the upper arm (C) or belly. Reproduced with permission from Medtronic GmbH, Germany.

While the focus for diabetes is on home use devices, parallel to this, clinical analyzers have matured, which include several biosensors together with ion-selective and oxygen sensors. Most of these sensors are relying on electrochemical measurement principles, while some use optical read-out based on absorption or fluorescence.

In addition to the blood glucose meter, many other point-of-care devices enable decentralized testing for specific substances or measurements at home, in emergencies, in doctor's offices, or directly at the bedside in hospitals. Enzymatic biosensors integrated with microfluidics allow multiparameter monitoring from tiny samples (Figure 1.7). Also, immunoassays are used in such point-of-care testing (POCT) systems, which utilize antibodies to detect the analyte selectively. Electrochemical methods allow the read-out and inherent signal amplification, as described in Section 10.1.7.

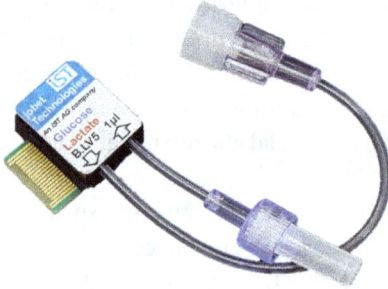

Figure 1.7: Enzymatic flow-through biosensor device: the cartridge contains sensors for glucose, lactate, glutamine, or glutamate. The low internal volume of 80 nl allows for samples' volumes below 1 μl. Reproduced with permission from Jobst Technologies GmbH, Germany.

Chemo and biosensors play an essential role in observing the metabolism in cells cultured artificially outside the organism [5], Section 10.1.8. The cells are either kept simply in culture or grown in chips comprising microfluidics to mimic even the function of organs (organ-on-chip). The goal of these approaches is to have a model with human cells in the lab, allowing drug tests to avoid animal experiments or even test several therapeutic strategies with a patient's cells to choose the most promising therapy for the individual patient. In the future, those approaches will be an essential part of personalized medicine.

1.2.2 Neuroscience, neurotechnology, and auditory nerve stimulation

Research in neuroscience requires information on the brain at many different levels. The realm spans from behavioral studies, investigation of the neural activity in particular functional units by magnetic resonance imaging, electrical signals recorded inside or outside the skull to detailed information about neurotransmitters on the molecular level. For the latter, carbon fibers, in combination with sophisticated electrochemical protocols act as electrochemical sensors. Fast scan cyclic voltammetry enables selective detection of dopamine with high temporal and spatial resolution (Sections 5.3.10 and 10.3.4). However, those approaches are limited to research in animals because of the vast invasiveness.

Neurotechnology enables recording the electrical signals and brain stimulation in humans by implants with noble metal electrodes on or in the neuronal tissue acting as a brain-computer-interface (BCI). Neurologists apply electrode arrays below the skull for intracranial electroencephalography or place stimulation electrodes into a specific brain region, even down to the cerebellum. This so-called deep brain stimulation (DBS) can help patients with Parkinson's diseases, epilepsy, Tourette syndrome, or severe cases of depression.

Figure 1.8 shows a commercial neurotechnological implant. It features electrodes to record signals, stimulation electrodes, and an electronics unit. The electronic units (1) connects from below the skin wirelessly to a control unit outside the patient, containing the power supply. The DBS electrodes (2) allow placement deep in the brain, e. g., in the thalamus or cerebellum. The array electrodes (3) deliver signals from their position on the cerebral cortex. Such electrodes are called electrocorticography (ECoG) array and can optionally stimulate. Depending on the application, the implants are dedicated for closed-loop operation, meaning the stimulation is autonomously applied based on the recorded signals. Commercial implants mainly rely on platinum or platinum-iridium alloys as the electrode material; many alternatives are discussed and evaluated in research (Section 10.3.3).

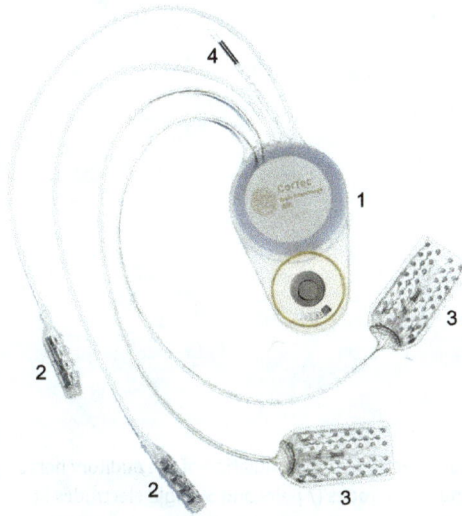

Figure 1.8: The Brain Interchange One system is a brain implant allowing the combination of various electrodes. The electronics unit (1) connects to DBS (2) and array electrodes (3). Reproduced with permission from CorTec GmbH, Germany.

The long-term stability of such electrodes is an essential prerequisite. Therefore, the corrosion behavior of such electrodes in the brain needs to be well understood and optimized. Here, electrochemical methods help better understand the processes at the tissue/electrode interface and allow for the investigation of possible corrosion mechanisms. Recent research points to the possibility of recruiting existing stimulation electrodes for electrochemical sensing, which even might allow for routine application in humans [6]. The next decade will show if such approaches are beneficial and safe enough to be applied with humans and eventually enter clinical routine. Section 10.3.5 discusses this topic in more details.

Another active implant is the cochlear implant (CI), allowing stimulation of the auditory nerve. After the cardiac stimulator, this is the second most frequently applied implant with electrical stimulation. For patients with severe hearing loss but intact auditory nerve, this implant stimulates the nerve with signals derived from an external microphone and converted by a speech processor.

Figure 1.9 shows a commercial CI with the electronic unit and the electrode array, which the surgeon inserts into the cochlea. The electrode array contains contacts for 12 channels, representing a specific frequency band in the hearing process. While patients mainly used a CI for orientation and speech understanding in the early days, current implants even enable listening to music.

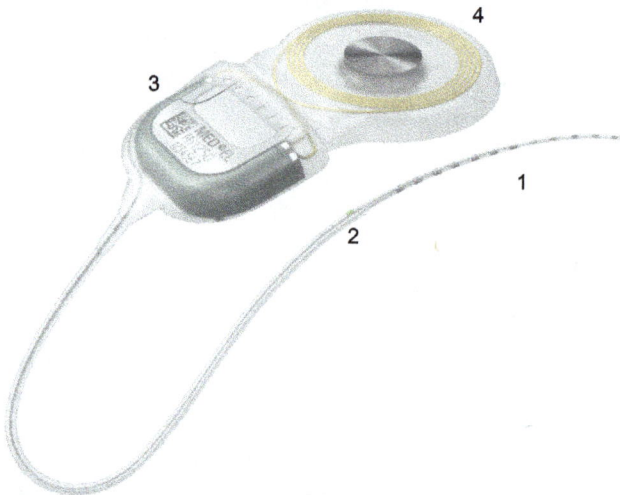

Figure 1.9: SYNCHRONY 2 Cochlear Implant with platinum electrodes for stimulation of the auditory nerve. The surgeon inserts the electrode array (1) with 19 platinum electrodes (7 pairs and 5 single electrodes at the tip) into the cochlea; the orientation marker (2) remains outside. The stimulator (3) with the coil (4) is located under the skin behind the ear. The coil of an audio processor attaches at the same position outside of the body. Reproduced with permission from MED-EL, Austria.

From an electrochemical perspective, the questions concerning the tissue-electrode interface are comparable to the situation with brain implants. Any discussion for brain implants on electrochemical methods helping understand the microenvironment, potential corrosions mechanism, and the electrochemical processes occurring in stimulation may similarly contribute to the field of auditory nerve stimulation and vice versa. Especially the long experience with CIs, including some patients wearing the electrodes over decades, may also help the brain implants, which have a younger clinical history and lower case numbers.

1.2.3 Microelectronics

Microelectronics and the semiconductor industry primarily benefit from electrochemical methods to electroplate copper as interconnects (Section 9.2.1). For a long, copper was the standard material for all interconnects, but further downscaling requires other materials less prone to migration than copper and, therefore, without the need of a barrier metal. Some fabs recently introduced cobalt for the tiniest structures in the 10 nm range close to the individual transistors. Another material candidate is ruthenium.

Besides, electrodeposition of various metals and alloys as well as electrochemical inline process control, such as for wet chemical etching, can be found in many fabs. While the large dimensions of the production lines require highly specific instrumentation devices and electrolyte handling, the underlying theory and concepts do not differ much from the other three fields described above.

1.2.4 Energy applications

Among the different fields of applications, it is the energy-related topic prospering most. In not much more than a decade, lithium-ion batteries became the essential electrochemical units, which are ubiquitous in our daily lives. We can find those rechargeable cells in a tiny hearing aid or earbuds, in smartwatches, in smartphones, in tablets, and in notebooks. Huger cells are part of cordless tools or store energy in electric vehicles. On an even larger scale, those batteries play an essential role in integrating renewable energies from solar and wind power into our power grid.

The Royal Swedish Academy of Science accounted for the importance of lithium-ion batteries by rewarding Stanley Whittingham, John Goodenough, and Akira Yoshino with the Nobel Prize in Chemistry 2019. In contrast to formerly developed batteries, the lithium-ion battery does not rely on electrochemical reactions of the electrode material but on lithium-ion movement between the electrodes in which the ions intercalate.

The human-made climate change and especially the latest public awareness of global warming directs many resources for research and business promotion to achieve a fossil fuel-free society, maybe even supported by negative carbon dioxide release. However, the drawback with any single-sides measures is the damage caused in other areas. For example, focusing on electrical mobility helps to avoid exhaust gas emissions from cars, but we also have to consider who, and under which circumstances, harvests lithium and especially cobalt. Researchers may provide solutions to avoid those metals in the long run, e. g., by using sodium instead or by relying on a completely different storage principle. Hydrogen gas, methane, *synthetic natural gas* (SNG), or synthetic fuels are possible alternatives to store energy with electrochemistry involved in the production and, in some cases, in the conversion back to electrical energy. Electrolysis allows the conversion of electrical power to gas, known as *power-to-gas* (P2G). Hydrogen forms directly from splitting water, processing to other fuels can even consume CO_2.

Among the complex relationships prescribing the climate on our planet, the common ground is that the atmospheric carbon oxide concentration increases rapidly (Figure 1.10), with a high likelihood leading to global warming. And, while there is controversy on the exact numbers, a substantial share of the CO_2 emission is human-made. I am sure that lowering or even reversing carbon dioxide emissions is essential for the future of the global population. I am also sure that electrochemical methods can play an important role here. On a detailed level, the exact direction is unclear. Many protagonists try to shape public awareness toward their own opinion either because of commercial interest or ideology, both in favor or against our environment. Any student, scientist, or engineer working in the energy field will face rational and emotional discussions. That is why I strongly recommend checking some basic facts continuously to find and adjust your own opinion and arguments. A starting point could be the web publication Our World in Data.[2]

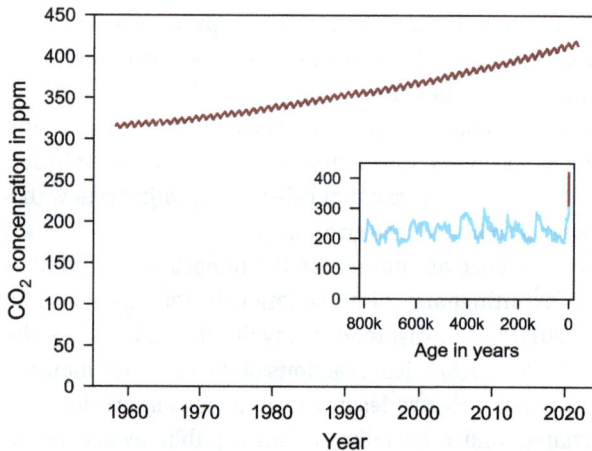

Figure 1.10: The atmospheric CO_2 concentration, measured since 1958 at the Mauna Loa Observatory, Hawaii, the so-called "Keeling Curve" [7]. Much more alarming is to see this curve in a larger context, in the inset combined with 800,000 years of ice-core data [8].

---

2 https://ourworldindata.org/co2-and-other-greenhouse-gas-emissions

1.3 Electrochemical cells

We now approach the arena of any electrochemical experiment. The arrangement of two electrodes in an electrolyte is called *electrochemical cell*. An electrolyte is an ion conductive phase, either liquid or solid. Formally, gases could also be considered as electrolytes, but ionized gas is normally considered to be part of the discipline of plasma physics rather than electrochemistry.

Most simple cases of an electrolyte are an aqueous solution of salt (such as potassium chloride) or an acid (such as sulfuric acid). In the electrolyte, current can flow due to ion conduction. In contrast, wires and circuitry through which currents can flow by electron conduction connect the electrodes. At the interface between electrode and electrolyte, redox processes take place, translating between ion and electron conduction.

The *anode* is the electrode at which the oxidation reaction occurs. The *cathode* is the electrode at which the reduction reaction takes place. While there are some mnemonic bridges to define the polarity of the anode for electronic parts, those hints can fail when considering electrochemical cells. To remember the terms anode and cathode correctly, one can think of **an ox** (imagine a male cattle ready to pull the plow) and a **red cat** (imagine a purring cat with red fur starring at you while you think about the cathode).

In general, two types of electrochemical cells can be distinguished (see Figure 1.11). In case a voltage is applied to the electrochemical cell (or a current source is connected), the arrangement is called *electrolytic cell*. At the positive terminal, the oxidation reaction occurs, and this electrode is accordingly called the *anode*. On the other side, at the negative terminal, the reduction reaction takes place, and it is therefore called the *cathode*. One example of the reactions in an electrolytic cell is the electrolysis of water, as indicated in Figure 1.11 (left). At the negative terminal, the reduction of protons produces hydrogen gas.

The International Union of Pure and Applied Chemistry (IUPAC)[3] recommends drawing the anode always as the left half-cell of an electrochemical cell. The polarity flips left or right, depending on the cell being an electrolytic or galvanic cell.

The other type of electrochemical cell is the *galvanic cell* or *voltaic cell* referring to the works of Luigi Galvani and Allesandro Volta. In contrast to the electrolytic cell, the electrical circuit acts as a consumer, and the electrochemical cell delivers charge. In the

---

3 The International Union of Pure and Applied Chemistry (IUPAC) is a worldwide organization that is part of the International Science Council, a nongovernmental organization uniting and representing scientists worldwide. One of the most considerable efforts of the IUPAC is the nomenclature of organic chemistry (published in the so-called "Blue Book") and inorganic chemistry ("Red Book"). In this textbook, the electrochemical terminology mainly follows the guidelines for analytical chemistry, the so-called "Orange Book" [9].

Electrolytic cell   **Galvanic cell**

$2H_2O \rightarrow$
$O_2 + 4H^+ + 4e^-$ | $2H^+ + 2e^- \rightarrow H_2$ $Zn \rightarrow Zn^{2+} + 2e^-$ | $Cu^{2+} + 2e^- \rightarrow Cu$
(oxidation) | (reduction)   (oxidation) | (reduction)

Anode   Cathode   Anode   Cathode

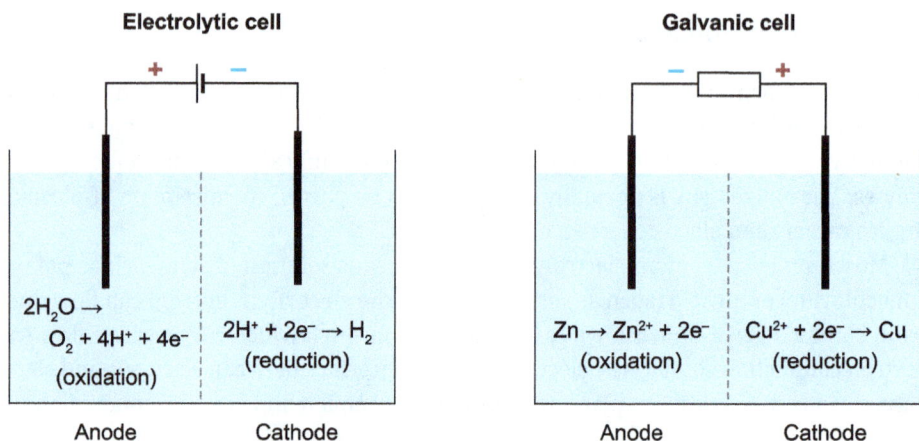

Figure 1.11: Electrochemical cells: the electrical circuit powers the electrolytic cell, the galvanic cell delivers charge to the electrical consumer. Please note the different polarities of the anode/cathode depending on the type of cell.

most simple case, a resistor connects the two electrodes. The anode, the electrode at which the oxidation occurs, is more negative than the electrode at which the reduction occurs. Thus the positive terminal of a galvanic cell is the cathode. A typical example of this type of cell is a battery getting discharged. The right cell in Figure 1.11 illustrates the situation in a Daniell cell (Section 1.1.3).

Both types of electrochemical cells deal with electrochemical processes at non-equilibrium. Energy either is brought into the electrochemical system from the electrical circuits (electrolytic cell), or an electrical circuit consumes the charge stored in the electrochemical cell (galvanic cell). The reactions in electrolytic cells do not occur spontaneously, while in a galvanic cell, the reactions run spontaneously once the electrical circuit is closed.

1.3.1 Primary and secondary cells

A nonrechargeable battery[4] is called a *primary cell* and is always a galvanic cell. A rechargeable battery is called a *secondary cell* and is an example of an electrochemical cell, which can be both electrolytic and galvanic cell. The well-known lead-acid battery, which is still prevalent in cars with combustion motors, is an example of what a secondary cell is. It is an electrochemical cell with two lead electrodes in an aqueous sulfuric acid solution as the electrolyte. The lead electrode at the positive terminal is

---

4 Strictly spoken, a *battery* is the technical realization of one or several connected primary or secondary cells in an appropriate package.

coated with a paste containing lead oxide (PbO_2). During supplying (galvanic cell), the negative terminal is the anode, and the reactions are

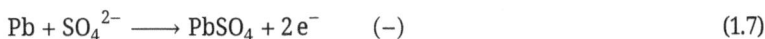

$$Pb + SO_4^{2-} \longrightarrow PbSO_4 + 2\,e^- \quad (-) \tag{1.7}$$

$$PbO_2 + 4\,H^+ + SO_4^{2-} + 2\,e^- \longrightarrow 2\,H_2O + PbSO_4 \quad (+) \tag{1.8}$$

In a discharged battery, both electrodes are covered with $PbSO_4$. During discharge, the SO_4^{2-} is removed from the electrolyte solution lowering the concentration of the sulfuric acid. As the density of an aqueous sulfuric solution strongly depends on its concentration, a hydrometer allows checking the charge status of a lead-acid battery.

While recharging the battery (electrolytic cell), the positive terminal is the anode, and the reactions are reversed:

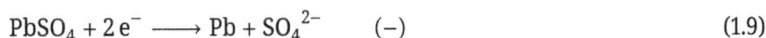

$$PbSO_4 + 2\,e^- \longrightarrow Pb + SO_4^{2-} \quad (-) \tag{1.9}$$

$$2\,H_2O + PbSO_4 \longrightarrow PbO_2 + 4\,H^+ + SO_4^{2-} + 2\,e^- \quad (+) \tag{1.10}$$

The nominal cell voltage of the lead-acid battery cell is 2.05 V. Therefore, a car battery contains six cells connected in series. Nowadays, commercial batteries contain grid electrodes instead of lead plates to increase the effective surface area but still feature sufficient current conduction. Repetitive discharge/charging cycles might cause dendrites' growth, eventually resulting in a short-circuit of the affected cell. Therefore, separators are introduced in between the two electrodes allowing ion conduction but avoiding direct contact of the two electrodes. Figure 1.12 shows a commercial lead-acid battery used in cars.

1.3.2 Half-cell

An electrochemical cell can be seen as two separate *half-cells* consisting of one electrode in an electrolyte each (Figure 1.13A). The theoretical treatment of an electrochemical cell usually considers two independent half-cells. With macroscopic electrodes in a large electrolyte vessel, this concept can be widely applied. For microscopic setups, the two involved electrodes often cannot be treated independently because of unwanted or desired cross-talk in-between (Section 10.1.7).

In case cross-talk between the two electrodes in an electrochemical cell should be avoided or different electrolytes for the two half-cells are required, an ion-permeable membrane can separate the half-cells (e. g., the separator in Figure 1.12) or a *salt-bridge* (Figure 1.13B). A salt-bridge connects the two half-cells by the same or a different ion-conducting medium. Practically glass or polymeric tubes are used, either with a tiny diameter or filled with a jellified salt solution to avoid flow between the two half-cell electrolytes. Depending on the salt solutions considerable diffusion potentials (Section 4.2.5) can develop.

Ion-permeable separator

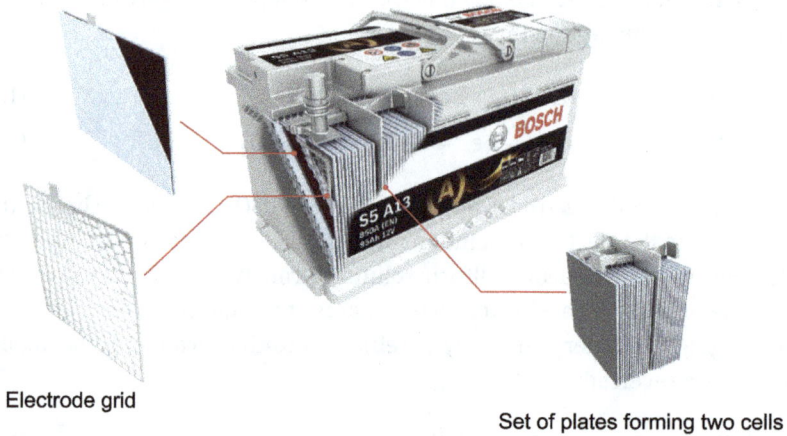

Electrode grid

Set of plates forming two cells

Figure 1.12: Commercial lead-acid battery for cars. Each of the six cells consists of several electrode plates connected in parallel. The grid electrodes' design allows for optimal current distribution; a lead alloy increases their corrosion resistance. Ion-permeable separators avoid direct contact between neighboring electrodes. Adapted with permission from Bosch.

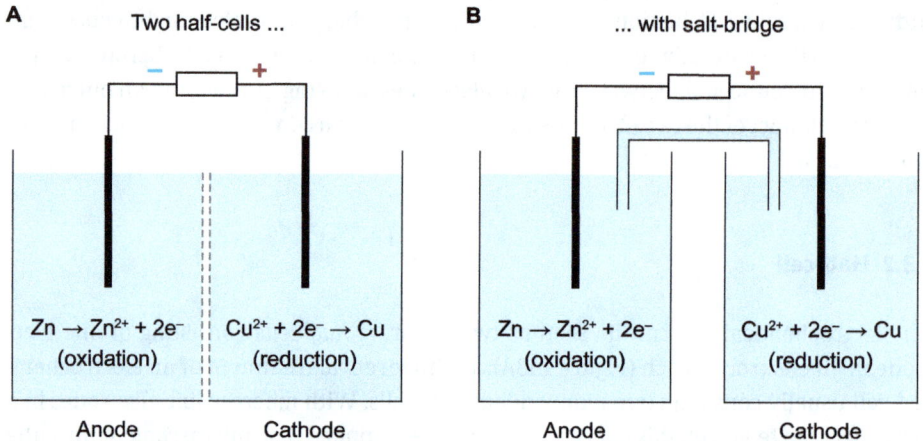

A Two half-cells ...

$$Zn \rightarrow Zn^{2+} + 2e^-$$
(oxidation)

$$Cu^{2+} + 2e^- \rightarrow Cu$$
(reduction)

Anode Cathode

B ... with salt-bridge

$$Zn \rightarrow Zn^{2+} + 2e^-$$
(oxidation)

$$Cu^{2+} + 2e^- \rightarrow Cu$$
(reduction)

Anode Cathode

Figure 1.13: An electrochemical cell (here a galvanic cell) can be considered as two half-cells (A). Also, in practical realization, an electrochemical cell can be arranged by two half-cells connected with a salt-bridge allowing ion conduction (B).

1.3.3 Electrochemical cell in equilibrium

In equilibrium, without external current flow, an electrochemical cell can exhibit a potential difference between its electrodes called *open circuit potential* (OCP) or more correctly *open circuit voltage* (OCV).[5] In the example depicted in Figure 1.14 two half-cells connected by a salt-bridge form the electrochemical cell. The electrical circuit is closed by a voltmeter (with ideally infinite internal resistance). Therefore, no current flow is possible, and the measured cell voltage E_{cell} is the difference between the OCPs of the two half-cells. In each half-cell, a reversible redox reaction takes place with no net current (the current of the backward reaction balances the current of the forward reaction). Because at each electrode, there is the oxidation and the reduction taking place, the terms anode and cathode are not meaningful.

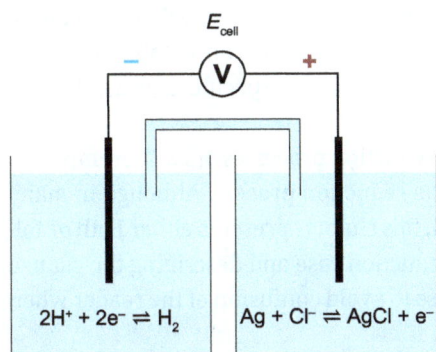

$$E_{cell}$$

$$2H^+ + 2e^- \rightleftharpoons H_2 \qquad Ag + Cl^- \rightleftharpoons AgCl + e^-$$

Figure 1.14: An electrochemical cell in equilibrium. The cell voltage E_{cell} is the difference between the open circuit potentials (OCPs) of the two half-cells. The left half-cell could be the standard hydrogen electrode (SHE) used to define the zero-point of the electrochemical potential scale.

In the example of Figure 1.14, the left half-cell is the hydrogen electrode. With platinum as electrode material, an electrode with unity activity of protons (pH 0) and a gas supply for hydrogen gas (normal pressure $p_N = 101.325\,\text{kPa}$) bubbling around the electrode the setup is called *standard hydrogen electrode* (SHE), and its potential is defined as 0 V (Section 4.2.1).

The right half-cell comprises an Ag/AgCl electrode. That is a silver electrode coated with silver salt in an electrolyte with Cl^- ions (Section 4.2.2). In this example, it is helpful to have the hydrogen gas in the electrolyte of the left half-cell only. Therefore, the two half-cells are connected by a salt-bridge, an aqueous, conducting phase, which hinders significant ion exchange between the two half-cells. Especially with reference electrodes, a bridge electrolyte separated by diaphragms is common (Section 4.2.5). In a microchip, a thin capillary filled with the salt solution has the same role.

---

5 In electrical engineering, the term *potential* is the level of a single terminal, while the term *voltage* is used for the difference between two potentials. In electrochemistry, the terms are sometimes not strictly separated, e. g., in some textbooks or research papers the term *open circuit potential* (OCP) is used for the measured cell voltage in equilibrium, although the term *open circuit voltage* (OCV) would be more appropriate.

2 Electrochemical theory

Textbooks can present electrochemical theory in at least two ways. The more formal way is to start with the equilibrium and progressing toward the general nonequilibrium case. The focus is on rigorous derivation from thermodynamics. The second possibility is to consider electrochemical processes in general and introducing the equilibrium as a special case. I believe this route is more in line with the practical approach to electrochemistry, therefore, this chapter goes that way.

The goal here is to learn all the essentials to understand and successfully apply electrochemical methods. Accordingly, the chapter aims at a condensed rather than all indetailed presentation. Deeper understanding and more details can be obtained from various other textbooks [1, 10–12]. For more information on specific aspects, the text sometimes refers to the original works.

Traditionally, electroanalysis took care for reduction processes only. Therefore, it is common to present the theory by examining the reduction process. Although in many fields of application oxidation processes prevail, this chapter presents either both or follows the classical approach of explaining the reduction case and describing the change required to apply the theory to the oxidation case to avoid confusion of the reader when working with different textbooks.

The traditional focus on the reduction process causes another confusion: an electrical current generated by a reduction has a negative sign in terms of the convention in physics, electrical engineering, or electronics. For long, scientists, especially in the context of polarography, displayed the reduction currents with positive numbers as these processes were the ones of interest. Nowadays, we face an inconsistent situation: some publications, particularly older ones or from the United States, prefer current-potential diagrams with positive currents for the reduction ("reduction up"). Simultaneously, the majority of the application-driven literature follows the more rational way ("oxidation up"). See also Section 5.3.1.

2.1 Conventions

The different disciplines contributing to the electrochemical theory in combination with the practical methods cause a mix up of notation and labeling of the variables. Therefore, this chapter starts with an introduction to the conventions used in this book. Besides Chapter 7 on electrochemical impedance spectroscopy and some briefly mentioned methods, it was possible to label all quantities uniquely.

Potential vs. energy
The electrical potential E in Volts (V) is referred to the potential of the standard hydrogen electrode (Section 4.2.1) if nothing else is specified. Depending on the context, es-

https://doi.org/10.1515/9783111488844-002

pecially when describing an experiment, the potential may be implicitly related to the reference electrode used. Reading the original research work asks to check the potential scale of a plot or value carefully. In contrast to the strict distinguishing in electrical engineering between a potential (referring to the vacuum level) and a voltage (the difference between two potentials) in electrochemistry, the term potential is used in both cases.

To not confuse with the potential, W (work) in J is used for the energy. One exception is the *Gibbs free energy G*. As it is the only form used in this book, the letter G symbolizes the molar Gibbs free energy in $J\,mol^{-1}$.

Current vs. current density and charge vs. charge density

While the current I in A is the measured or applied quantity, its normalization by the electrode area, the current density i in $A\,m^{-2}$, is more helpful when presenting and comparing data. Important to note, that for normalization the geometric (projected) electrode area is considered instead of the entire surface area due to roughness. Similarly, the charge Q in A s needs to be distinguished from the charge density q in $A\,s\,m^{-2}$.

The concept using the geometrical rather than the total surface area allows for a simple identification of the influence of electrode roughness (see Figure 8.16 for an example).

Flux vs. flux density

Along with the current and current density, the species' transfer involved in an electrochemical reaction can be expressed by the species flux J in $mol\,s^{-1}$ and its flux density j in $mol\,s^{-1}\,m^{-2}$. For mass transfer to or from an electrode, the flux is normalized by the electrode's geometrical area. In general, the area for the flux density can be any cross-section depending on the context.

Concentration

The concentration c always refers to the number of species (typically in mol) per volume of solvent. For easy to read representation of equations, the concentration of species X is written as [X] instead of c_X.

2.2 Faradaic processes

A *faradaic process* is a redox process, a chemical reaction involving an electron transfer (oxidation or reduction). Accordingly, a *faradaic current* is a current due to a change in redox state. A common confusion is mistakenly thinking of capacitive currents as "Faraday currents." However, a capacitive effect does not include a change in redox state and is, therefore, a non-faradaic process.

In the presentation of electrochemical theory and methods, the fictive species O (oxidized form) and R (reduced form) are used to describe prototypic n-electron processes:

$$\text{Reduction:} \quad O + ne^- \longrightarrow R \tag{2.1}$$

$$\text{Oxidation:} \quad R \longrightarrow O + ne^- \tag{2.2}$$

2.2.1 Faraday's law

The *Faraday's law* connects the number of species involved in a redox reaction with the required or released charge. Considering the mass of the species (e. g., mass of deposited layer in case of electrodeposition) the law can be formulated as

$$m = \frac{M}{zF} \cdot Q \tag{2.3}$$

with m the mass, M the molar mass, z the valency, and Q the charge. Historically, the Faraday constant F was obtained experimentally, while it is nowadays expressed in terms of the elementary charge q_e and the Avogadro constant N_A:

$$F = q_e \cdot N_A = 96\,485 \,\text{A s mol}^{-1} \tag{2.4}$$

! It is essential to clearly distinguish between the number of electrons n involved in a reaction and the valency z, both as dimensionless numbers. $n = 1, 2, 3 \ldots$ is a natural number, while the valency $z = \pm 1, \pm 2, \pm 3 \ldots$ contains a sign. This textbook carefully considers the difference between n and z. Generally, there are many inconsistencies around n and z in textbooks and the scientific literature.

Alternatively, Faraday's law can be more catchily written with the species flux density j instead of the mass and the current density i instead of the charge:

$$i = zF \cdot j \tag{2.5}$$

This form of Faraday's law allows linking a measured current to a species concentration in a diffusion-controlled scenario or can connect the normal species flux into an electrode with the current when running simulations of the electrode's microenvironment.

---

i ■■□ Task 2.1: You have two different metal plating baths for the same metal. In one plating bath, the deposition follows a two-electron process; in the other one, a one-electron process. If you want to achieve the same mass of deposited metal within the same time, for which bath you need to apply a higher current density? Use Faraday's law to support your argument.

2.2.2 Electron transfer

When we want to understand what is going on in a Faradaic process, we should focus on electron transfer. There are several aspects and different levels of abstraction. In the end, each approach targets to express the complex situation by a simple quantity relating to experimentally accessible effects. From a practical perspective most important is to be aware of the assumption made during the simplification. That is why you find in this book so many hints and warnings about the assumptions.

The chemist's perspective illustrates the electron transfer by the energy diagram of the situation at the electrode/solution interface (Figure 2.1). For a metal, the electrode's energy diagram reduces to the energy level of electrons. In the solution, we need to consider the substances' molecular nature. Looking at the energy diagram, we need to consider the highest occupied molecule orbital (HOMO) and the lowest unoccupied molecule orbital (LUMO). Electron transfer occurs when an electron from the electrode fills into the LUMO (reduction) or escapes from the HOMO to the electrode (oxidation).

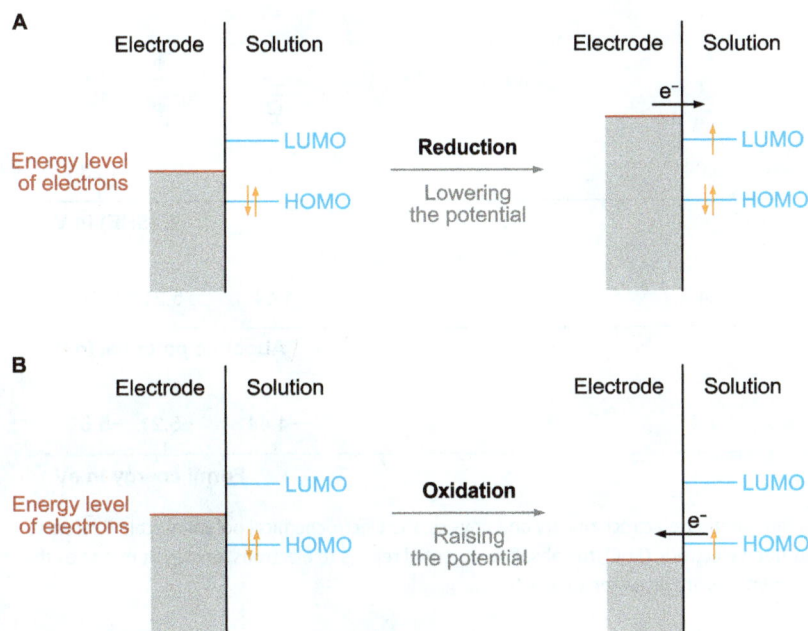

Figure 2.1: Electron transfer at a metal electrode: energy diagrams for the situation at the electrode/solution interface. Lowering the electrode potential (raising the electron's energy level) enables the reduction process (A); raising the electrode potential (lowering the electron's energy level) drives the oxidation (B).

An electron can pass along the electrode/solution interface when it is energetically favored (after the process, it is on a lower energy level). Driving the reduction reaction requires a higher energy level of the electrons in the electrode. The electrical potential E and the energy ΔG are linked inversely:

$$\Delta G = -nFE \tag{2.6}$$

Lowering the electrical potential of the electrode causes a rise in the energy level. With high enough energy (low enough potential), the reduction process (transfer of an electron from the electrode) can occur. The oxidation process runs the opposite way: Increasing the electrode's electrical potential (lowering the electron's energy level) causes the oxidation process (transfer of an electron to the electrode).

It is not only about the relation between energy and potential; also, there are different potential scales in electrochemistry and physics. The electrochemical scale refers to the standard hydrogen electrode (SHE), while in physics, the vacuum level (electrons at rest) was chosen as the reference point. Figure 2.2 illustrates the relation.

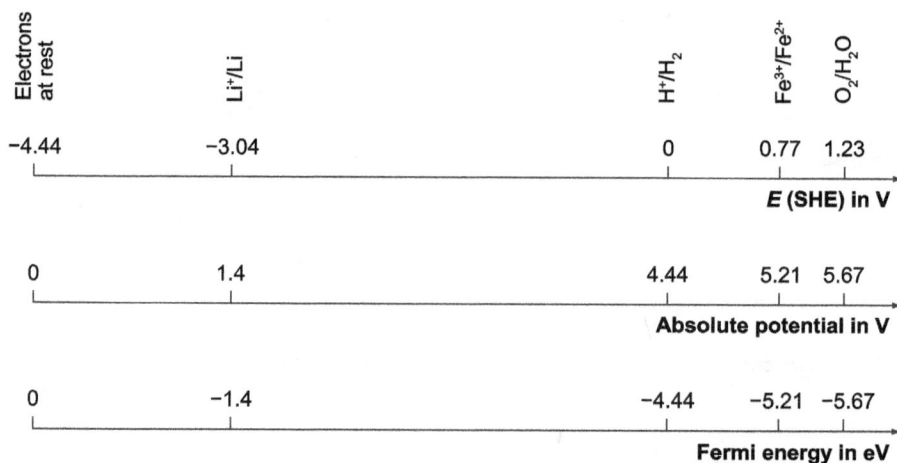

Figure 2.2: Bringing together electrochemistry and physics: the electrochemical potential scale E refers to the standard hydrogen electrode (SHE); the absolute potential relates to electrons' energy at rest so as the energy scale. The numbers are given for 25 °C.

All potentials E in this book refer to SHE if not mentioned explicitly. Accordingly, the standard potential of the H^+/H_2 reaction is at zero. In an aqueous system, the two gas-evolving reactions (H_2 and O_2) defines the possible domain. Another essential hallmark is iron (Fe^{3+}/Fe^{2+}) at 0.77 V. The most left-positioned reaction is with lithium, indicating its high reactivity and enormous potential for energy storage.

The exact interrelation between the scales depends on the Fermi energy at the SHE, which is temperature-dependent. The value −4.44 eV is for 25 °C [13]. For real electrodes, more effects may play a role when linking the different scales [14].

2.2.3 Faradaic processes in equilibrium – Nernst equation

We now assume that our fictive species O and R are a *redox couple*, which means they are linked by an *n*-electron redox process

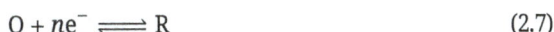

$$O + ne^- \rightleftharpoons R \qquad (2.7)$$

in which the forward and backward reaction is possible. Practical realizations of commonly used redox couples are discussed in Section 4.6.

When talking about O and R in a real scenario, the concentration c, which is the number of species per solvent volume, is not a proper measure to discuss a faradaic process as we cannot be sure that all species will take part in the reaction. The more appropriate quantity is the activity a, which considers only the number of species per solvent volume, which are active in a reaction. For higher concentrations, the species might hinder each other so the activity is lower than the concentration. The dimensionless activity coefficient y accounts for those effects connecting the activity and concentration:

$$a = y \cdot c \qquad (2.8)$$

The activity coefficient y depends on the concentration of all the different species in the solution as well as the solvent. For strongly diluted solutions, it approaches 1. Practically each species/electrolyte combination needs empirical values provided in tables or calculated by software.

Discussing a Faradaic process in equilibrium, we can rely on thermodynamics to calculate the change in the free energy depending on the involved species' activity:

$$\Delta G = \Delta G^0 + RT \ln \frac{a_R}{a_O} \qquad (2.9)$$

ΔG^0 relates to the free energy at standard conditions, R is the (molar) gas constant.

Introducing the link between energy and potential (equation (2.6)) leads to the Nernst equation with the electrode's actual potential E and the standard potential E^0:

$$E = E^0 + \frac{RT}{nF} \cdot \ln \frac{a_O}{a_R} \qquad (2.10)$$

Practically, one would prefer the Nernst equation in terms of concentration rather than activity. With equation (2.8) rearrangement leads to

$$E = E^0 + \frac{RT}{nF} \cdot \ln \frac{\gamma_O c_O}{\gamma_R c_R} = E^0 + \frac{RT}{nF} \cdot \ln \frac{\gamma_O}{\gamma_R} + \frac{RT}{nF} \cdot \ln \frac{[O]}{[R]}. \tag{2.11}$$

The standard potential E^0 together with the term containing the activity coefficients forms the so-called *formal potential*:

$$E^{0\prime} = E^0 + \frac{RT}{nF} \cdot \ln \frac{\gamma_O}{\gamma_R}, \tag{2.12}$$

and allows to rewrite the Nernst equation in terms of concentration:

$$E = E^{0\prime} + \frac{RT}{nF} \cdot \ln \frac{[O]}{[R]} \tag{2.13}$$

Please note that the formal potential is concentration-dependent, and assuming it as constant is valid only for reasonable small changes in concentration.

Another approximation $E^0 \approx E^{0\prime}$ is possible for diluted solutions ($c \ll 1\,\text{mM}$):

$$E \approx E^0 + \frac{RT}{nF} \cdot \ln \frac{[O]}{[R]} \tag{2.14}$$

---

i ■□□ Task 2.2: Rewrite the Nernst equation (equation (2.10)) with the decadic logarithm (lg) instead of the natural logarithm (ln). How much is the factor before the decadic logarithm for a one-electron process at room temperature?

For reactions with multiple involved species, one can enhance the redox equation from the simple redox couple to

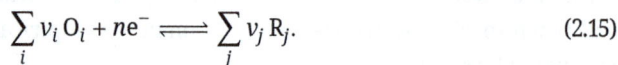

$$\sum_i v_i O_i + ne^- \rightleftharpoons \sum_j v_j R_j. \tag{2.15}$$

Here, v_i and v_j denotes the stoichiometric number of all species i and j. Such a situation requires to formulate the Nernst equation in its general form:

$$E = E^0 + \frac{RT}{nF} \cdot \ln \frac{\prod_i a_{O,i}^{v_i}}{\prod_j a_{R,j}^{v_j}} \tag{2.16}$$

More application-oriented variants of the Nernst equation are the Nikolsky–Eisenman equation in the context of ion-selective electrodes (Section 5.1.3) and the concept of mixed-potentials in corrosion (Section 8.1.1). Related topics are the Donnan potential (Section 5.1.2) at membranes and the Goldman equation when discussing membrane potentials in physiology.

2.2.4 General electrode reaction

While the electron transfer is essential in all cases, it may not be the rate-determining step for an electrode reaction. Zooming out from the electrode/solution interface, we need to consider how O and R come to and away from the interface. Figure 2.3 summarizes the different steps involved in the reduction process. The oxidized species (O) somewhere in the bulk of the electrolyte needs to reach the electrode's microenvironment. Either diffusion, convection, or migration dominates this mass transfer (Section 2.4). In any case, diffusion governs the final distance toward the electrode.

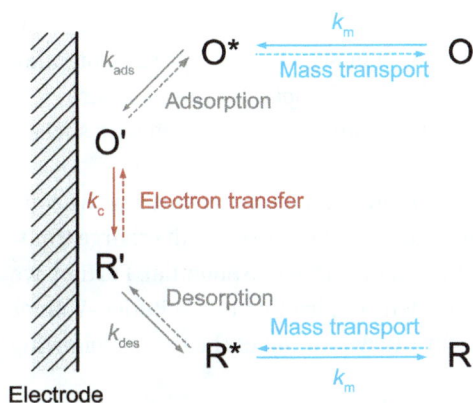

Figure 2.3: General electrode reaction: the solid lines represent the steps of a reduction reaction at the cathode with the charge transfer rate constant k_c. The dashed lines illustrate an oxidation reaction for which the electrode would be the anode. Here, the charge transfer rate constant would be k_a.

The mass transfer coefficient k_m represents the mass transfer from the bulk situation (species O) to the near vicinity of the electrode (species denoted as O*). Before the electron transfer can happen, the O* has to find its way to the electrode surface where it adsorbs labeled with O'. The associated rate constant is k_{ads}. The charge transfer rate constant k_c quantifies the reduction reaction from O' to R'. Afterward, R' desorbs to R* and leaves the electrode's vicinity to the bulk (denoted as R). Figure 2.3 shows the pathway of the reduction process in solid. The dashed lines indicate the oxidation process with the same steps but in reverse order. Here, the rate constant would be k_a.

Often the electron transfer is the rate-determining step compared to the adsorption. Therefore, the typical discussion is about whether the mass transfer or the electron transfer limits the reaction. In the latter case, the Butler–Volmer equation presented in the next section allows the charge transfer description. In some cases, more complex models are required, for example, when solvation plays a role, as in Marcus theory. Both rate constant k_{ads} and k_c (for oxidation k_a) include several substeps itself. From an experimental perspective, those effects are normally not distinguishable, and it is sufficient to express them together as an overall rate constant. This approach might be too coarse when dealing with the exact nature of the interfacial region between electrode and electrolyte (Section 2.6) – the same applies when describing electrode reactions with ab-initio molecular dynamics simulations or for research in catalysis.

2.3 Faradaic processes: kinetic control – electron transfer

So far, we discussed the electron transfer in terms of thermodynamics, what is the general prerequisite (Figure 2.1), and how this is linked to the equilibrium potential apart from standard conditions (Section 2.2.3). Thermodynamics provides us with the information of whether a process can occur at a certain energy level (electrode potential) but cannot predict how fast the process runs and if it happens at all. In Figures 5.25 and 5.26, we see practical examples for this complex interplay interwoven with surface reactions of the electrode material.

Now we come to the kinetic perspective of the electron transfer following the commonly used activated complex theory. With this theory, we can express the charge transfer rate constants k_c and k_a in their dependency on the electrode potential. Further, we will reformulate the rate constants as current (density), leading to the Butler–Volmer equation. We discuss the reduction pathway for the derivation of the theory following the typical approach in the textbooks:

$$O + ne^- \longrightarrow R \tag{2.17}$$

Afterward, we transfer the results to the oxidation and formulate the Butler–Volmer equation comprising both oxidation and reduction.

2.3.1 Activated complex theory

The activated complex or transition state theory assumes that the reduction reaction proceed from the reactant (O) to the product (R) via a transition state, the activated complex. The activated complex represents an energy barrier, and the reaction rate varies depending on how easily it can be overcome. Figure 2.4 shows the Gibbs free energy G along the reaction coordinate. The energy barrier $\Delta G_c^{\ddagger}$ for the reduction (the forward reaction) is lower than the barrier $\Delta G_a^{\ddagger}$ for the oxidation (the backward reaction).

The frequency of overcoming the energy barrier $\Delta G^{\ddagger}$, the rate constant k in $m\,s^{-1}$, follows the Arrhenius equation with a frequency factor A:

$$k = A \cdot \exp -\frac{\Delta G^{\ddagger}}{RT} \tag{2.18}$$

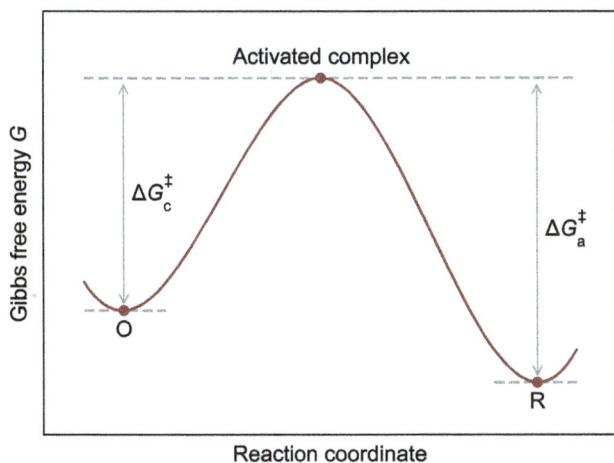

Figure 2.4: Activated complex theory: energy barrier between the reactants and products.

A potential E applied to the electrode changes the Gibbs free energy of O by $-nFE$. To describe the impact of the applied potential on the rate constant requires calculating the energy of the activated complex. Linearization of the energy diagram facilitates the calculation (Figure 2.5). The relationship between a change x in the free energy of O causes a change of $a_c x$ at the activated complex. The factor a_c depends on the symmetry of the energy barrier and is therefore called the *symmetry factor* or *transfer coefficient*.

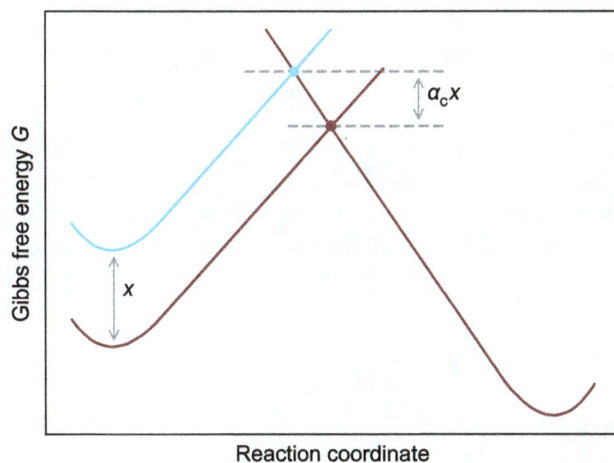

Figure 2.5: Activated complex theory: linearization of the energy barrier around the activated complex. An applied potential causes a change of the reactant's energy by x resulting in a change of the energy of the activated complex by $a_c x$.

How much the energy barrier $\Delta G_c^{\ddagger}$ deviates from the energy barrier in equilibrium $\Delta G_{c,0}^{\ddagger}$ depends on how much the potential E differs from the equilibrium potential E^0:

$$\Delta G_c^{\ddagger} = \Delta G_{c,0}^{\ddagger} + \alpha_c nF(E - E^0) \tag{2.19}$$

For the oxidation reaction the expression for the energy barrier is alike, but please note the difference in the sign:

$$\Delta G_a^{\ddagger} = \Delta G_{a,0}^{\ddagger} - \alpha_a nF(E - E^0) \tag{2.20}$$

α_c is the symmetry factor or transfer coefficient with values between 0 and 1. $\alpha_c = 0.5$ as in Figure 2.5 indicates symmetry of the energy barrier. Higher value of α_c means the activated complex is on the reaction coordinate closer to the reactants.

Insertion of equations (2.19) or (2.20) in equation (2.18) leads to the rate constants for the reduction or oxidation reaction in terms of the applied potential:

$$k_c = k_0 \cdot \exp \frac{-\alpha_c nF(E - E^0)}{RT} \tag{2.21}$$

$$k_a = k_0 \cdot \exp \frac{\alpha_a nF(E - E^0)}{RT} \tag{2.22}$$

The frequency factor and the energy barrier in equilibrium $\Delta G_0^{\ddagger}$ are summarized by k_0. Because in equilibrium ($E = E^0$) the rate constants k_c and k_a are identical, k_0 needs to be the same in both equations.

2.3.2 Butler–Volmer equation

For electrode reactions, the reaction rate v links the species concentration (at the electrode) with the rate constant k:

$$v = kc \tag{2.23}$$

The reaction rate can also be understood as species flux into or from the electrode, allowing to apply Faraday's law (equation (2.5)) to link the reaction rates to the current density:

$$i_c = -nFk_0[O^*] \cdot \exp \frac{-\alpha_c nF(E - E^0)}{RT} \tag{2.24}$$

$$i_a = nFk_0[R^*] \cdot \exp \frac{\alpha_a nF(E - E^0)}{RT} \tag{2.25}$$

The total current of both the reduction and oxidation together

$$i = i_a + i_c \tag{2.26}$$

leads to one variant of the *Butler–Volmer equation* assuming the electron transfer is the rate-determining step for the electrode reaction:

$$i = nFk_0\left([R^*] \cdot \exp\frac{a_a nF(E - E^0)}{RT} - [O^*] \cdot \exp\frac{-a_c nF(E - E^0)}{RT} \right) \tag{2.27}$$

Formally, equation (2.27) should depend on O' and R' (the adsorbed species) rather than O* and R* (Figure 2.3). As the Butler–Volmer equation's derivation does not consider adsorption, this section follows the usual approach to write O* and R* instead.

Several modifications allow for a more handsome expression. In the case of a one-step electron transfer, the transfer coefficients add up to

$$a_a + a_c = 1. \tag{2.28}$$

This relation allows simplifying by defining a single a. Here, a accounts for the reduction; some textbooks, describe it vice versa:

$$a = a_c \quad \text{and} \quad a = 1 - a_a \tag{2.29}$$

For most metals, $a \approx 0.5$ (a symmetrical energy barrier) is a good approximation if the exact value is unknown. For semiconductors electrodes, a approaches 1 (n-type semiconductor) or 0 (p-type semiconductor).

The overpotential η can be understood as the deviation of the electrode potential from the equilibrium:

$$\eta = E - E^0 \tag{2.30}$$

Strictly speaking, the overpotential here is the activation overpotential. For a more detailed discussion of overpotentials, see Section 2.5.2.

The *exchange current density* i_0 combines the concentrations [R] and [O] with the other proportional constants:

$$i_0 = nFk_0[R]^a[O]^{1-a} \tag{2.31}$$

Because of the assumption that the electron transfer is the rate-determining step, the influence by mass transfer or adsorption can be neglected, so [R*] = [R] and [O*] = [O] (compare Figure 2.3 for the definition of the concentrations). The exponents for the concentrations come from the Nernst equation used considering equation (2.27) in equilibrium and bring together the concentrations before the bracket.

The exchange current density is a measure for the ongoing forward and backward reactions, which compensate in equilibrium to obtain a zero net current. Besides its role as a scaling factor in the Butler–Volmer equation, the exchange current density helps to choose an optimal redox system in a reference electrode (Section 4.2) and represents the corrosion rate (Section 8.1.4).

With the modifications above, the Butler–Volmer equation is

$$i = i_0 \cdot \left(\exp \frac{(1-\alpha)nF \cdot \eta}{RT} - \exp \frac{-\alpha nF \cdot \eta}{RT} \right) \tag{2.32}$$

The Butler–Volmer equation is sometimes called the "diode equation of electrochemistry" because of its similar appearance. Figure 2.6 shows plots with a variation of the energy barrier's symmetry (A) and exchange current densities (B).

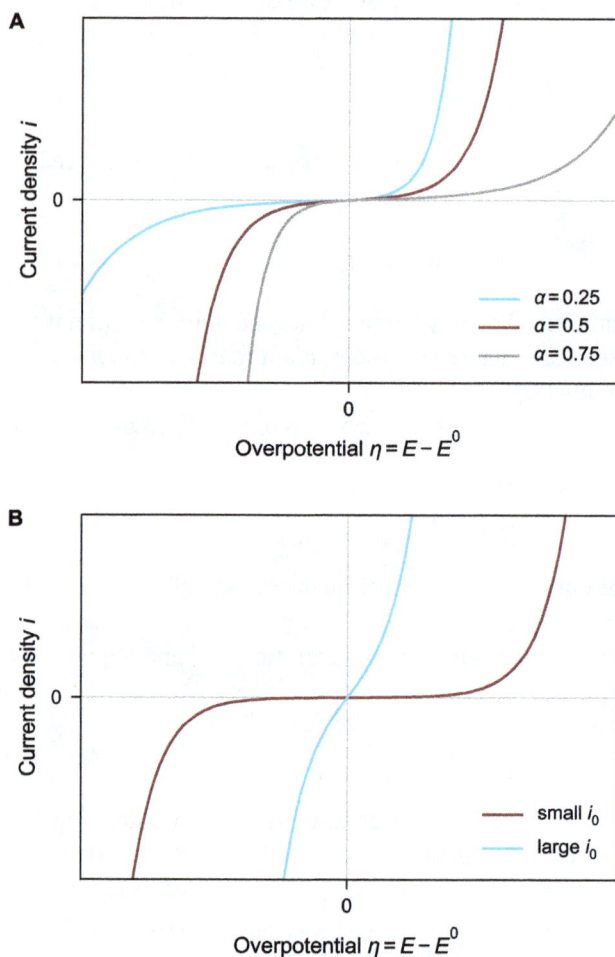

Figure 2.6: Butler–Volmer equation: the symmetry factor (α) prescribes the prevalence of oxidation or reduction reaction (A). The exchange current density (i_0) is a measure of how difficult the electrode gets polarized (B).

Table 2.1 shows some typical exchange current density values. The data indicates the strong dependency on the electrode material. For example, hydrogen hardly gets formed on mercury, while platinum strongly catalyzes the reaction.

Table 2.1: Exchange current density i_0 at the given concentration and the standard exchange current density i_{00} referring to $[O] = [R] = 1$ M. All values are to room temperature (25 °C) [12].

Redox couple	Electrolyte	Electrode	i_0 mA cm^{-2}	i_{00} in mA cm^{-2}
5 mM Fe^{3+}/Fe^{2+}	1 M HClO$_4$	Pt	2	4×10^2
10 mM Fe(CN)$_6$$^{3-}$/Fe(CN)$_6$$^{4-}$	0.5 M K$_2$SO$_4$	Pt	50	5×10^3
Ag/1 mM Ag$^+$	1 M HClO$_4$	Ag	150	13×10^3
H$_2$/H$^+$	1 M H$_2$SO$_4$	Hg	1×10^{-9}	1×10^{-9}
H$_2$/H$^+$	1 M H$_2$SO$_4$	Pt	1	1
H$_2$/OH$^-$	1 M KOH	Pt	1	1
O$_2$/H$^+$	1 M H$_2$SO$_4$	Pt	1×10^{-3}	1×10^{-3}
O$_2$/OH$^-$	1 M KOH	Pt	1×10^{-3}	1×10^{-3}

2.3.3 Charge-transfer close to equilibrium: linearization

The Butler–Volmer equation allows for further simplifications. Close to the equilibrium, that is low overpotential, linearization is possible:

$$i = i_0 \frac{nF}{RT} \cdot \eta \tag{2.33}$$

The current-potential relationship is now similar to Ohm's law; this suggests defining a charge-transfer resistance R_{ct} (called polarization resistance in context of corrosion analysis) accordingly:

$$R_{ct} = \frac{RT}{nFi_0} \tag{2.34}$$

The resistance becomes smaller for higher exchange current density. Chapter 7 provides a more rigorous discussion on applying a resistor to model electrochemical reactions by equivalent circuits.

■□□ Task 2.3: Derive the approximation for the charge transfer close to equilibrium (equation (2.34)). Start with the Butler–Volmer equation (equation (2.32)) and consider the relationship $\exp x \approx 1 + x$ for small x.

2.3.4 Charge-transfer far from equilibrium: Tafel plot

Another approximation is for large overpotentials. At strongly negative overpotentials, the reduction is dominant and equation (2.32) reduces to

$$i = -i_0 \cdot \exp \frac{-\alpha nF \cdot \eta}{RT} \qquad (2.35)$$

At very high overpotentials, oxidation prevails, and equation (2.32) reduces to

$$i = i_0 \cdot \exp \frac{(1 - \alpha)nF \cdot \eta}{RT} \qquad (2.36)$$

Rewriting the above equations and taking account of their exponential nature leads to the *Tafel equations*:

$$\ln |i| = \frac{-\alpha nF}{RT} \cdot \eta + \ln i_0 \qquad (2.37)$$

$$\ln |i| = \frac{(1 - \alpha)nF}{RT} \cdot \eta + \ln i_0 \qquad (2.38)$$

The absolute value avoids the logarithm of a negative number as it would be the case for a reduction current. The equations already suggest plotting the Butler–Volmer equation in a $\ln |i|$ versus η diagram, the so-called *Tafel plot* (Figure 2.7). The red curve represents the complete Butler–Volmer equation. The dashed grey curves are the Tafel equations which intersect at $\ln i_0$.

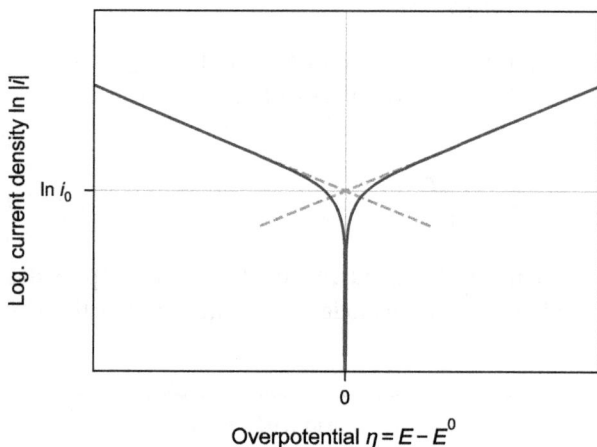

Figure 2.7: Tafel plot with the overpotential on the *x*-axis. For potentials far from equilibrium, the current in logarithmic presentation becomes linear. Extrapolation of the two linear regions leads to the exchange current density.

Tafel plots became an essential tool in corrosion analysis. Even several electrochemical instrumentation programs allow showing the measured data live as a Tafel plot.

Initially, Julius Tafel (1862–1918) used a presentation with the overpotential versus the current density and formulated an empirical equation with the decadic logarithm to interpret his data [15]:

$$\eta = \text{const} + b \lg |i| \tag{2.39}$$

Figure 2.8 shows a Tafel plot with the overpotential on the y-axis. b is what we today call the *Tafel slope*.

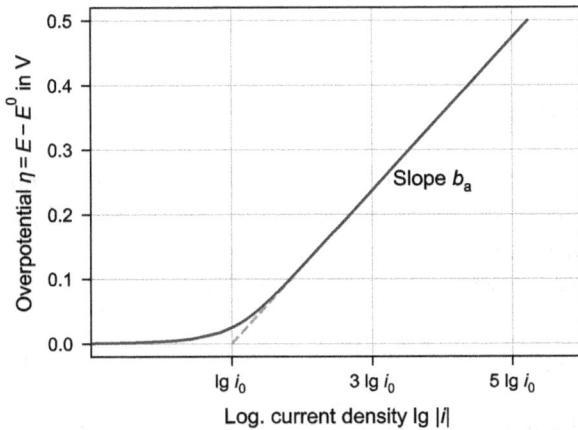

Figure 2.8: Tafel plot with the overpotential on the y-axis for an oxidation process. The voltage scales is for a one-electron process with $\alpha_a = 0.5$ at room temperature.

The Tafel slopes for the reduction and oxidation side are

$$b_c = -\frac{2.303RT}{\alpha_c nF} \quad \text{and} \quad b_a = \frac{2.303RT}{\alpha_a nF} \tag{2.40}$$

In the case of a one-electron process and $\alpha_a = 0.5$, the Tafel slope for the oxidation reaction is $b_a = 118\,\text{mV}$ per decade change in current density at room temperature. It is essential to always clarify whether the decadic or natural logarithm was used when reporting the slopes.

2.4 Faradaic processes: mass transfer control

While in the previous section, the focus was on electron transfer, this section deals with the mass transfer of the electroactive species. Three different mechanisms may play a role:

– Diffusion, the movement of species along a concentration gradient
– Migration, the movement of charged species in an electrical field (potential gradient)
– Convection, the movement of species by fluid flow

2.4.1 Diffusion

A system with different local concentrations of species tends to reach the state of equal distribution. The driving force due to a concentration gradient superimposes the arbitrary thermal movement of any particle. This process is called diffusion. Under stationary conditions (time-invariance), *Fick's first law* describes the resulting species flux, here written for a one-dimensional situation:

$$j(x) = -D \cdot \frac{\partial c(x)}{\partial x}$$

(2.41)

The species flux density j is inverse proportional to the concentration gradient. The minus sign is because of the tendency of the system to reach equal distribution. D, the proportional constant, is the diffusion constant.

A time-variant situation requires description by *Fick's second law*:

$$\frac{\partial c(x,t)}{\partial t} = D \cdot \frac{\partial^2 c(x,t)}{\partial x^2}$$

(2.42)

Fick's second law is a second-order linear partial differential equation, a form of equation called the Poisson equation. There is no general analytical solution available asking for simulation tools to describe the diffusion situation in front of an electrode (any FEM software can do this). For certain boundary conditions, analytic solutions exist, e. g., leading to the Cottrell equation (equation (2.56), task 2.4).

The diffusion constant can be approximated by the Stokes–Einstein equation assuming spherical particles (radius r):

$$D = \frac{k_B T}{6\pi \eta r}$$

(2.43)

The diffusion constant not only depends on the species itself but also the medium's viscosity η, and temperature. k_B is the Boltzmann constant.

In general, the approximation of spherical species is not always appropriate, especially when discussing molecules. Here, empirical data plays an essential role. Sections 5.2.2, 6.1.1, and 6.2.1 describe methods to measure diffusion constants.

2.4.2 Migration

Electrostatic forces on charged species in an electrical field cause a species flux depending on the species' charge and the properties of the surrounding medium:

$$j(x) = -\frac{zFDc}{RT} \cdot \frac{\partial \phi(x)}{\partial x} \tag{2.44}$$

The valency z accounts for the species' charge; the diffusion constant D describes the species' ability to move in the specific medium. Please note the usage of the electrostatic potential ϕ in the electrical field's description to not confuse with the electrode potential E.

Typically, the electroneutrality approximation is used here, which means no significant charge separation occurs. Focus on tiny dimensions (nm) or small timescales (ns) may require a more rigorous treatment by numerical simulation methods.

The *Debye length* helps to judge the applicability of the electroneutrality approximation. The Deybe length λ_D is an estimate for the distance when the electrolyte screens a charged species' electrical field:

$$\lambda_D = \sqrt{\frac{\epsilon_r \epsilon_0 k_B T}{N_A q_e^2 \sum_i z_i^2 c_i}} \tag{2.45}$$

All charged species in the electrolyte contribute with their valency z_i and c_i. ϵ_r is the relative permittivity of the solvent, ϵ_0 the permittivity of free space. For example, at room temperature, the Debye length of an aqueous 0.1 M KCl solution is 1 nm. That means for distances larger than 1 nm electroneutrality can be assumed.

2.4.3 Convection

Convection means the movement of the species of interest together with the surrounding medium. That is in contrast to diffusion and migration, for which the species is moving within the medium. Generally, the species flux due to convection in a fluid with flow velocity $v(x)$ is

$$j(x) = c \cdot v(x) \tag{2.46}$$

The flow velocity has the same dimension as the mass transfer constant and is equivalent to it in mass transfer dominated by convection.

Convection can have two origins: in thermal convection, fluid moves driven by density gradients caused by thermal gradients; applying external fluid flow results in forced convection. Thermal convection is typically unwanted and preferably avoided. In microsystems, thermal convection usually is negligible, while the diffusion situation in macroscopic measurement cells often suffers from it. Forced convection provides sophisticated mass transfer control and enables several electrochemical methods (Section 6.1). However, for all those systems, the transfer directly at the electrode is governed by diffusion only.

2.4.4 Nernst–Planck equation

The Nernst–Planck equation describes the combination of the different transport mechanisms (diffusion, migration, and convection). For simplicity, we consider the flux of a species i in a one-dimensional case assuming stationary conditions:

$$j_i(x) = \underbrace{-D_i \cdot \frac{\partial c_i(x)}{\partial x}}_{\text{Diffusion}} \underbrace{- \frac{z_i F D_i c_i}{RT} \cdot \frac{\partial \phi(x)}{\partial x}}_{\text{Migration}} + \underbrace{c_i \cdot v(x)}_{\text{Convection}} \tag{2.47}$$

The formulation in terms of species flux density can be linked to the mass transfer coefficient $k_{m\,i}$ as depicted in Figure 2.3 by normalization with the species' concentration:

$$k_{m\,i} = \frac{j_i}{c_i} \tag{2.48}$$

Most electrochemical experiments aim to focus on only one mechanism of mass transfer. Stirring, fluid flow at a rotating electrode (Section 6.1.1) or flow in a microchannel causes domination of convection with the species flux depending on the fluid's velocity only.

Small electrolyte volumes, no mechanical disturbance of the setup, or even gelification of the electrolyte can help avoid convection. Thermal equilibration in a temperature box helps to reduce thermal convection but at the same time might introduce forced convection by vibration due to a fan.

Lowering the electrolyte's resistance reduces the electrical field strength and, accordingly, migration. The electrolyte solution's concentration should be much greater than the electroactive species' one, as is the case in many analytic applications (millimolar and below analyte concentration in an electrolyte with up to a molar salt concentration). An example with high concentrations of the electroactive species in which migration and diffusion plays an important role is electroplating (Figure 2.9).

2.4.5 Diffusion: constant polarization

A common situation is diffusion toward an electrode under constant polarization. Microsensors often have a diffusion-limiting membrane in front of the electrode, and the electrode is strongly polarized so that the electrode reaction consumes all analyte

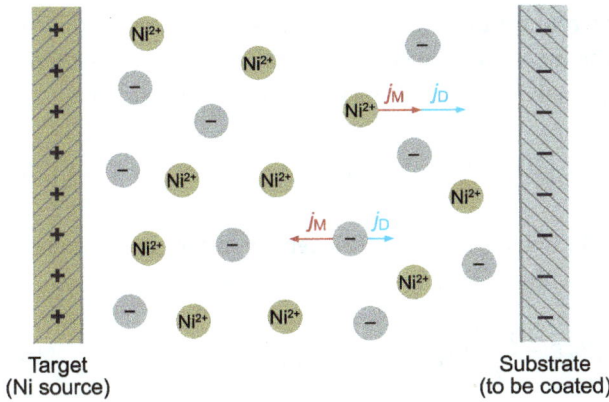

Figure 2.9: Example of an unstirred nickel plating bath. The substrate is the cathode at which nickel deposits from Ni^{2+} ions in the solution. At the nickel anode (the so-called nickel target), Ni^{2+} ions dissolve and refresh the plating bath. The Ni^{2+} ions move toward the substrate by both migration j_M (the negative electrode attracts positive particles) and diffusion j_D (the deposition causes a depletion in front of the electrode and, therefore, a concentration gradient).

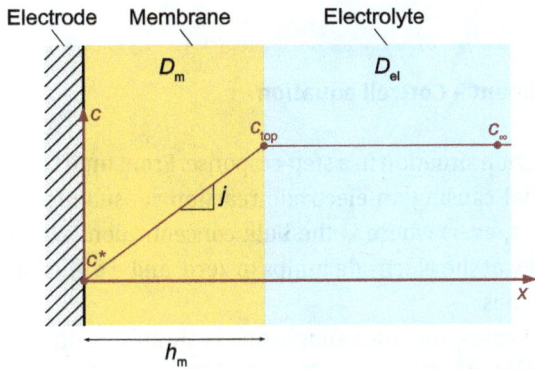

Figure 2.10: Diffusion situation in front of a membrane-covered electrode, assuming strong enough polarization that the electrode reaction consumes all the considered species. The slope of the concentration drop within the membrane is the species flux density j.

molecules. Figure 2.10 illustrates this situation: a membrane (layer height h_m) covers the electrode.

Fick's first law (equation (2.41)) allows to describe the stationary diffusion situation:

$$j = -D_m \cdot \frac{c_{top} - c^*}{h_m} \tag{2.49}$$

The analyte concentration c^* at the electrode is zero assuming strong enough polarization. The diffusion constant of the analyte in the membrane (D_m) is much lower than the diffusion constant in the surrounding electrolyte (D_{el}). Therefore, most of the concentration gradient is within the membrane layer, and the concentration at the top of the membrane (c_{top}) is approximately the same as the bulk concentration (c_∞):

$$j = -D_m \cdot \frac{c_\infty}{h_m} \tag{2.50}$$

Faraday's law (equation (2.5)) links the species flux density to the current density leading to a simple linear relationship:

$$i = zF(-j) = \frac{zFD_m}{h_m} c_\infty = \text{const.} \cdot c_\infty \tag{2.51}$$

The species flux appears with a negative sign as species leave the system (are consumed by the electrode reaction). Equation (2.51) is the transfer function when such a membrane-covered electrode is part of an amperometric microsensor (Section 5.2). The sensor current depends on the membrane properties only. Additionally, such an arrangement is not bothered by convection in the electrolyte and, e. g., allows for flow-rate independent sensors.

2.4.6 Diffusion: potential step experiment – Cottrell equation

Another important scenario is the diffusion situation in a step-response. From time zero, the electrode is polarized at a potential causing an electrode reaction consuming all the considered species ($c^* = 0$). Before, everywhere is the bulk concentration. When switching on, the concentration directly at the electrode jumps to zero, and a diffusion profile in front of the electrode establishes.

Figure 2.11 shows the three typical cases: the most simple one is an infinite planar electrode (A), at which a linear diffusion profile forms. The hemispherical electrode (B) causes a spherical diffusion profile, and a disk electrode (C) results in a situation

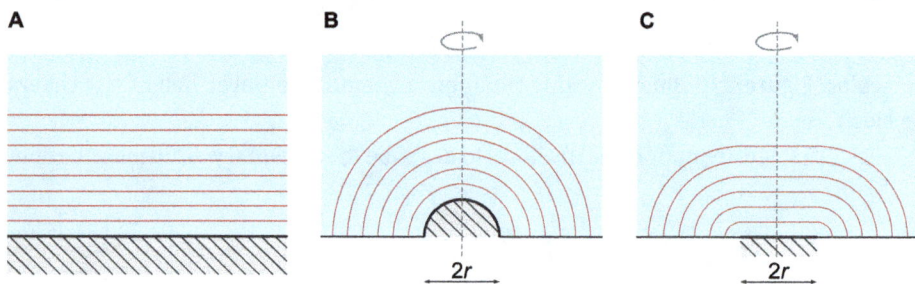

Figure 2.11: Diffusion profiles: planar electrode (A), hemispherical electrode (B), and disk electrode (C). The red lines are levels of equal concentration dropping toward the electrode.

in between. For large times or small diameter, the disc electrode's diffusion profile approaches the hemispherical case.

Infinite planar electrode
An infinite planar electrode is the most straightforward case. The diffusion occurs in one dimension only, along an axis x as the distance from the electrode. The time-dependent diffusion situation requires a solution to Fick's second law. The initial and boundary conditions are:

-   Initially, there is equal distribution of the species of interest with its bulk concentration c_∞.
-   After switching on the electrode potential, the concentration directly a the electrode c^* is zero.
-   Far from the electrode, there is an exhaustless source for the species with its bulk concentration c_∞.

Solving equation (2.42) (task 2.4) leads to the time-dependent concentration distribution:

$$c(x, t) = c_\infty \mathrm{erf}\left(\frac{x}{2\sqrt{Dt}} \right) \tag{2.52}$$

Figure 2.12 visualizes the solution with typical values. Plot A shows the decay of the concentration after switching on at specific distances from the electrode surface. In plot B, the concentration directly at the electrode is zero, while the gradient is propagating with time from the electrode.

Please note how fast the major part of the concentration drop moves away from the electrode surface. While in macroscopic experiments, a few hundred micrometers are not "far," in a microsystem, it can be more than the total electrolyte space. That means the assumption of a large electrolyte volume in front of the infinite planar electrode is often not applicable in microsystems.

The diffusion layer thickness δ is a way to characterize the diffusion distribution. It is understood as the zone in the electrode's neighborhood where the diffusion gradient drops, while the outside is assumed to be undisturbed. This zone is also called the *Nernst diffusion layer*. Technically, the diffusion layer thickness is defined by

$$\delta = \sqrt{Dt}. \tag{2.53}$$

From equation (2.52), one can see that for $x = \delta$ the concentration drops to 52 % of the bulk concentration. So the idea that most of the diffusion gradient drops within the diffusion layer is misleading. That is more realistic for $x = 3\delta$ where the concentration is 97 % of c_∞. Figure 2.13 visualizes the diffusion layer thickness for different diffusion constant. Typical diffusion constants for gases in water are between the two values.

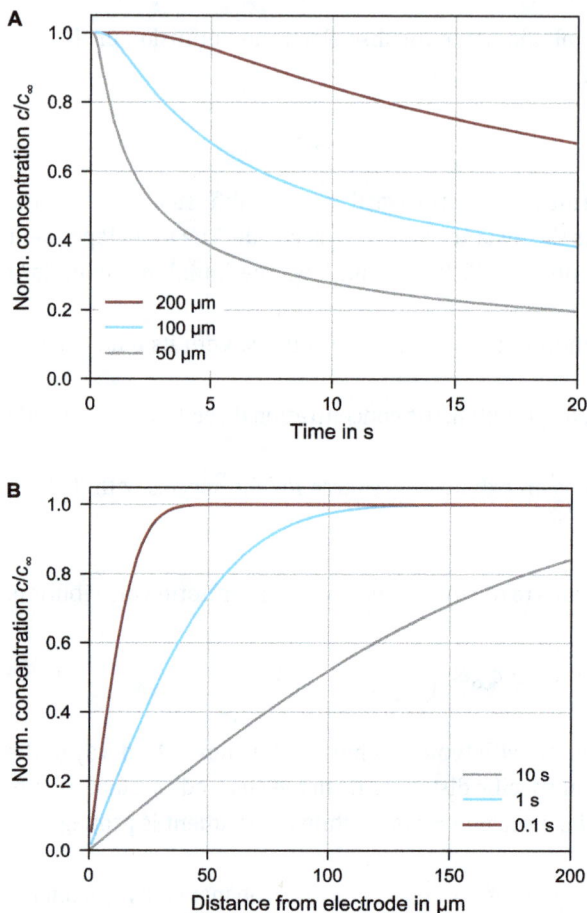

Figure 2.12: Time-dependent concentration distribution for a step-response at an infinite planar electrode. The concentration is normalized to the bulk concentration c_∞. The numbers are for $D = 1 \times 10^{-5}\,\mathrm{cm^2\,s^{-1}}$.

The species flux "into" the electrode, that means the species consumed by the electrode reaction, is the derivative of the concentration distribution equation (2.52) at the electrode ($x = 0$):

$$j(t) = D\frac{\partial c(x,t)}{\partial x}\bigg|_{x=0} \tag{2.54}$$

Calculating the derivation and insertion into Faraday's law equation (2.5) leads to the Cottrell equation:

$$i(t) = zFc_\infty\sqrt{\frac{D}{\pi t}} \tag{2.55}$$

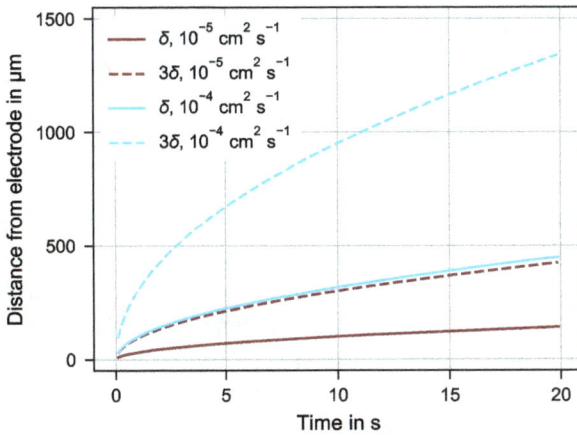

Figure 2.13: Diffusion layer thickness δ for a step-response at an infinite planar electrode. The concentration is 52 % of the bulk concentration at δ and 97 % at 3δ. The colors are for different diffusion constants.

The current density drops with the inverse of the square root of time. For longer times, the current approaches zero.

■■■ Task 2.4: Derive the Cottrell equation (equation (2.55)) for infinite planar electrodes (linear diffusion).
Use the coordinate x as distance from the electrode.
- Initial condition: $c(x, 0) = c_\infty$
- Boundary condition at the electrode: $c(0, t) = 0$
- Boundary condition (infinite source): $x \rightarrow \infty : c(x, t) = c_\infty$

The derivation requires two steps (the first one is more tricky):
1. Use Fick's second law (equation (2.42)) and find a solution for $c(x, t)$. Preferably use the Laplace transformation to solve the differential equation. Hint: it is important to find an appropriate substitution for $c(x, t)$. You should end up with equation (2.52).
2. Use Faradays law (equation (2.5)) and derive the Cottrell equation. Take into account the species flux into the electrode (equation (2.54)).

Hemispherical electrode
The situation at a hemispherical electrode deviates substantially from the planar one, as shown when comparing Figure 2.11(A) and (B). With progressing time, the newly accessed diffusion plane's surface is the same as the electrode area in the case of a planar electrode, while the spherical diffusion profile results in a continuous increase. An additional radial term in the Cottrell equation for the hemispherical electrode reflects this increase:

$$i(t) = zFc_\infty \left(\sqrt{\frac{D}{\pi t}} + \frac{D}{r} \right) \tag{2.56}$$

In contrast to the planar situation, the current density reaches steady-state for longer times ($t \gg \frac{r^2}{\pi D}$):

$$i_{ss} = \frac{zFc_{\infty}D}{r} \tag{2.57}$$

Figure 2.14 compares the current density response for an infinite planar and a hemispherical electrode. The numbers are for a two-electron oxidation process with the species' diffusion constant of 1×10^{-5} cm^2 s^{-1}. The diameter of the hemispherical electrode is 50 µm. The hemispherical solution is the planar solution offset by the steady-state current density. For a larger diameter, the hemispherical curve would resemble more the values of the planar one.

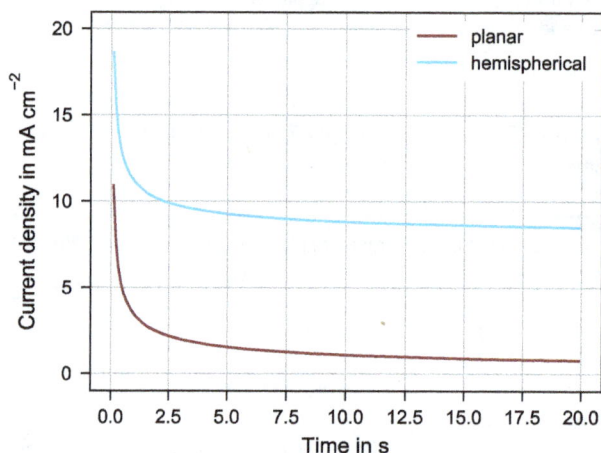

Figure 2.14: Cottrell equation for an infinite planar and a hemispherical electrode ($D = 1 \times 10^{-5}$ cm^2 s^{-1}, $z = 2$, $c = 1$ mM, for the hemispherical electrode: $r = 25$ µm).

Disk-shaped electrode

For disk-shaped electrodes, no exact analytical solution exists. Shoup and Szabo [16] suggested a widely accepted semiempirical equation that matches very well the experimental data:

$$i(t) = zFc_{\infty}\frac{4D}{\pi r} \cdot f(\mathcal{T}) \tag{2.58}$$

The approximation function is formulated using the dimensionless time $\mathcal{T}$:

$$f(\mathcal{T}) = 0.7854 + \frac{0.8862}{\sqrt{\mathcal{T}}} + 0.2146 \cdot \exp\left(\frac{-0.7823}{\sqrt{\mathcal{T}}}\right) \quad \text{with } \mathcal{T} = \frac{4Dt}{r^2} \tag{2.59}$$

In the steady-state, the current density i_{ss} resembles that of hemispherical electrodes:

$$i_{ss} = \frac{zFc_\infty D}{r} \tag{2.60}$$

However, conversion to steady-state current (not current density) results in different equations as the disk's electrode area is πr^2 compared to $2\pi r^2$ for the hemispherical electrode.

Usage of piecewise-defined equations allow for even higher accuracy, as discussed by Mahon and Oldham [17].

■■□ Task 2.5: Diffusion constant measurements by a step-response (Cottrell equation):
- You have an infinite planar electrode and record the current response after switching on to a potential at which you can oxidize two fictive substances A and B.
- You do this experiment for both substances. You wait 2 seconds after switching on and read the current value. For substance B, you get double the amount of current as for A.
- The diffusion constant of A is 1×10^{-5} cm^2 s^{-1}. What is the diffusion constant of B?

2.5 Faradaic processes: combined kinetic and mass transfer control

In the previous sections, the theory appears as electrochemical systems follow either kinetic limitations (the electron transfer is the rate-determining step) or mass transfer limitation. Unfortunately, most real situations are somewhere in between – for small overpotential following the Butler–Volmer equation (equation (2.32)) while reaching a limiting current i_l when the overpotentials are high enough that mass-transfer limitation kicks in. An enhancement of the Butler–Volmer equation considers this effect:

$$i = i_0 \cdot \left(\left(1 - \frac{i}{i_{l,a}}\right) \cdot \exp \frac{(1-\alpha)nF \cdot \eta}{RT} - \left(1 - \frac{i}{i_{l,c}}\right) \cdot \exp \frac{-\alpha nF \cdot \eta}{RT} \right) \tag{2.61}$$

Figure 2.15 shows different cases for this equation comparing different ratios between limiting and exchange current density. Please note that the limiting current density is used to normalize all curves. Figure 2.16 brings the case of equation (2.61) together with the situation without mass transfer limitation (equation (2.32)).

2.5.1 Reversibility

The term reversibility has a different meaning depending on the subject. Unfortunately, most deviate from the common sense understanding based on common language.

Thermodynamic reversibility requires thermodynamic equilibrium with its surrounding. When reversed, an ideal thermodynamically reversible process would leave

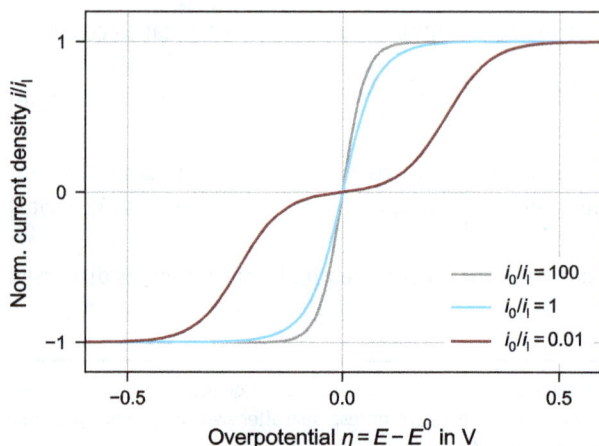

Figure 2.15: Butler–Volmer equation considering the influence of the mass transfer: current density normalized by the limiting current density i_l for different ratios between limiting and exchange current density. Assumptions: $a = 0.5$, a one-electron process, $i_l = i_{l,a} = -i_{l,c}$, and room temperature.

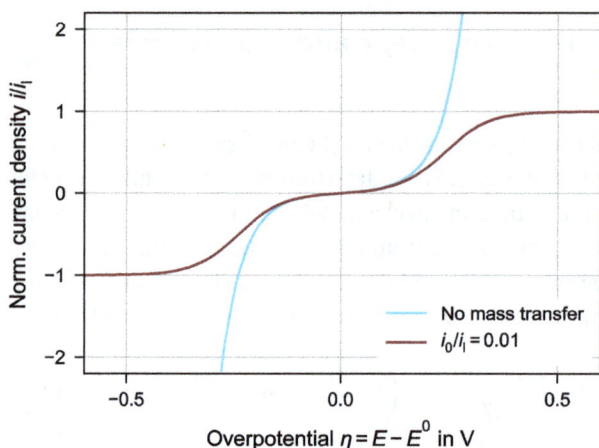

Figure 2.16: Butler–Volmer equation considering the influence of the mass transfer and comparison to the situation without mass-transfer limitation (B). Assumptions: $a = 0.5$, a one-electron process, $i_l = i_{l,a} = -i_{l,c}$, and room temperature.

its environment without any change. Complete thermodynamic reversibility exists only as an approximation. In chemistry, a reaction is reversible if the forward and reverse pathways can proceed simultaneously. Chemical reversibility applied to a redox equation means that the involved species form a redox couple. Electrochemical reversibility considers the rate of electron transfer compared to the mass transfer.

Electrochemical reversible systems

The electron transfer rate is high (a low energy barrier for the activated complex, Figure 2.4), resulting in equilibrium directly a the electrode. In terms of rate constants, a reversible system is characterized by $k_0 \gg k_m$. The standard rate constant k_0 representing the electron transfer is high compared to the rate constant for the mass transfer k_m, which means the electrode reaction is much faster than the mass transfer.

Electrochemical irreversible systems

Here, the electron transfer rate is low (a high energy barrier for the activated complex). Only by applying high overpotentials, the reaction can reach mass transfer control. In terms of rate constants, an irreversible system is characterized by $k_0 \ll k_m$.

We have a more rigorous discussion on electrochemical reversibility and experimental possibilities to determine it with cyclic voltammetry in Section 5.3.2.

2.5.2 Overpotential

In Section 2.3.2, we used the overpotential η as a shortcut for $E - E^0$ (equation (2.30)), the deviation from the thermodynamic equilibrium. In a more general understanding, the overpotential describes the difference between the potential at which the reaction takes place experimentally and the thermodynamic value. Overpotentials are an alternative approach to formulate both kinetic and mass transfer control. Additionally, overpotentials can represent several factors appearing as a loss to the applied voltage.

Activation overpotential

The Butler–Volmer equation describes the current density of a system under kinetic control as a function of the applied overpotential. Inversion of equation (2.36) leads to the *activation overpotential* (for an oxidation process):

$$\eta_{act} = \frac{RT}{(1-\alpha)nF} \cdot \ln\frac{i}{i_0} \tag{2.62}$$

This equation also resembles the Tafel equation (equation (2.39)) and represents the applied overpotential required for the reaction's activation.

Concentration overpotential

A system influenced by mass transfer control requires more driving voltage as the closer the current reaches the limiting current. Taking the oxidation side of the enhanced

Butler–Volmer equation (equation (2.61)) and inversion leads to

$$\eta = \frac{RT}{(1-\alpha)nF} \cdot \ln\frac{i}{i_0} + \frac{RT}{(1-\alpha)nF} \cdot \ln\left(1-\frac{i}{i_1}\right) \tag{2.63}$$

Additional to the activation overpotential, a second term representing the mass transfer shows up and is often called *concentration overpotential*:

$$\eta_{\text{conc}} = \frac{RT}{(1-\alpha)nF} \cdot \ln\left(1-\frac{i}{i_1}\right) \tag{2.64}$$

Resistive overpotential

Additionally, ohmic losses, mainly due to the electrolyte resistivity, reduce the effect of the applied potential. This loss is called *resistive overpotential* or IR drop, depending on the resistance R and the current I. Later on (Section 3.2.4), we see that this resistance in a three-electrode setup is the uncompensated resistance R_{u}.

The overall overpotential taking into account the charge transfer, the mass transfer, and the IR drop is

$$\eta = \eta_{\text{act}} + \eta_{\text{conc}} + IR \tag{2.65}$$

Figure 2.17 shows the combined overpotentials occurring in an electrolytic cell following equation (2.65). In a galvanic cell, the overpotentials appear as losses lowering the cell voltage predicted by thermodynamics (Figure 2.18).

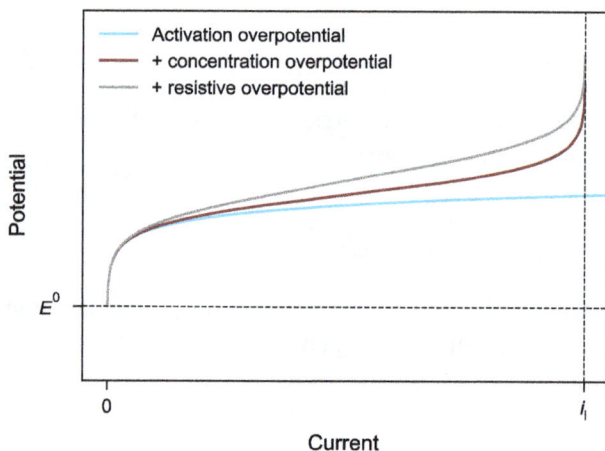

Figure 2.17: Overpotentials for an oxidation process in an electrolytic cell. The plot is similar to the oxidation side of Figure 2.16 with the axes exchanged.

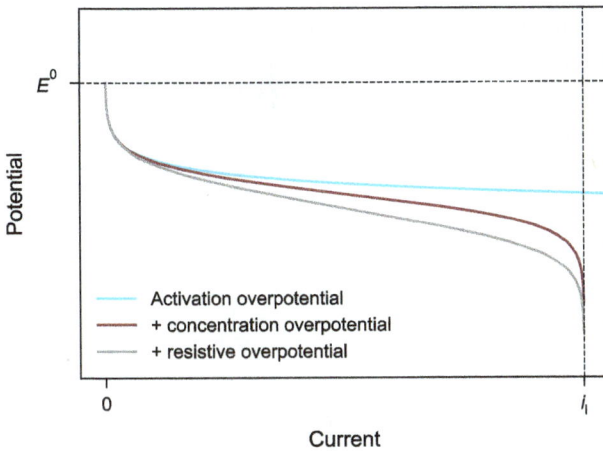

Figure 2.18: Overpotentials for an oxidation process in a galvanic cell.

2.6 Interfacial region

Until here, we concentrated on global quantities describing the electrochemical phenomena in terms of thermodynamics, reaction kinetics, and mass transfer. Often this perspective is most beneficial to understand and analyze electrochemical methods. However, some effects are hard to explain by this bigger picture, especially when discussing micro- and nanosystems. This section zooms closer into the interface between electrode and electrolyte down to the molecular level.

Figure 2.19 illustrates the regions of interest for the effects at the interface. In the electrode, this region is the space charge region; in the electrolyte, it is the electrical double layer (EDL). For a metal electrode, the space-charge region is, for most cases, of low relevance, while in semiconductor electrodes, it is the key to understand its function. The EDL is always essential to consider.

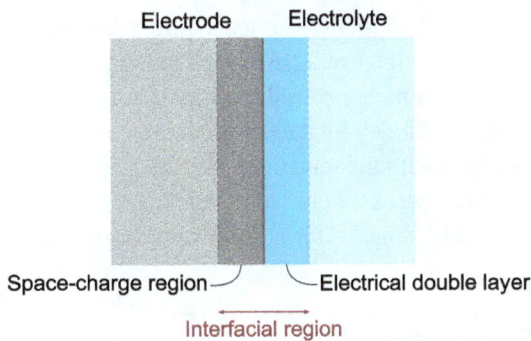

Figure 2.19: Regions of interest for effects at the electrode-electrolyte interface.

The potential curve along the interface is, in a rough approximation, a voltage step (Figure 2.20). The potential on the electrode side is called the *Galvani potential* ϕ (sometimes: inner potential). In the electrolyte, next to the interface, it is the *Volta potential* ψ (sometimes: outer potential). The difference between the Galvani and Volta potential is called the *surface potential* χ.

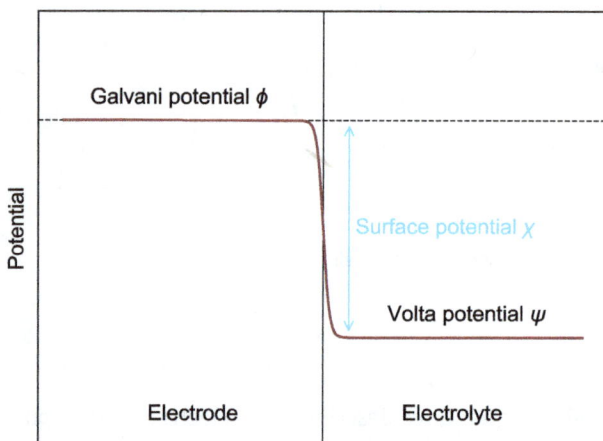

Figure 2.20: The potential curve at the electrode-electrolyte interface for a positively charged electrode. Please be aware that some other textbooks define the potentials differently or even contradictory. Always double-check the exact understanding of the terms that occur in a scientific publication.

2.6.1 Electrical double layer

The term electrical double layer goes back to the early model of Helmholtz. He assumed a fixed layer of adsorbed ions with an opposite charge in front of the electrode (Figure 2.21A). With the electrode, this layer forms a parallel-plate capacitor. Its capacity is independent of the electrode potential and electrolyte concentration.

More elaborate models came up: Gouy and Chapman assumed a diffuse layer comprising potential and concentration dependency (Figure 2.21B). Stern fused the initial Helmholtz model with the Gouy–Chapman (GC) model considering both a fixed layer directly at the electrode and a diffuse layer. Grahame's further improvements were the distinguishment of specifically adsorbed ions without solvation, the inner Helmholtz plane (IHP), and solvated ions with opposite charge as the electrode, the outer Helmholtz plane (OHP). Figure 2.21C illustrates the combined model for a negatively charged electrode.

The IHP and OHP assume a compact layer with point charges. To get an idea of the dimensions, we can look at a negatively charged electrode in saltwater: The center of the IHP is one ion radius away from the electrode (0.1 nm for sodium). In this example,

A Helmholtz   **B** Gouy-Chapman   **C** Stern / Grahame

IHP OHP

⊕ Cation   ⊖ Anion   ◯ Water molecule

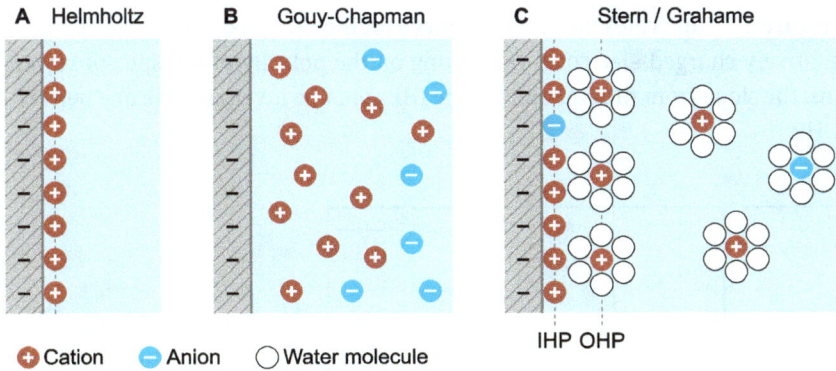

Figure 2.21: Electrical double layer models: the Helmholtz model (A) with a compact layer and the Gouy–Chapman model (B) with a diffuse layer were brought together by Stern and Grahame (C). The specifically adsorbed ions form the inner Helmholtz layer (IHP), and solvated ions form the outer Helmholtz layer (OHP). A diffuse layer of solvated ions follows the Helmholtz layers.

the OHP would be around 0.6 nm from the electrode (the hydrated sodium ion's radius is 0.4 nm). Typically, the compact layer is below 1 nm, followed by a diffuse layer in the range of up to 10 nm (depending on the ion strength).

As a rule of thumb, using the Debye length (equation (2.45)), the thickness of the electrical double layer is $1.5 \cdot \Lambda_D$. Figure 2.22 shows the typical range in an aqueous KCl solution.

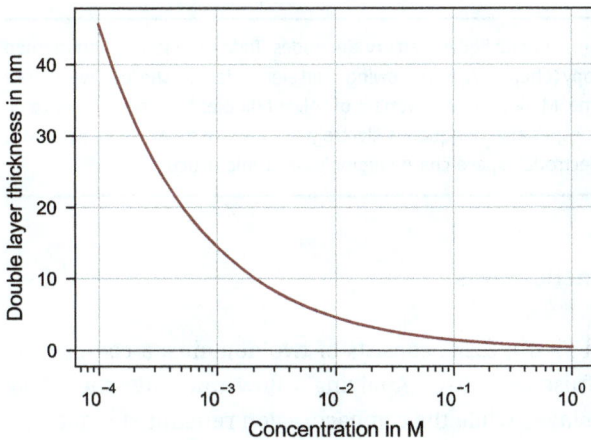

Figure 2.22: Thickness of the electrical double layer, approximated using the Debye length ($1.5 \cdot \Lambda_D$, $n = 1$, ϵ_r based on KCl solutions).

The potential profile through the interfacial region strongly depends on the applied model. In reality, the "true" curve shape is highly decisive for a reaction mechanism's specifics. Figure 2.23 illustrates the potential curve for the EDL model as in Figure 2.21(C) with a negatively charged electrode. Depending on the polarity of the specifically adsorbed ions, the slope from the electrode to the IHP could be inverse to the one between IHP and OHP.

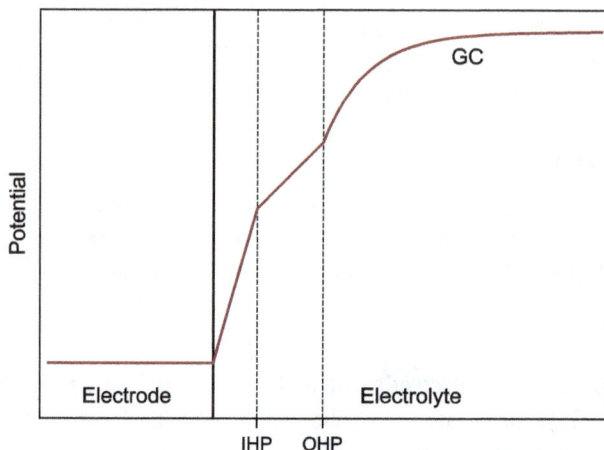

Figure 2.23: Potential in an EDL model as in Figure 2.21C with a negatively charged electrode. The diffuse layer, according to Gouy–Chapman (GC), follows the inner Helmholtz plane (IHP) and the outer Helmholtz plane (OHP).

ℹ️ All of these models originate in measurements with liquid mercury electrodes. Today, the state-of-art method is electrochemical impedance spectroscopy (Chapter 7), also allowing solid electrodes. On the level we discuss here, the overall picture remains the same. More precise treatment of solvent dipoles and their influence on the electrical field improves the models, especially for aqueous electrolytes. In more detail, solid electrodes may also require consideration of the electrode's space-charge region's electronic structure.

2.6.2 Zeta potential – electrokinetic effects

The EDL model, as presented in Figure 2.23C, consists of two domains: a compact region (IHP and OHP) and the diffuse layer (GC). Applying a flow can move the whole electrolyte, including the diffuse layer, while the compact region remains stationary. A shear plane separates the two areas. As the solvated ions' center defines the OHP, the shear plane is slightly further away from the electrode than the OHP (Figure 2.24). The potential at this shear plane is called the *zeta potential* ζ.

The zeta potential plays an essential role in electrokinetic effects. Superimposing a flow causes charge separation and, therefore, the formation of the so-called streaming potential. Inversely, applying an electrical field to the electrolyte (vertical in Figure 2.24)

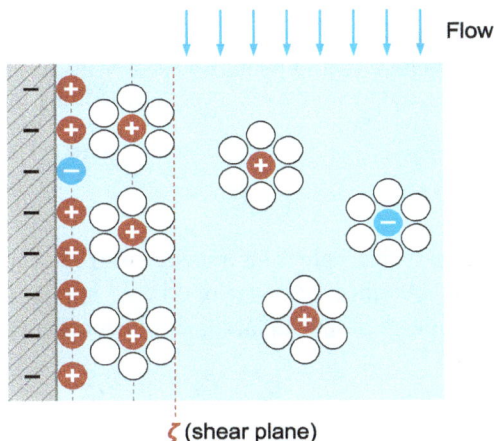

Figure 2.24: A flow perpendicular to the electrode's surface causes electrolyte movement separated from the stationary layers next to the electrode by a shear plane. The potential at this plane is the zeta potential ζ.

causes fluid movement (*electroosmosis*). The streaming potential principle (or depending on the measurement setup: streaming current) allows measurement of the solid material's surface properties. Those zeta potential measurements also allow measurements with colloids providing an indicator for the stability of suspension and emulsions. Section 6.1.4 discusses the application in the characterization of material surfaces.

2.7 Non-faradaic processes

Compared to the faradaic processes (redox reactions), the non-faradaic processes causing a current without a redox process are less spectacular. Primarily, they are capacitive currents due to the EDL's charge-up and pseudo-capacitive effects originating in adsorption of or intercalation with charged species.

2.7.1 Capacitive currents

Treatment of the capacitive currents due to the EDL's charge-up often assumes the double layer as an ideal capacitor and reduces the complex models, as discussed in Section 2.6.1, to a voltage-independent capacity. Typical values for the double layer capacity density at metal electrodes in aqueous electrolytes range from 10 to 100 $\mu F\ cm^{-2}$.

The capacitive current depends linearly on the double layer capacity C_{dl} and the change in potential:

$$I = C_{dl} \cdot \frac{dE}{dt} \qquad (2.66)$$

Linear scan experiment

In a linear scan experiment, the potential is ramped up or down by a fixed scanrate of $v = dE/dt$. Accordingly, a constant capacitive current shows up, its sign depending on the scan direction:

$$I = \pm C_{dl} \cdot v \qquad (2.67)$$

For a $1\,cm^2$ electrode, with the above given typical values for a metal electrode, the double layer capacity ranges from 10 to 100 µF. A typical scan rate of $0.1\,V\,s^{-1}$ implies a constant capacitive current from 1 to 10 µA, superimposing all faradaic contributions.

Potential step experiment

Another typical case in which the double layer capacity is visible and its current is essential to know is the potential step experiment described while deriving the Cottrell equation (Section 2.4.6). Here, the double layer capacity forms with the cell's resistance (more precisely, the uncompensated resistance R_u, see Section 3.2.4) an RC circuit. This circuit's charging current is

$$I = \frac{\Delta E}{R_u} \cdot \exp\left(-\frac{t}{\tau}\right) \quad \text{with } \tau = R_u C_{dl} \qquad (2.68)$$

in response to a potential step of ΔE at $t = 0$.

For a $1\,cm^2$ electrode and an uncompensated resistance of $10\,\Omega$, the time constant τ ranges for our typical double layer capacity density values from 0.1 to 1 ms only. Those values suggest that the capacitive contribution is minor for the typical time scales of 20 ms and above[1] in potential step experiments. However, several factors can cause the capacitive currents to be more prominent compared to the faradaic signals:

1. The finite current to the instrumentation limits that can flow in response to a potential step cause delayed charging up of the double layer and, therefore, a capacitive contribution for more extended periods.
2. For mass transfer limitation of the faradaic process, a microrough electrode surface does not increase the faradaic current but the double layer capacity, which depends on the real electrode surface. The roughness factor Rf relates the real surface area to the electrode's footprint (Section 8.4.2).

Figure 2.25 shows an example of the double layer capacity's influence on the Cottrell response for a planar electrode. The grey curve is similar to Figure 2.14. The red curve additionally contains the capacitive current assuming a microrough electrode (Rf = 10).

---

[1] The 20 ms comes from the mains frequency of 50 Hz. Preferable acquisition times are at least as long as one power cycle to reduce the mains frequency's influence. Additionally, most modern potentiostats specifically filters on 50 Hz (or 60 Hz depending on the country) and its harmonics.

A

B

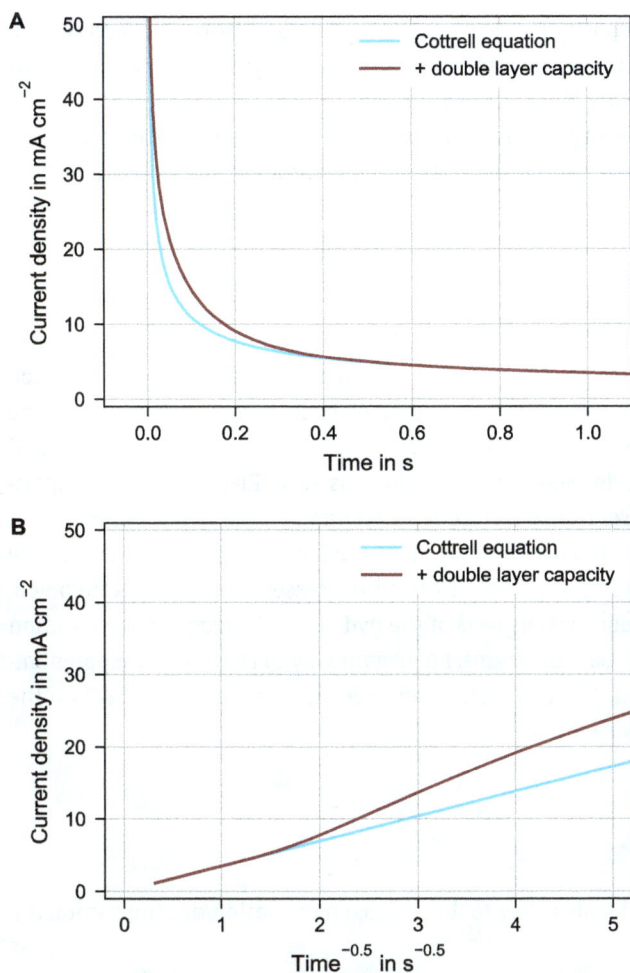

Figure 2.25: Cottrell equation for a microrough $1\,cm^2$ planar electrode with superimposed capacitive current ($D = 1 \times 10^{-5}\,cm^2\,s^{-1}$, $z = 2$, $c = 1\,mM$, double layer capacity density $100\,\mu F\,cm^{-2}$, Rf = 10).

This effect is more drastically visible in plot B with the inverse of the square root of time on the x-axis as commonly used to determine diffusion constants (Section 5.2.2).

2.7.2 Pseudo-capacitive effects

Apart from the double layer capacity, adsorption of ions or intercalation with charged species can cause nonfaradic currents. Ions can adsorb at a given potential up to or a fraction of a monolayer on the electrode. Usually, those processes are fast compared to redox reactions. Examples are proton adsorption and desorption on platinum (Section 8.4) or underpotential deposition (Section 9.2.4). While for adsorption, ions remain at the surface, intercalation is the reversible inclusion into layered materials such as graphite. The main intercalation application is charge storage.

Recently, the term pseudo-capacity is coined mainly by the field of supercapacitors. Here, pseudo-capacitive effects can originate from both faradaic and nonfaradic processes (Section 11.2.3).

2.8 Potential scales

One of the most severe causes of misunderstandings when switching between disciplines is the scales to which a potential is defined. In electrical engineering, the voltage relates to a zero-point within the circuit. The electrochemical theory uses the potential of the hydrogen electrode under standard conditions as zero. Electrochemical experiments measure the applied reference electrode's potential, which is not necessarily a hydrogen reference electrode. In an aqueous system, we have to consider the influence of the pH value on the meaning of potential for all processes with protons involved. This section considers the theoretical aspects of the hydrogen electrode and its relation to the pH. Afterward, the Pourbaix diagrams, an elegant way to represent the potential-pH landscape, are introduced. The discussion continues in Sections 3.2 and 4.2 on the instrumentation and reference electrodes.

2.8.1 Influence of pH

The electrochemical potential scale refers to the hydrogen formation/decomposition under standard conditions:

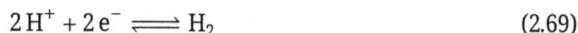

$$2\,H^+ + 2\,e^- \rightleftharpoons H_2 \tag{2.69}$$

With the Nernst equation (equation (2.16)), the definition $E^0_{H^+/H_2} = 0$ and unit partial pressure of hydrogen ($p_{H_2} = 1$), the potential of the hydrogen electrode is

$$E_{H^+/H_2} = \frac{RT}{2F}\ln a^2_{H^+} = \frac{RT}{F}\ln a_{H^+} \tag{2.70}$$

or reformulated with the decadic logarithm

$$E_{H^+/H_2} = 2.303\frac{RT}{F}\lg a_{H^+} = -2.303\frac{RT}{F}\cdot pH \tag{2.71}$$

The prefactor of the logarithm with $\ln 10 = 2.303$ is 59 mV at room temperature. The pH comes in using its definition $pH = -\lg a_{H^+}$.

The potential scale related to $E^0_{H^+/H_2}$ is called the standard hydrogen electrode (SHE) scale. Consideration of the pH leads to the reversible hydrogen electrode (RHE) scale. Both scales can be seen as theoretical concepts but also exist as practical electrodes. With equation (2.71), the relation between the two potential scales is

$$E_{RHE} = E_{SHE} + 2.303\frac{RT}{F} \cdot pH \tag{2.72}$$

Published results leave the reader often with confusion about what a reported potential means. Therefore, **!** it is essential to clearly state the used reference electrode, and as we have seen from the discussion above, the pH value. If there is no strict convention in your subject, I strongly recommend converting all potentials to the RHE scale. Besides more straightforward comparability between different works, this also allows for a transparent presentation of the pH values' influence. See Daubinger et al. [18] as an example.

2.8.2 Stability of water

Similarly, to the discussion for the hydrogen in the previous section, the pH dependency of the oxygen formation/decomposition

$$O_2 + 4H^+ + 4e^- \rightleftharpoons 2H_2O \tag{2.73}$$

can be formulated:

$$E_{O_2/H_2O} = E^0_{O_2/H_2O} - 2.303\frac{RT}{F} \cdot pH \tag{2.74}$$

The reaction's potential at standard conditions is $E^0_{O_2/H_2O} = 1.23\,V$.

Equations (2.71) and (2.73) help illustrate water stability in a potential-pH diagram (Figure 2.26). From this diagram, one can quickly identify the region of water stability. The lower line also represents the zero-point of the RHE potential scale.

2.8.3 Pourbaix diagrams

Potentials-pH diagrams are a mighty tool to present thermodynamic data of electrochemical reactions. Initially, Marcel Pourbaix (1904–1998) suggested those diagrams while collecting data for his "Atlas of Electrochemical Equilibria," initially published in French in 1963.

Figure 2.27A shows the (simplified) Pourbaix diagram of zinc in water. Three phases (Zn, Zn^{2+}, ZnO) are connected by three different boundary lines, each being exemplary for the three possible classes of processes:

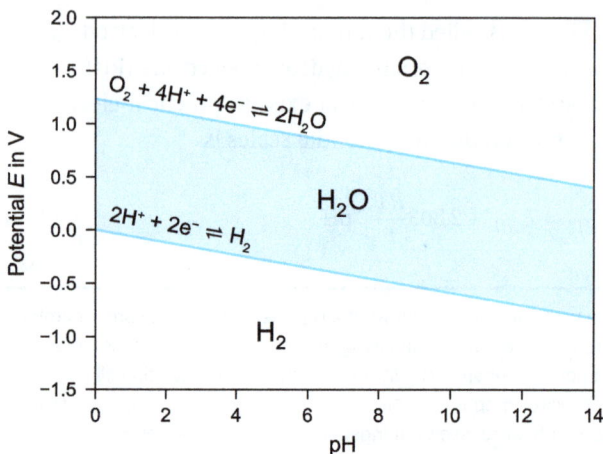

Figure 2.26: Stability of water: the two lines represent the equilibrium potentials at different pH for the decomposition reactions. In between the two lines, water is stable, above oxygen and below hydrogen is formed.

Horizontal boundary line

The horizontal line represents a redox reaction independent of pH (no H^+/OH^-, only electrons are involved):

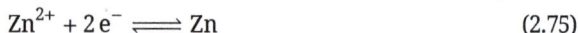

$$Zn^{2+} + 2\,e^- \rightleftharpoons Zn \tag{2.75}$$

Typically, the metal dissolution is an example of a pH-independent process.

Vertical boundary line

The vertical line represents a chemical equilibrium (an acid-base reaction; only H^+/OH^-, no electrons are involved):

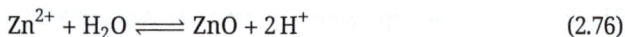

$$Zn^{2+} + H_2O \rightleftharpoons ZnO + 2\,H^+ \tag{2.76}$$

The chemical equilibrium is independent of the electrode potential.

Sloped boundary lines

Redox reactions dependent on pH (both H^+/OH^- and electrons are involved) appear as a sloped line:

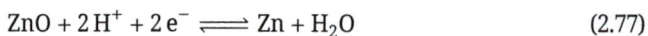

$$ZnO + 2\,H^+ + 2\,e^- \rightleftharpoons Zn + H_2O \tag{2.77}$$

The formation of oxides or hydroxides is a typical example. The slope represents the ratio between H^+/OH^- and electrons in the reaction equation. For a ratio 1 : 1, the

A

B

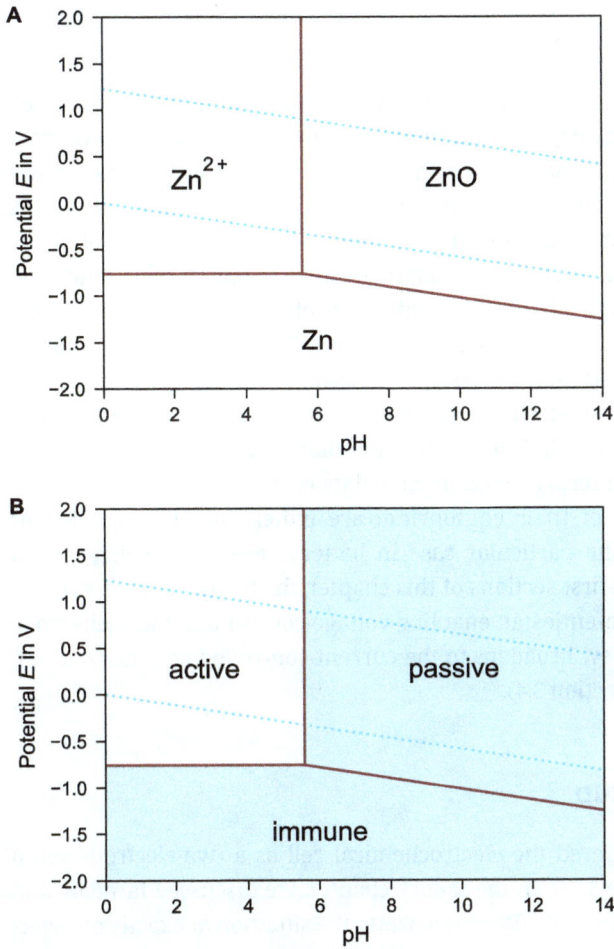

Figure 2.27: Simplified Pourbaix diagram of zinc in water (A) and its stable regions (B). The blue dotted lines indicate the stability of water.

slope is −59 mV at room temperature. In this case, the lines are parallel to the dotted lines in the plot, which indicate the stability of water (compare Figure 2.26).

Often Pourbaix diagrams are used to indicate the stability of each phase (Figure 2.27B). All metals are stable (immune) at sufficiently low potentials and may undergo corrosion (active regions) or passivation at higher potential. Often at high potentials, the passive layer loses its stability, and further dissolution occurs (trans-passivity). A detailed discussion of Pourbaix diagrams' meaning in corrosion science is in Section 8.1.3. Important to note that the thermodynamic information describes which processes may occur. If at all, and at which rate a process takes place requires further description by kinetic analysis.

3 Instrumentation

This chapter on instrumentation discusses the setup around an electrochemical cell from the perspective of electrical engineering. In case you are not familiar with electronics, namely operational amplifier circuits, you might benefit from checking the primer in Appendix B. While chemists sometimes have trouble relating effects observed during experiments to instrumentation details, electrical engineers struggle with the reference electrode potentials and their relation to the electrochemical potential scale. Our modern instrumentation device, typically fully controlled by software, simplifies its operation tremendously. At the same time, this is a tempting pitfall to forget about the circuit and data acquisition details. That said, this chapter seeks to bridge the two perspectives by discussing electronics and the electrochemical cell together. Although most of you will not build a potentiostat yourself, I urge you to familiarize yourself with the inner life of at least a simple one to interpret experimental data correctly.

Electrochemical cells out of their equilibrium are either voltage-controlled or current-controlled (ignoring the particular case in battery research dealing with a constant resistive load). In the first sections of this chapter, the focus is on the voltage-controlled situation and the potentiostat, enabling voltage-control in a three-electrode setup. In the latter part, the view broadens to the current-controlled case (Section 3.3) and potential measurement (Section 3.4).

3.1 Three-electrode setup

In the Introduction, we considered the electrochemical cell as a two-electrode setup based on two half-cells (Figure 1.13). In the theory chapter, we discussed faradaic and non-faradaic currents at one electrode. We often prefer the situation at exactly one electrode to describe or use an electrochemical effect; the second half-cell is only needed to close the circuit. In a voltage-controlled case, we face the issue that the same current (with opposite sign) needs to pass through both electrodes of the electrochemical cell while the applied voltage allows the control of the potential difference between the two but not precisely of the one electrode of interest. Figure 3.1 illustrates the dilemma in a simplified current-potential diagram. At the electrode of interest, called working electrode (WE), other faradaic or non-faradaic processes than at the second electrode can occur. An applied voltage (E_{cell}) causes a current at the WE, which with the opposite sign, needs to pass through the second electrode. However, to control the electrochemical process, we need to define the WE potential E_{WE}, while the connection between E_{cell} and E_{WE} largely depends on the electrode materials, the cell geometry, and most critically, the cell current.

https://doi.org/10.1515/9783111488844-003

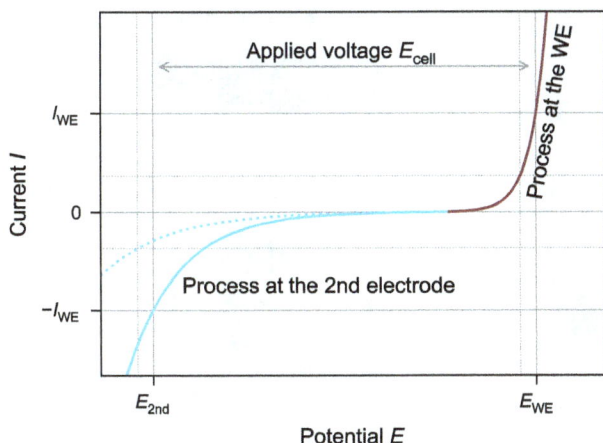

Figure 3.1: Electrochemical cell: simplified current-potential situation in a two-electrode setup. The electrode of interest is called the working electrode (WE). A variation in the process at the 2nd electrode (dashed lines) causes for same applied E_{cell} an entirely different situation resulting in a much lower current.

One strategy to focus more on the WE and predominantly control E_{WE} is to use a substantially larger second electrode or combine a polarizable WE with a non-polarizable second electrode (Section 4.1.2).

The better and most applied approach is the separation of the second electrode into a counter electrode (CE), allowing the passage of the current and a reference electrode (RE), which is kept current-less and ensures a stable potential independent of the WE's current. Figure 3.2 illustrates this three-electrode approach in comparison to the two-electrode situation. Here, the applied voltage allows direct control of E_{WE} (given that the potential of the used reference electrode is known). The device ensuring that the cell voltage (E_{cell}) is the desired set value (E_{set}) is called potentiostat, in Figure 3.2 still as a black box. The following section will look inside potentiostats aiming to not think of a black box anymore.

3.2 Potentiostat

The role of the potentiostat is to ensure appropriate conditions in a voltage-controlled three-electrode setup and provide measurements of the cell current. The potentiostat requires a set value (E_{set}) for the cell voltage (E_{cell}), which is the potential at the WE minus the potential at the RE. A potentiostat circuit outputs the current measurement as a voltage proportional to the cell current.

A potentiostat with a single control amplifier is the most simple circuit (Figure 3.3). In the circuit diagram, the blue area is the electrochemical cell with its three electrodes. A shunt resistor connects the WE to the circuit's ground. The resistance should be small

Two-electrode setup

Three-electrode setup

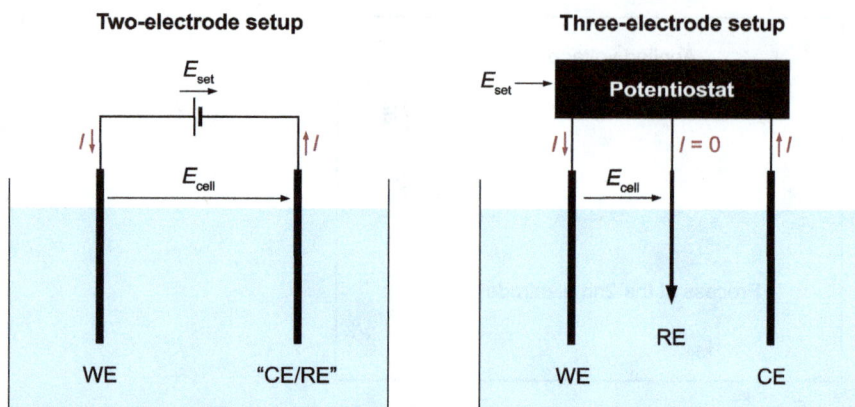

Figure 3.2: Electrochemical cell: two- versus three-electrode setup. The device controlling and measuring the current so that the cell voltage E_{cell} matches the set voltage E_{set} is a potentiostat.

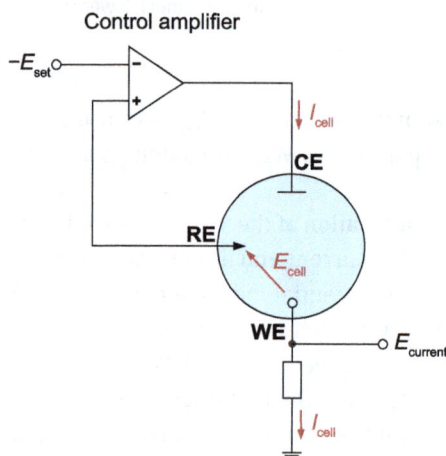

Figure 3.3: Basic potentiostat with a single control amplifier. The WE is on the GND side connected by a shunt resistor to measure the cell current. The control amplifier drives the voltage at the CE.

enough that the voltage drop is negligible, allowing to treat the WE like directly connected to GND. The cell voltage (E_{cell}) expresses how much more positive the WE is than the RE. The WE is at zero in the circuit's potential scale, the RE accordingly at $-E_{cell}$. At the positive input of the amplifier is the RE; the set value is at its negative terminal. In this circuit, the set value needs to be applied inverted ($-E_{set}$).

The electrochemical cell is in the feedback loop with the CE at the output of the amplifier. When the amplifier is operating within its specifications, the difference between its two inputs is zero. Accordingly, $-E_{cell} - (-E_{set}) = 0$, what is our desired condition ($E_{cell} = E_{set}$). The control amplifier ensures this by driving an appropriate output voltage causing a current through the CE and WE. No current flows through the RE, assuming no current leaks at the amplifier's input.

The voltage drop at the shunt resistor allows for the current measurement (the output is a voltage $E_{current}$ proportional to the current with the resistance as the scaling factor). However, a small value of the shunt resistor causes a low output voltage $E_{current}$, while a larger resistance causes a voltage drop causing an error in the condition $E_{cell} = E_{set}$ as the WE is not precisely on zero. Besides, this simple potentiostat works in its practical realization on a breadboard as long as the currents are not too tiny and the electrochemical cell has a predominant faradaic process reducing the influence of the cell's capacity.

Another shortcoming of this simple potentiostat is using the same operational amplifier for the control and the measurement of the RE potential. Actual components always need to compromise in one or the other aspect. There are types with low input leakage currents, different ones with high bandwidth, or the ability to drive high currents.

3.2.1 Adder potentiostat

The adder potentiostat is an advanced circuit overcoming several shortcomings of the basic potentiostat discussed in the previous section. This circuit consists of at least three operational amplifiers (Figure 3.4). The WE connects to virtual ground (VGND, Section B.2 in the Appendix) at a current follower and is therefore at zero potential in the circuit's scale. The RE connects to the input of a voltage follower. In the circuit's potential scale, it is at $-E_{cell}$, so is the output of the voltage follower.

The name of the adder potentiostat comes from the inverting adder, which serves as the control amplifier. At the adder's inputs, one or more set values are added together and compared to zero (the difference between the two operational amplifier's inputs is zero and the noninverting input connects to GND):

$$E_{set,1} + E_{set,2} + E_{set,3} - E_{cell} = 0 \tag{3.1}$$

The inverting adder ensures by driving a voltage at the CE that E_{cell} equals the sum of the set values. The different set values can be in older setups from a DC voltage source and, e.g., a ramp generator or a pulse generator. In modern potentiostats, a set value is often from a digital-analog converter and others from analog signal generators for high-speed signals.

The current follower outputs a voltage ($-E_{current}$) proportional to the cell current. Depending on the current range, one or more amplifier stages may be at the output of the current follower to provide reasonable voltage levels to the digital-analog converter.

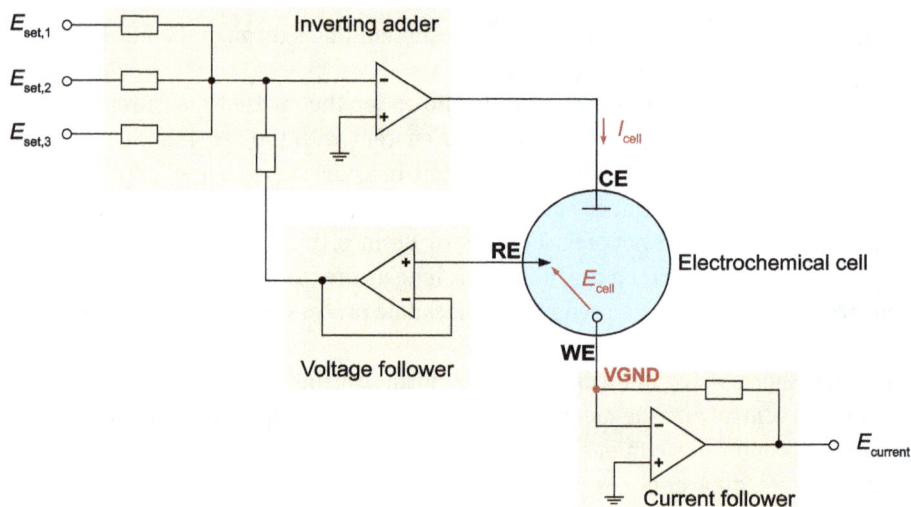

Figure 3.4: Adder potentiostat: the WE connects to a current follower for the current measurement and is accordingly on virtual ground (VGND). A voltage follower buffers the RE potential, and an inverting adder serves as the control amplifier driving the voltage at the CE.

The circuit, as in Figure 3.4, works well in actual experiments. It might be improved by some filtering to ensure stability and eliminate noise at the output of the current follower. Section 3.2.3 discusses the impact of such filtering.

A critical shortcoming of the adder potentiostat is its inability to drive multiple working electrodes sharing a single reference and counter electrode. With some circuit enhancements to a so-called bipotentiostat, it is possible to operate a second working electrode in the same electrochemical cell. Beyond two working electrodes, another circuit, preferable with the counter electrode on GND, is needed. Besides, many commercial potentiostat devices use an adder potentiostat circuit.

3.2.2 Signal generation and data acquisition

DC sources and analog ramp generators traditionally generated the set values for the potentiostat. Today's potentiostats generate the set values by digital-analog converters, which is in many aspects a big step forward. However, a severe downside is the inherent discretization of the applied potential into small steps, which is especially critical when applying a ramp. Figure 3.5(A) illustrates this situation by comparing the ideal voltage ramp and the applied signal in the form of a staircase due to the digital-analog conversion. Figure 3.5(B) shows the current response to such a staircase.

Commercial potentiostats provide several solutions: additionally to the standard signal generation by a digital-analog converter, an analog ramp generator allows for

A

B

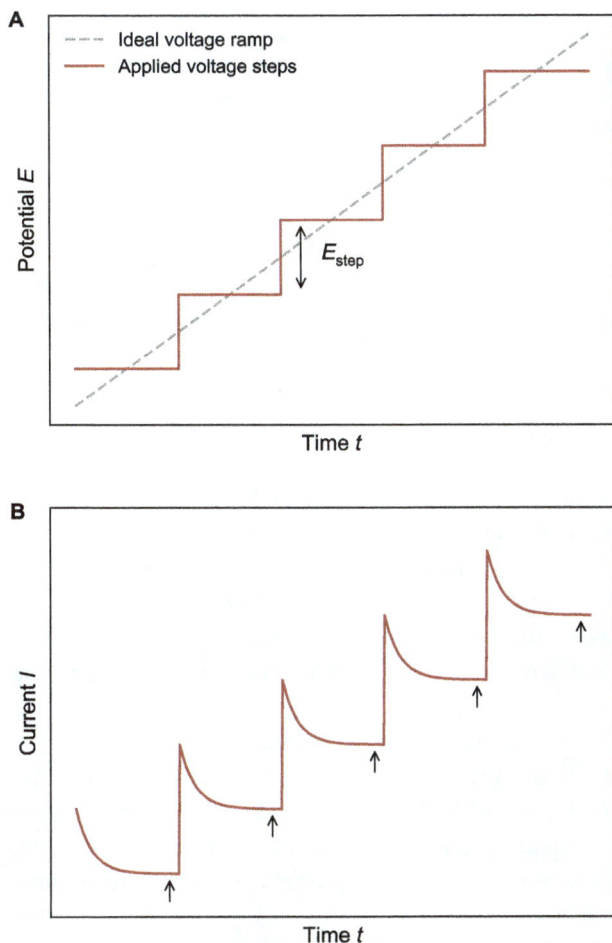

Figure 3.5: Digital signal generation leads to staircase voltage ramps (A), causing a sequence of small step responses in the current (B). Sampling the current at the end of each step (indicated with the arrows) results in a severe information loss as fast processes and the capacitive response may be missed out.

smooth ramps (depending on the manufacturer called "true linear scan generator" or similar). The more simple approach is to combine a staircase signal generation with an appropriate data acquisition scheme. Instead of sampling at the end (or any other single point) of each step's current response, the devices integrate the current over the whole step by either an analog integrator or by using the average of many samples. The manufacturers call the latter sampling method "current averaging" or "surface mode" in contrast to the "standard mode" or "fast mode."

! Understanding what happens in staircase ramps, how fast the voltage rise for step responses, and other effects like noise or stability, as discussed in the next section, often require awareness of the raw signals. Many potentiostats make the analog signals of the cell voltage and the (amplified) output of the current follower available at auxiliary connections. Using a two-channel oscilloscope (one channel for the voltage, one for the current signal) allows monitoring the raw values on different time scales. In my opinion, an oscilloscope should be part of the standard equipment at an electrochemical setup.

Figure 3.6 demonstrates the implications for a cyclic voltammogram of a platinum electrode (Section 8.4 discusses platinum electrochemistry in more detail) and how an oscilloscope helps demystify the observed results. The ramp signal with a slope of $100\,\mathrm{mV\,s^{-1}}$ was digitally generated as a staircase (step height 10 mV). That corresponds to a step duration of 100 ms, a reasonable value to avoid interference with the 20 ms from the 50 Hz mains frequency (16.7 ms for 60 Hz).[1]

The current was measured once per step ("staircase") and also read out by an integrator summing up the current over the whole step duration. Figure 3.6(A) shows the tremendous difference between the two acquisition methods. All typical platinum features on the left side of the curve miss in the curve without integration, and any consideration of the charge enclosed by the curves would result in totally wrong values. At $-0.3\,\mathrm{V}$ in the upward scan (indicated by the points), the current reading even has opposite sign.

Connecting an oscilloscope to the current monitor output of the potentiostat reveals what happened. Figure 3.6(B) with the current's raw signal over time shows the exponential decay within each step's duration. The integration of the whole step results in positive values due to the charge-up of the double layer capacity. At the specific potential, the current without capacitive contribution, the oxygen reduction with its negative current dominates.

Measurements at the end of each step in staircase ramps cause severe information loss and might even hide features, especially when using cyclic voltammetry to investigate materials. This point cannot be stressed enough: you find many cyclic voltammograms suffering from this effect even in peer-reviewed scientific literature. Before publishing cyclic voltammetry data on novel or differently deposited materials, you should ensure that the setup can reproduce well-known curves, such as from a platinum electrode.

While the staircase ramp with a sampling of the current at a specific point seems at first glance like a bug only, it is also a valuable method to, e. g., suppress capacitive effects by taking the current readings at later timepoint in each step. Some devices even

---

1 Experimental details: 3 mm Pt disk electrode in aerated PBS, potentials refer to Ag/AgCl with 3 M KCl inner filling; scanrate $100\,\mathrm{mV\,s^{-1}}$, potential step 10 mV; Autolab PGSTAT128N with FI20 analog filter and integrator module; software NOVA 2, which automatically readouts and discharge the integrator at each potential step.

A

B

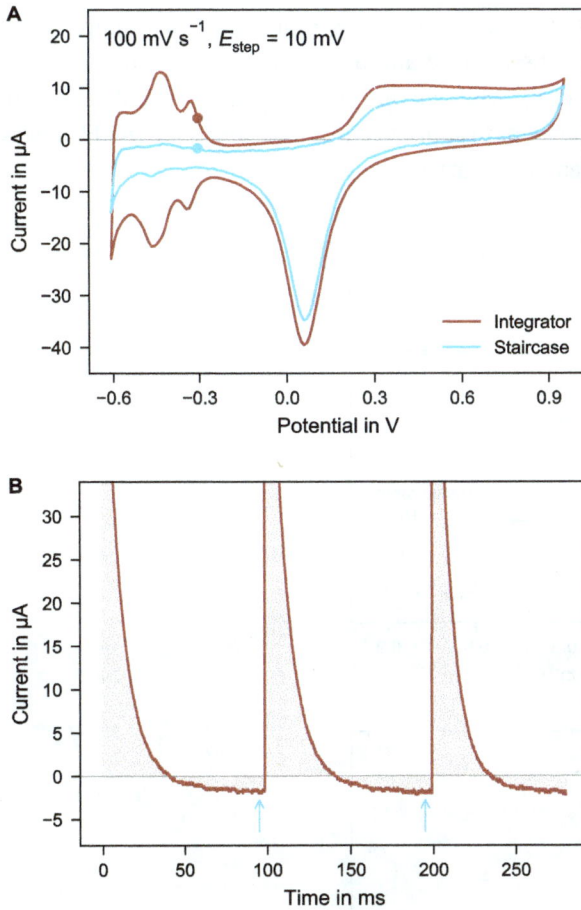

Figure 3.6: The practical implication of the digital signal generation on cyclic voltammograms of a platinum electrode: different sampling methods cause completely different curves (A), raw data accessed by an oscilloscope (B) reveals the situation at −0.3 V in the upward scan (at the points in A), arrows indicate the acquisition in case of staircase.

allow the selection of the fraction of the step at which the current is sampled (i. e., the positions of the arrows in Figure 3.5(B) within in the steps).

Different signal generation/data acquisition combinations are possible for linear sweeps or cyclic voltammograms, depending on the potentiostat model:

1. Analog voltage ramp, current measurement at discrete steps
2. Staircase voltage ramp, current measurement at the end of each step
3. Staircase voltage ramp, current integration using an analog integrator discharged after each step
4. Staircase voltage ramp, continuous sampling of the current and calculating the average

Figure 3.7 compares the different possible combinations with a practical example.[2] The analog ramp is expected to be the best choice and used as a reference in both plots. Figure 3.7(A) compares staircase ramps with current sampling at the end of each step. While decreasing the step duration helps to approach the ideal curve, one would always underestimate the charge enclosed in the curve.

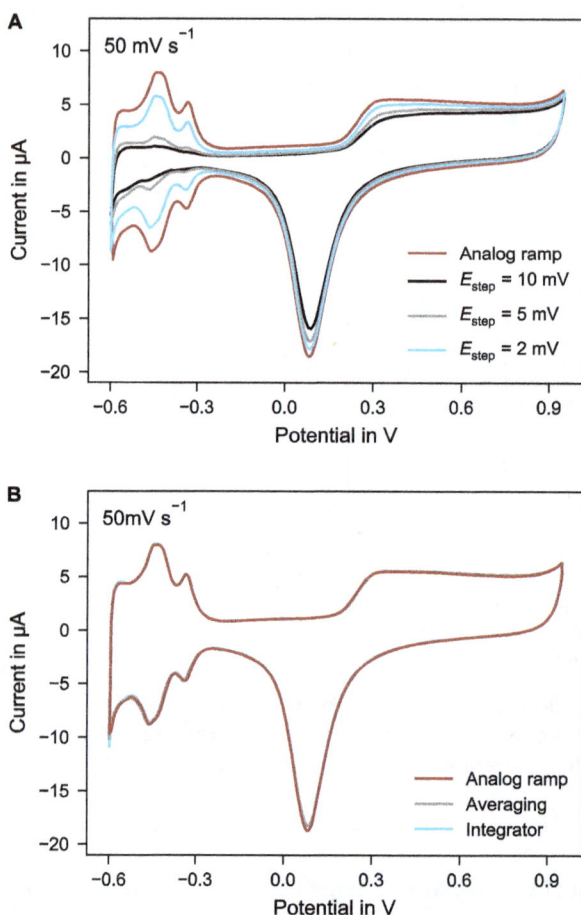

Figure 3.7: Different signal generation/data acquisition combinations at the example of a platinum cyclic voltammogram: staircase ramps with current sampling a the end of each step result in severe information loss compared to an analog ramp (A). Using an analog integrator or current averaging makes it possible to get results with a staircase voltage ramp that resembles the curve generated by an analog ramp (B).

---

2 Experimental details: 3 mm Pt disk electrode in deoxygenated PBS, potentials refer to Ag/AgCl with 3 M KCl inner filling; scanrate 50 mV s^{-1}; Autolab PGSTAT128N, SCAN250 true linear scan generator module, FI20 analog filter and integrator module; IVIUM CompactStat for the current averaging and staircase curves.

Using an analog integrator or taking the average from continuous samples during the whole step duration in staircase generated voltage ramps resembles the cyclic voltammogram with an analog ramp generator (Figure 3.7(A)). Additionally, the integrator and the current averaging may help to reduce noise.

However, in both methods, it is essential to use a sufficiently large current range (larger than the displayed values) to avoid clipping of the initially huger value in each step. If no analog ramp generator is available, one can check the validity of the measurement by a larger current range and ensure that the integrated or averaged current does not increase. Additionally, one should pay attention to the voltage step E_{step} and ensure that it does not cause a too large integration or averaging period (depending on the scanrate). In case of doubt, cross-check by lowering the E_{step}, which should not change the overall cyclic voltammogram appearance or enclosed charge.

3.2.3 Stability, noise, and compliance voltage

Three other aspects occurring during experiments are issues with stability, noise, and compliance voltage. Debugging a setup might get tricky when these issues are interlinked. We will identify all three within the adder potentiostat circuit; however, the concepts apply to other potentiostats as well.

Stability
The control loop in the adder potentiostat is from the output of the inverting adder (control amplifier), the electrochemical cell, the voltage follower, and back into the input of the inverting adder (illustrated by the blue loop in Figure 3.8). Depending on the nature of the electrochemical cell, the potentiostat's control loop might get unstable, and even oscillations might occur. Especially with cells showing strongly capacitive behavior, the risk is high. Therefore, the adder circuit often comprises damping elements limiting the bandwidth of the control loop. The tradeoff is between higher stability and high speed for, e. g., ensuring steep voltage steps or allowing for fast ramps.

Most commercial potentiostats allow changing the bandwidth in two or three steps, often with names referring to "high stability" and "high speed" rather than numeric values of the corner frequencies. A setting for stability limits the bandwidth, e. g., to 10 kHz. For some applications, e. g., with electrochemical impedance spectroscopy (Chapter 7) at high frequencies, circuits without damping are a better choice.

High-end devices may comprise an "oscillation protection" switching the cell off when an oscillating instability occurs to protect the cell and, in extreme cases, the device itself. Deactivating those features requires a good knowledge about the actual situation and preferable observing the potentiostats analog monitors with an oscilloscope.

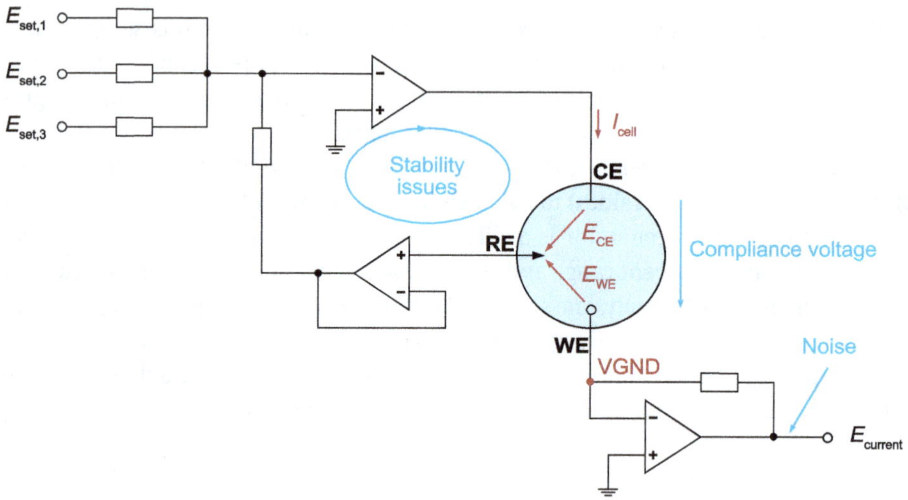

Figure 3.8: Practical considerations illustrated in the circuit of an adder potentiostat: It is important to distinguish between *stability issues*, which affect the electrochemical cells, and *noise*, which affects the output only. The *compliance voltage* is the maximal potential difference between CE and WE the potentiostat can drive to control the electrochemical cell.

Noise

Electrical noise influencing the cell current can be an issue, especially when dealing with low currents (pA and nA range). While digital filters can eliminate the noise in the measured data, severe noise might bring the amplification stages after the current follower or the analog-digital converter to their limits. Therefore, analog filters directly after the current follower help cancel the noise before its amplification or conversion.

! Often commercial potentiostat software offers automatic settings for stability and filters to eliminate noise. In many cases, this allows for appropriate settings working with most standard applications. When dealing with electrochemistry in microsystems, the risk is high to leave the realm of "standard applications." A minor change in other instrument parameters may cause the automatic selection to choose different damping or filters, causing different and maybe awkward behavior of the whole setup.

Compliance voltage

In the adder potentiostat, polarizing the CE potential establishes the cell voltage according to the set values. When the inverting adder (control amplifier) reaches its maximum or minimum output voltage, the circuit fails to ensure $E_{cell} = E_{set}$. As the WE electrode is on VGND, the compliance voltage is the same as the maximal or minimum output voltage of the inverting adder (Figure 3.8).

Some potentiostat software has an indicator for voltage compliance issues, or when the device records the actual cell voltage deviation from the set value is a warning. Elimination of the matter is often easily possible by choosing a different or larger counter electrode, changing the electrode placement, or increasing the electrolyte conductivity (compare the discussion to Figure 3.9).

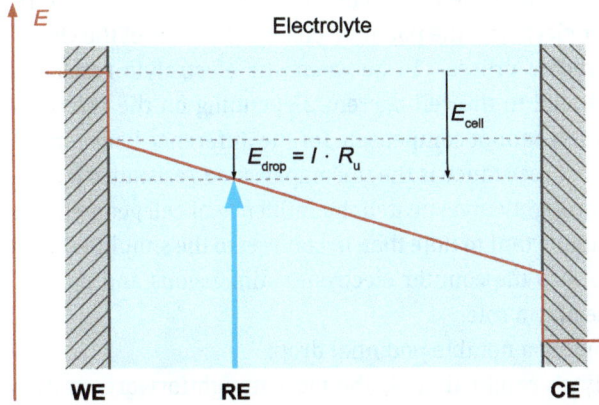

Figure 3.9: Illustration of the potential drop (IR drop) occurring in a three-electrode setup: The electrolyte resistance causes a potential gradient between working and counter electrode.

In practice, two very challenging scenarios might occur:

1. The filter for the noise hides instability issues. While the current signal may look unsuspicious, the measured values could be unrelated to the electrochemical processes. One can solve this situation by manually setting stability and filters for the noise to observe the damping's influence and check if a less filtered signal still shows the same trend as the nicely filtered one.

2. The voltage compliance reaches the potentiostat's maximum hindering the cell voltage from reaching the set value. Interwoven with stability issues, only transiently getting to the voltage compliance limit makes this effect complicated to debug. One indication for such a situation is when changing the counter electrode dimensions or its position shifts the system's behavior.

In both cases, observation of the analog monitor outputs of the potentiostat with an oscilloscope reveals the details immediately to avoid cumbersome indirect debugging.

3.2.4 Potential drop (IR drop) and its compensation

The terms potential drop or IR drop refer to the part of the cell voltage lost because of the electrical resistance of the electrolyte or cables. Apart from battery research, in most experiments, the electrolyte resistance dominates possible contribution from the wires and electrical connections. The introduction of a current-less reference electrode in the three-electrode setup limits the relevant resistance to the one between the working and reference electrode. The potentiostat, by principle, cannot compensate any resistance here without knowledge of its value, and this resistance is therefore called the uncompensated resistance R_u.

Figure 3.9 illustrates the situation of the potentials in an electrochemical cell. Directly at the working and counter electrode, the potential changes because of the charge transfer process, typically a faradaic process. In between, the electrolyte resistance causes a potential drop proportional to the cell current. Depending on the reference electrode's position, the potentiostat cannot compensate for the difference between WE and RE, called E_{drop}. The associated resistance is the uncompensated resistance R_u.

Myland and Oldham [19] thoroughly investigated the influence of cell geometry on uncompensated resistance. It is important to note that, in contrast to the simplified view in Figure 3.9, in practical setups also the counter electrode's dimensions and position relative to the working electrode play a role.

Several strategies allow avoiding a notable potential drop:

1. The increase of the electrolyte's conductivity is the most straightforward strategy to avoid a high potential drop. Besides appropriate electrolyte selection, addition of conducting salts may help.
2. Placement of the reference electrode in close vicinity of the working electrode reduces the influence of the electrolyte resistance. The Luggin capillary (Figure 4.13) helps to access the potential in the electrolyte very close to the working electrode.
3. Depending on the application, e. g., for electrodeposition, lowering the cell current is an appropriate strategy to minimize the potential drop.

Some applications hinder an independent selection of the electrolyte's conductivity. Additionally, an electrode featuring micro or nanostructures or generally a substantial electrode roughness can cause the potential drop to occur between the structures, making it impossible to bring the reference electrode close enough. Therefore, measurement and compensation of the uncompensated resistance is another possibility.

Measurement of the uncompensated resistance

An equivalent circuit model of the electrochemical cell helps to understand the methods to measure the uncompensated resistance. Figure 3.10 shows a typical model. Resistors R_{ct} represent the charge transfer at the working and counter electrode. In parallel, there are capacitors for the electrical double layer forming a simple Randles circuit (Section 7.2.3). Between WE and CE, the cell current I passes through the electrolyte resistance, split into two components R_u and R by the RE position. If available, electrochemical impedance spectroscopy (Chapter 7) allows precise the determination of the uncompensated resistance.

An alternate method to measure the uncompensated resistance is the current interrupt technique. During potentiostatic control, the cell current is interrupted for a short period (e. g., in the millisecond range), and the open-circuit potential of the cell allows

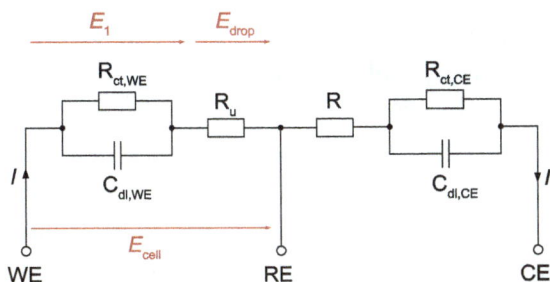

Figure 3.10: Equivalent circuit model to explain the principles for IR drop measurement: the RC parallel circuits represent the charge transfer processes and the electrolytic double layer at the working and counter electrode. The two resistors in the middle are the electrolyte resistance, with the uncompensated resistance R_u the fraction between working and reference electrode.

the determination of the potential drop. Figure 3.11 illustrates such an measurement. At time zero, the measurement device interrupts the current flow, and the potential drops instantaneously by E_{drop} to E_1. The charge transfer resistor slowly discharges the double layer capacity, and the potential drops exponentially starting from E_1. Measurement of the potential decay and extrapolation to time zero allows accessing E_{drop} by a simple procedure. Knowing the cell current I just before the interrupt leads to the uncompensated resistance $R_u = E_{drop}/I$.

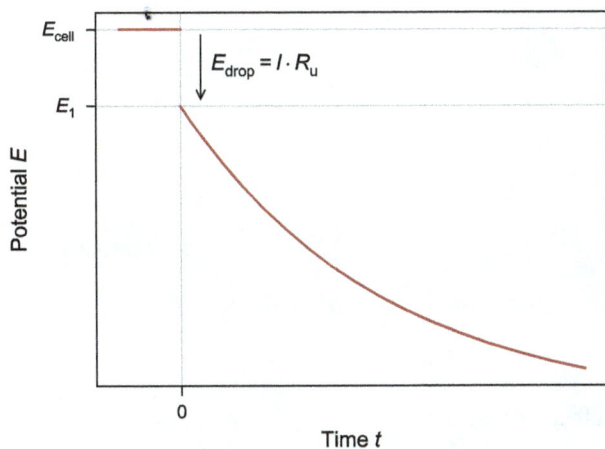

Figure 3.11: Current interrupt technique: on interruption of the cell current I at time zero, the cell voltage instantaneously drops by E_{drop} to the value E_1, buffered by the double layer capacity.

The current interrupt technique allows for intermittent measurement of the uncompensated resistance and can account for changes in the electrolyte composition during an experiment.

A third option to estimate the uncompensated resistance is to use the compensation circuit discussed in the next section and find by trial-and-error the point of instability. This approach is only possible with robust electrodes (i. e., no nanofeatured electrode).

Potential drop compensation

Once the value of the uncompensated resistance is known, an enhancement of the adder potentiostat allows for its compensation (Figure 3.12). A signal proportional to the cell current is linked back to the summation point. The adjustable resistor controls the magnitude of the feedback to match different values of R_u. The positive feedback always implies the risk of overcompensation and, therefore, instability of the potentiostat circuit. In practical experiments, compensation of around 90 % from R_u is optimal.

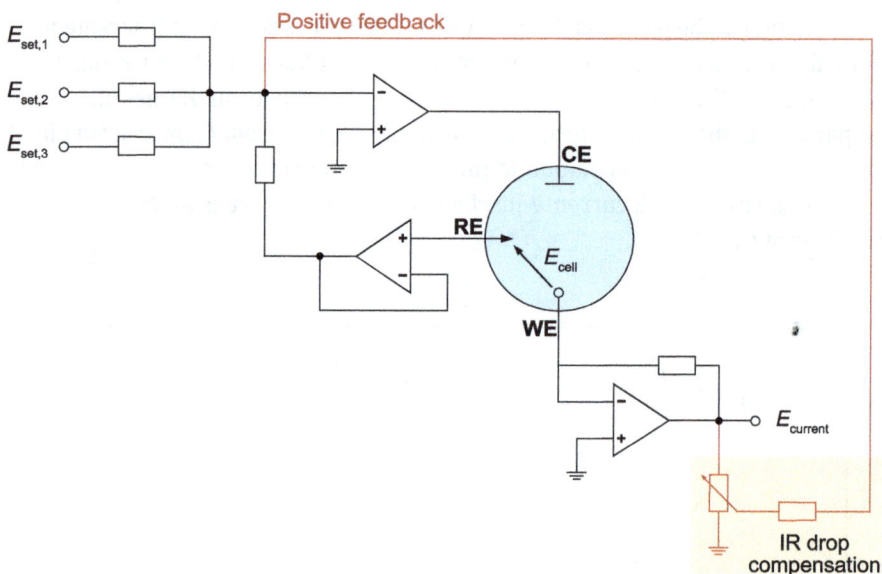

Figure 3.12: Adder potentiostat: potential drop compensation by current feedback.

3.2.5 Potential of the counter electrode

The process at the counter electrode and its potential not only matters to stay within the compliance voltage. Especially in microsystems, the WE and CE are often close enough that changes in the electrode's microenvironment (pH, reaction products) may influence the other electrodes' reaction. This cross-talk allows even a feedback mode between WE and CE as, e. g., used in Clark-type oxygen sensors following the Ross principle; see Section 10.1.2.

If the potentiostat device does not provide this functionality per se, measurement of the CE potential is quite challenging. Especially with low cell currents, the leakage currents through the potential measurement may disturb the potentiostat's functionality. An option is using a voltmeter with high internal impedance or an electrometer device (Section 3.4) between CE and WE. In the case of an adder potentiostat, it is even better to measure between CE and AGND to not compromise the potentiostat's current measurement at all. However, if more than one of the devices have an earth connection, it is crucial to ensure that all GND levels are the same. Additionally, external links to the electrochemical cell may increase the noise or even cause stability issues.

3.2.6 Commercial devices

A potentiostat as a device typically contains the potentiostat circuit itself and sophisticated circuits for signal generation. Today most instruments include digital components for control and data acquisition too. Commercial potentiostat devices may also comprise other functionality such as a galvanostat or a frequency analyzer.

Figure 3.13 illustrates the variety of commercially available options. Depending on the application, selection criteria should be, e. g., portability versus flexibility. By now, there is not a universally optimal potentiostat; each of the available options comes with some limitations. All devices are capable of successfully running (chrono)amperometric methods assuming the current ranges are suitable. Dealing with, e. g., cyclic voltammetry of noble metals requires more sophisticated data acquisition schemes or analog ramp generation (Section 3.2.2), which may not be available at all or only with an optional module. More advanced voltammetry methods and electrochemical impedance spectroscopy are available in several devices but not with the same functionality and accuracy. I strongly recommend getting a demonstration or obtaining a test device before buying an expensive model. When dealing with electrochemistry at the micro- and nanoscale, demonstration or testing should focus on the desired current range rather than a macroscopic cell. In any case, you should always prefer a device with analog monitor outputs to connect an oscilloscope over one without.

Many commercial devices use an internal circuit resembling the adder potentiostat as introduced above. Figure 3.14 shows the block diagram of a Metrohm Autolab PGSTAT128N device. The right part of the diagram is similar to Figure 3.4. A current follower (CF) connects the working electrode to virtual ground. A voltage follower (VF) provides the cell voltage as the difference between a reference (re) and sense (s) input. If the sense input connects to the working electrode, the configuration matches the situation discussed for the three-electrode setup (Figure 3.2).

The control amplifier (CA) with its summation point (Σ) forms an inverting adder. At the summation point, the cell voltage compares to set values from various input sources, such as a digital-analog converter (DAC164), the frequency analyzer module (FRA32M), or an analog ramp generator (SCAN250). Several modules use the output voltage of the

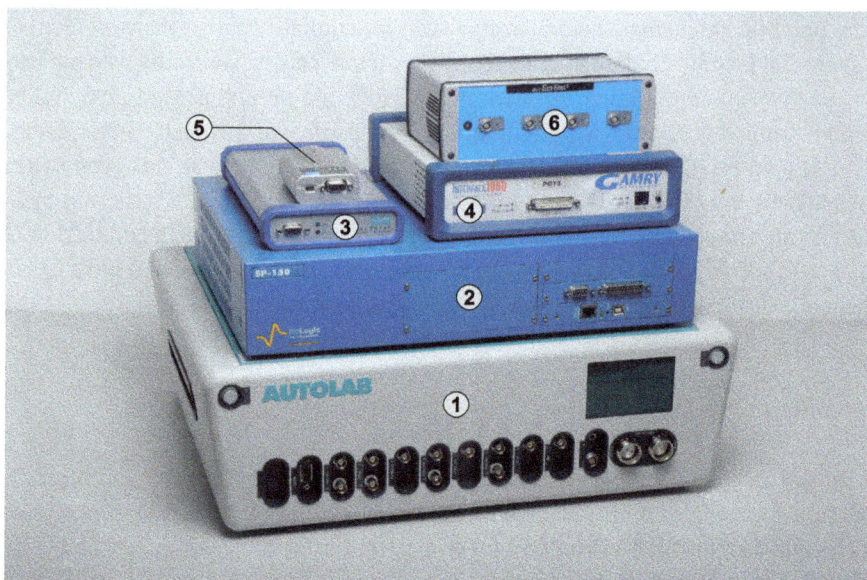

Figure 3.13: Different commercial potentiostats in various form factors: Metrohm Autolab PGSTAT128N (1), BioLogic SP-150 (2), Ivium CompactStat (3), Gamry Interface 1000 (4), PalmSens EmStat3 Blue (5), and PalmSens MultiEmStat3 (6). The collection illustrates the variety of commercial potentiostats and is not a comprehensive representation of the available products.

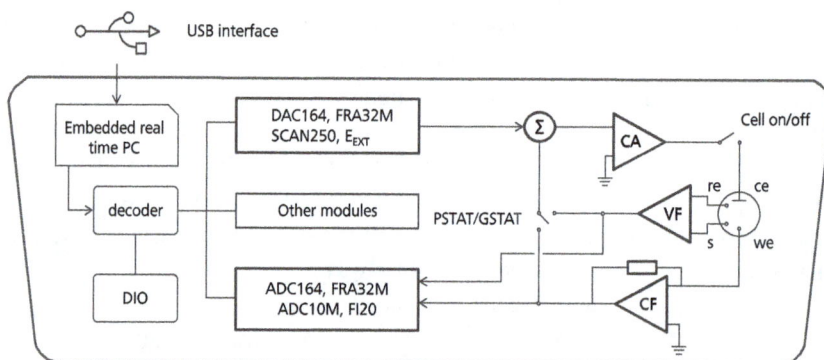

Figure 3.14: Block diagram of the Metrohm Autolab PGSTAT128N device revealing its internal circuit as an adder potentiostat. Reproduced with permission from Metrohm Autolab, The Netherlands.

current follower: an analog-digital converter ADC164 or ADC10M, the frequency analyzer module FRA32M, or the filter and analog integrator module FI20.

The use of an embedded real-time computer allows for exact timing what is essential for most electrochemical methods beyond potentiometry and amperometry. Additionally, the device works as a galvanostat (Section 3.3) depending on the "PSTAT/GSTAT" switch.

3.2.7 Integrated circuits, embedded potentiostats

Several potentiostats as an integrated circuit (IC) are on the market. While application-specific ICs seemed to be the way to go a decade back, several general-purpose ICs with and without integrated microcontrollers are now available as standard components. The book's webpage provides links to different embedded potentiostat projects giving schematics and firmware for the different potentiostat ICs.

An *embedded system* is usually understood as a specialized computer system, often realized with a microcontroller, in a single integrated chip. Typically, the focus is on physical interaction through sensors or actuators. In contrast to general-purpose computer systems, it is dedicated to a specific task, resulting in optimal resource efficiency, e. g., low power consumption, small form factor, or low price. Its software (firmware) is not intended for end-users. There is no strict definition of an embedded system, mainly because no single authority covers the different applications and technical realizations, in contrast to, e. g., the IUPAC for (electro-)chemistry.

   An *embedded potentiostat* is, accordingly, an embedded system that provides the functionality of a potentiostat. An embedded potentiostat's software and hardware interfaces are designed for manufacturers to build the potentiostat into their device instead of being ready for end-user applications. Therefore, embedded potentiostats are sometimes called OEM potentiostats, referring to the *original equipment manufacturer (OEM)* that produces parts and equipment that could be included in another manufacturer's end-user product.

Embedded potentiostats can achieve analytical performance comparable to bench-top devices depending on the firmware implementation of the electrochemical methods. However, two limitations usually occur:

1. The current range is limited, typically in the range of mA or below, which is often sufficient with sensors but not when applied in battery research or when working with fuel cells or electrolyzers.
2. The applied potentials are limited by the restrictions of the circuit's voltage levels. The limitation touches both the set value of the cell potential (between WE and RE) and the compliance voltage (between WE and CE). The latter can be especially critical, asking for optimal CE material and size selection.

Texas Instruments LMP91000

For more than a decade, the LMP91000 from Texas Instruments allows for amperometry in a three-electrode setup. The chip features a potentiostat circuit with a digital signal generation or optional analog input for the set value and an analog output for a voltage representing the cell current. The analog-digital conversion requires a separate IC. An I^2C interface allows the control of the LMP91000.

   The block diagram (Figure 3.15) reveals a potentiostat circuit similar to the adder potentiostat shown in Figure 3.4. The control amplifier (A1) compares the potential of the RE to the set value ("variable bias"). The WE connects to the current follower, here called the transimpedance amplifier (TIA). The digitally adjustable gain resistor of the

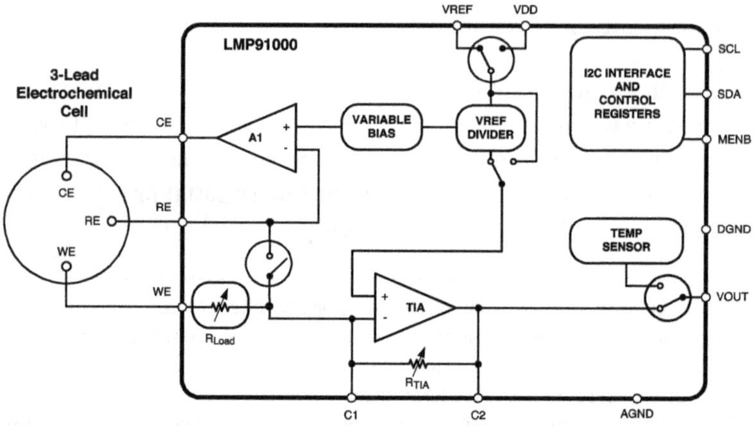

Figure 3.15: LMP91000: potentiostat IC with a circuit similar to the adder potentiostat, well suited for (chrono-)amperometry in low-power application scenarios. Courtesy of Texas Instruments.

TIA allows for current ranges from 5 μA to 750 μA full scale. Additionally, the circuit provides a cell conditioning current up to 10 mA. The zero-level at the noninverting input of the TIA is either defined by an external reference voltage or programmed to fractions of the supply voltage.

In principle, the LMP91000 would allow for potential ramps by itself, but it is best suited for (chrono)amperometric measurements (Section 5.2). Other methods such as cyclic voltammetry (Section 5.3.2), square wave voltammetry (Section 5.3.7), and normal pulse voltammetry (Section 5.3.6) are possible with an external generation of the set value as demonstrated by the KickStat project [20]. Depending on the configuration, the IC enables low-power applications below 100 μW total power consumption even with a three-electrode cell. The input bias current at the RE connection is below 100 pA at room temperature but may reach nearly up to 1 nA over the specified temperature range (−40 °C to 85 °C), possibly critical for microfabricated reference electrodes. With a price of less than $2 (October 2024, quantities > 1000), this IC allows for cost-effective designs.

Analog Devices AD5940/AD5941

More recently, Analog Devices came up with the AD5940/AD5941.[3] This IC features two analog front-end (AFE) loops with potentiostatic circuits combined with sophisticated signal generation and data acquisition possibilities (Figure 3.16). The difference between the two AFEs is the bandwidth, one for lower frequencies up to 200 Hz, and the high bandwidth AFE loop for higher frequencies up to 200 kHz and impedance analysis.

---

3 The difference between the two ICs is the package. The AD5940 is ready for chip-scale packaging technology with 3.6 mm × 4.2 mm outlines, allowing tiny designs. Prototyping or non-professional users may prefer the AD5941 with its Lead Frame Chip Scale Package (LFCSP), permitting more simple building.

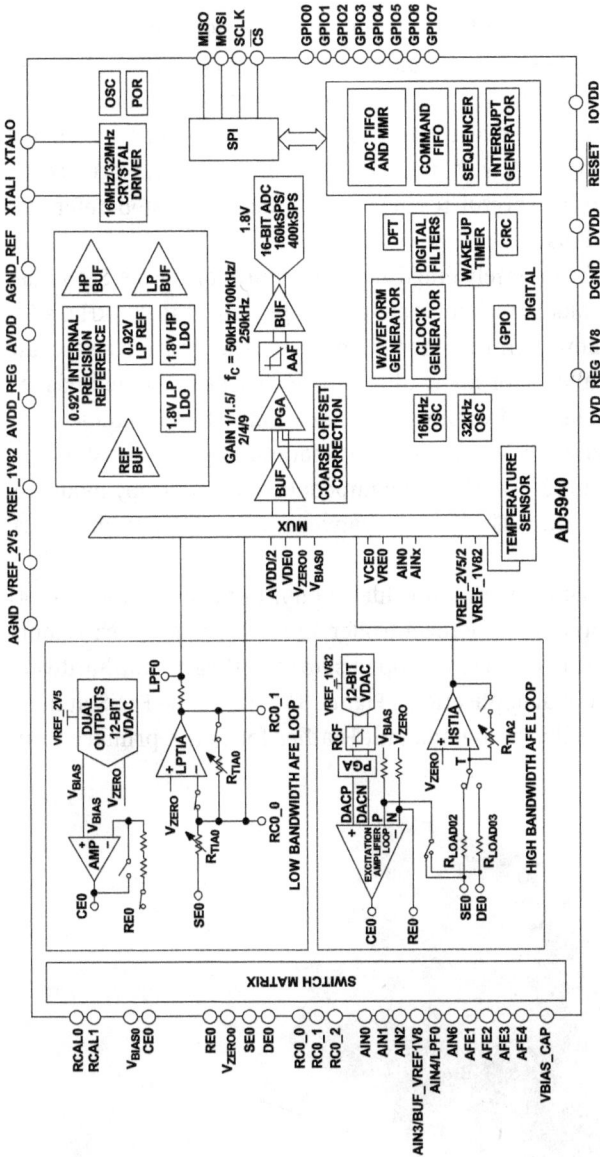

Figure 3.16: AD5940: potentiostat IC with separate different analog front-end (AFE) for low and high band-width operation. The IC allows measurements with many electrochemical methods, including impedance spectroscopy. All images, icons, and marks in this figure are owned by Analog Devices, Inc. ("ADI"), copyright © 2021. All Rights Reserved. These images, icons, and marks are reproduced with permission by ADI. No unauthorized reproduction, distribution, or usage is permitted without ADI's written consent.

A switch matrix allow configuring the different front-ends and signal paths. The primary circuit in each AFE resembles the circuit in Figure 3.4 with the current follower called transimpedance amplifier (TIA). "SE0" is the working electrode connection. The potentiostat can handle currents in between 50 pA and 3 mA. The input bias current at the RE connections is typically 20 pA with a maximum of 150 pA, which are acceptable values in general but require careful treatment for microfabricated reference electrodes in long-term applications.

A built-in sequencer allows for precisely timed electrochemical protocols. Set values with different waveforms (sinusoidal, trapezoid) are available with up to 200 kHz. A Serial Peripheral Interface (SPI) allows control of the IC and data readout from a separate microcontroller. The hardware allows for many different electrochemical methods, including chronoamperometry, all kinds of voltammetry, and electrochemical impedance spectroscopy. With the low-bandwidth AFE, low-power operation below 100 µW is possible, while using advanced features, especially the impedance spectroscopy mode, requires supply currents beyond 10 mA. The price is between $5 and $6 (October 2024, quantities > 1000).

Freiburg's Potentiostat (FreiStat) is an embedded potentiostat using the AD5941 and a software framework comprising microcontroller firmware and a corresponding Python library.[4] The FreiStat is provided as open-source software and hardware allowing free use, also in commercial applications. Figure 3.17 shows the FreiStat main board stacked onto a microcontroller in the Adafruit Feather format. A primary focus

50.8 mm

Figure 3.17: Freiburg's Potentiostat (FreiStat) with an AD5941. The FreiStat main board is stacked onto an Adafruit Feather M0 WiFi microcontroller running the FreiStat firmware.

---

4 https://github.com/IMTEK-FreiStat

in the firmware development was on the analytical performance of the electrochemical methods implementation, e. g., following the instrumentation aspects discussed in Section 3.2.2. Measurements with voltammetric and amperometric methods were compared to commercial bench-top devices, and cloud-based corrosion monitoring demonstrated the possibility of an Internet-of-Things (IoT) application [21].

ADuCM355

The ADuCM355 is a mixed-signal microcontroller including an Arm Cortex-M3 processor and potentiostat circuitry. The analog part features two low bandwidth AFE loops and one high bandwidth AFE loop like the ones of the AD5940 with similar specifications being able to handle currents in between 50 pA and 3 mA. The input bias current at the RE connections is typically 20 pA with a maximum of 150 pA. The two low bandwidth AFEs allow operation in parallel as bipotentiostat, and the high bandwidth AFE can connect either of the two input channels.

The wide variety of possible electrochemical methods combined with the data processing capabilities of a mixed-signal microcontroller makes the ADuCM355 well suited in combination with an electrochemical sensor as an edge device for IoT applications. The downside of its enormous complexity is the high initial learning effort when programming the device. The IC's price around $10 (October 2024, quantities > 1000), providing both the potentiostat and the microcontroller, indicates that the hardware costs play a minor role in comparison to the development effort for most applications.

The EmStat Pico by PalmSens is an OEM potentiostat module based on the ADuCM355 (Figure 3.18). Additionally, the reference electrode inputs connect through high impedance buffer amplifiers. The module's firmware offers several electrochemical methods for easy access from end-user software. Apart from OEM, the module is attractive for teaching, e. g., bringing embedded-systems students in contact with electrochemical and biosensors.

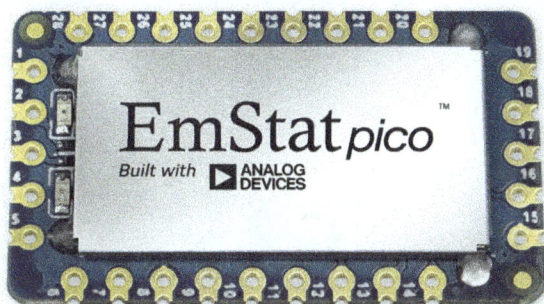

Figure 3.18: EmStat Pico module with an ADuCM355. A serial interface (UART) with a convenient communication protocol allows control and data readout. Reproduced with permission from PalmSens, The Netherlands.

3.3 Galvanostat

In contrast to the voltage-controlled situation by a potentiostat, the galvanostat controls the cell current. The galvanostatic setup (Figure 3.19) features a current source between two electrodes. Typically, E_{cell} denotes the voltage between the two electrodes and depends on both the electrodes' processes. A separate reference electrode (RE) allows for potential measurements to refer the electrode of interest, the working electrode (WE), to the electrochemical potential scale. The galvanostatic setup is well suited for electrodeposition because of the simple control of the amount deposited material using the deposition time. Also, in corrosion analysis, current-controlled methods are commonly used.

In essence, the galvanostat is a current source. A practical realization is an inverting amplifier with the electrochemical cell in the feedback loop (Figure 3.20). The set voltage controls the cell current with the resistor at the inverting input defining the current

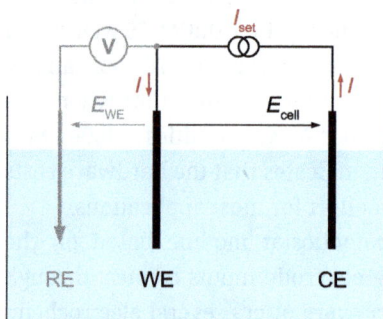

Figure 3.19: Galvanostatic setup: a current source controls the current between two electrodes (WE, CE) with an optional readout of the WE's potential vs. a reference electrode (RE).

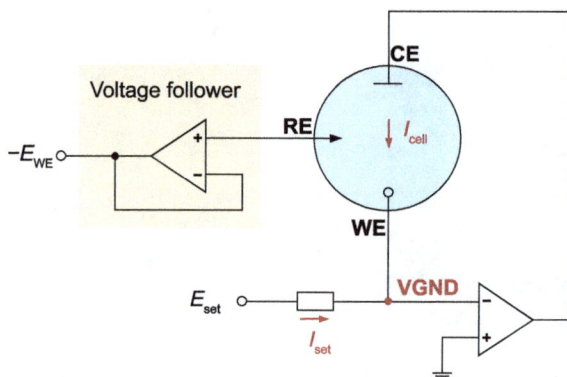

Figure 3.20: Simple galvanostat with the electrochemical cell in the feedback loop of an inverting amplifier. The third electrode (RE) and the voltage follower are optional.

range. The output voltage range of the operational amplifier determines the maximum compliance voltage of the galvanostat. Optionally a voltage follower reads the potential of the reference electrode. As the WE is on the virtual ground (VGND), the voltage follower's output is $-E_{WE}$.

3.4 Potential measurement

Potential measurements for potentiometry (Section 5.1) require voltmeters with very high input impedance or low input bias currents. An ordinary handheld multimeter offers a far too low input resistance (e. g., 10 MΩ or 100 MΩ), while a pH meter, which allows displaying the measured voltage directly, is a good choice.

Specific potentiometry instrumentation can rely on the outstanding instrumentation amplifier INA116[5] from Texas Instruments. Most prominently is the low input bias current of 3 fA typically and up to 25 fA maximum at 85 °C. The actual bias current depends on temperature and input voltage. The INA116 features guard outputs, which buffer the measured input voltages and can be used to polarize guard rings on the PCB or shields of the cabling to eliminate electromagnetic interference to the high impedance inputs.

The low input bias of the INA116 allows for potential measurements at the counter electrode in a potentiostatic setup without disturbing even lower currents typically occurring in microsensors [22]. In such applications, it is essential to avoid a mismatch of the different ground levels.

3.5 Electrical shielding

Electrochemical experiments can be sensitive to disturbance from electromagnetic waves, most critically with potentiometry and low-current applications. For some applications (e. g., amperometry), severe analog and digital filtering allow for adequate noise reduction. In general, appropriate shielding is the best option.

A faradaic cage made of solid metal or a fine metal mesh is the most effective way to shield a measurement cell from electromagnetic waves causing electrical noise issues. The cage must provide a well-conducting enclosure all around the cell (Figure 3.21). Loose connection, e. g., at the door, can lower overall efficiency. Especially for the lower frequency range, e. g., the 50 Hz or 60 Hz from the power net, it is advisable to use ferromagnetic materials. Holes or even a mesh are tolerable depending on the frequency

---

5 The manufacturer does not recommend the INA116 for new designs. Unfortunately, by now, there are no adequate replacement parts that feature comparably low input bias currents. Besides relying on INA116 stocks, the best possible solution is to build an electrometer amplifier circuit using discrete operational amplifiers with a low input bias current.

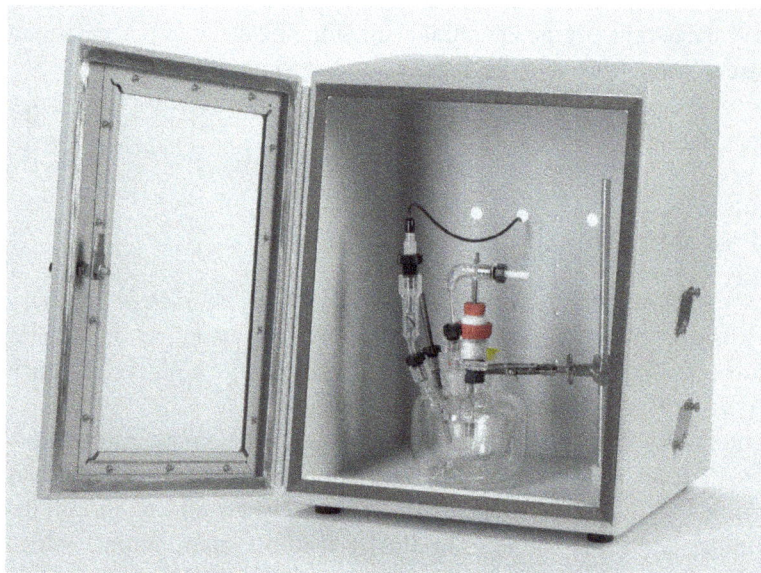

Figure 3.21: Faraday cage with a conductive glass window. Reproduced with permission of Gamry Instruments, USA.

range of the electromagnetic wave (the holes should be substantially smaller than the wavelength, which, e. g., reaches 6 cm for a Wi-Fi signal in the 5 GHz band). All cables to the electrochemical cell, especially outside the Faraday cage, should run close to each other, avoiding loops collecting disturbances. If possible, shorter cables are always the better option.

An important aspect is grounding. The Faraday cage is preferable on the GND level of the potentiostat (see Section B.2 for a more detailed discussion on different ground levels). If all connected devices are floating, the GND connections and an earth connection should meet at one point at the Faraday cage forming a star topology. If one measurement device connects to the earth, no further earth connection should exist to avoid loop-like topology ("ground loops"). Besides those rules, any specific arrangement might behave differently and sometimes unexpectedly. For each new setup, plan some time to check for noise and try different grounding methods. A battery-powered computer and a USB-powered potentiostat may help to identify grounding and earth connection issues.

4 Electrochemical laboratory

Following the chapters about the electrochemical theory and instrumentation, this chapter on the electrochemical laboratory discusses the different varieties of electrodes and their practical realization, labware, and electrolytes. A strong emphasis is on the reference electrodes and applying the electrochemical theory to describe their potential. The chapter comprises tables with properties of electrodes, electrolytes, and redox couples.

Assuming you know how to work with chemicals, this textbook does not provide many safety-related hints. If you are unfamiliar with chemistry and lab work, spend time on a proper introduction from specialists. Use appropriate protective measures (safety glasses, fumehood, gloves matching your chemicals, etc.), but do not forget common sense and blindly trust the protection. Carefully study the chemical's safety datasheets before starting any lab work.

4.1 Classification of electrodes

Different electrodes can be classified by the nature of the electrode reaction or by its polarizability. Knowing those terms not necessarily helps to understand and successfully apply electrochemical methods but allows to decipher electrochemical original works, which sometimes extensively employ those terms. It also might help to understand which other electrodes could be used as alternative by looking at the same class of electrodes. That is especially useful in case of reference electrodes.

4.1.1 Classes of electrodes

The IUPAC recommends for metallic electrodes the nomenclature using classes 0 to 3 [9]. Some other textbooks use similar groupings as electrodes of the first kind, second kind, etc., or type 1, type 2, etc.

Class 1

Metal electrodes in contact with their cations form class 1. A typical example is the reactions in a copper plating bath:

$$Cu^{2+} + 2\,e^- \rightleftharpoons Cu \quad E^0 = 0.34\,V \tag{4.1}$$

Class 2

Mostly used as reference electrodes, class 2 electrodes are metallic electrodes covered with their sparingly soluble salt in a solution with the anion. The most prominent ex-

https://doi.org/10.1515/9783111488844-004

ample is the Ag/AgCl reference electrode:

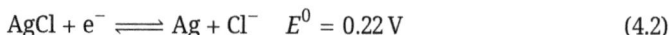

$$AgCl + e^- \rightleftharpoons Ag + Cl^- \quad E^0 = 0.22\,V \tag{4.2}$$

Class 3

The third class of electrodes is rather exotic. It groups metal electrodes with their salt (or soluble complex), a second salt (or complex), and a cation.

Class 0

Less common is the usage of the term class 0, which refers to inert electrodes. Ideally, inert electrodes are not involved in the redox reaction they enable. The term overlaps with the ideal polarizable electrode, as introduced in the following section.

4.1.2 Polarizability and potential window

The term *polarizability* is used to express the ability of an electrode to accept a potential without causing a current flow. Especially in biomedical engineering this description is useful to chose an appropriate electrode for stimulation (preferable high polarizability) or recording (preferable low polarizability). The quantity to describe the polarizability of an electrode is the exchange current density (equation (2.31)). High exchange current density means low polarizability.

Ideal polarizable electrode

At an *ideal polarizable electrode (IPE)*, no redox process occurs. Consequently, in equilibrium its potential is undefined. Upon polarization, the electrode can be at any potential level without a current flow due to a redox process (Figure 4.1(A)). The only current flowing while changing the electrode's potential is due to charge up of the double layer (Section 2.6.1). Therefore, the IPE is often considered as a pure capacitive element with its potential depending on picked up charges only.

Practical realization of a polarizable electrode are electrode/electrolyte combinations in which for a given potential window no redox process occurs. In aqueous systems, the potential window without a redox processes from decomposition (electrolysis) of water is called the *water window*. The water windows strongly depend on the electrode material, for example, it is around 2.2 V in case of glassy carbon or 1.4 V in case of gold. The exact number also depends on the electrolyte's pH [23]. Additionally, nanostructured electrode surfaces show a larger water window compared to macroscopic planar electrodes [18].

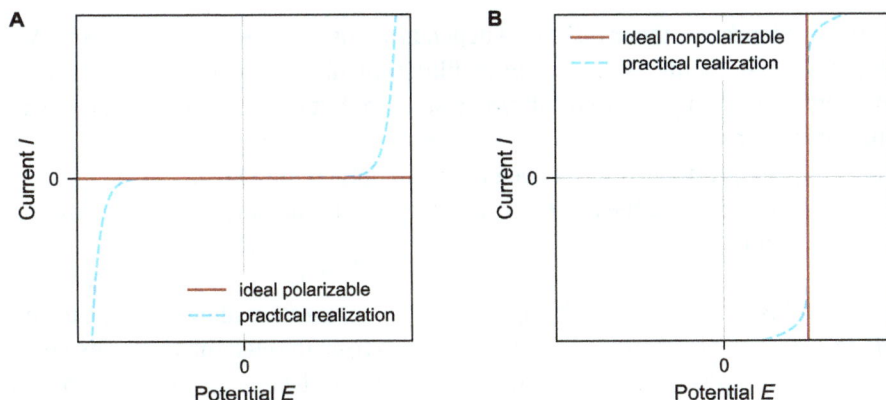

Figure 4.1: Ideal polarizable electrode (A) and ideal nonpolarizable electrode (B) with their practical realizations.

Ideal nonpolarizable electrode

The opposite case is the nonpolarizable electrode (Figure 4.1B). Ideally, the electrode keeps its potential independent of the current flow. A practical realization is an Ag/AgCl reference electrode. For a low current, the electrode keeps its potential independent of the current. However, current flow would degrade an Ag/AgCl electrode over time.

4.2 Reference electrodes

Reference electrodes (RE) are the key to reproducible experiments by defining the potential axis and, thereby, the position on the energy scale. While it is practically possible, though not advisable, to use a reference electrode in a two-electrode setup out of equilibrium, the regular operation is by ensuring zero current flow into the reference electrode. Potentiometry needs an RE in its two-electrode configuration and amperometry/voltammetry in the three-electrode setup. Many practical issues link back to an improper reference electrode.

Depending on the application, there are different requirements for the RE:

1. An equilibrium process with a high exchange current density: the exchange current density quantifies the rate of the two compensating reactions in equilibrium. For higher values, the equilibrium potential is less sensitive to external electrical disturbances.
2. An inner filling defines the electrolyte's equilibrium potential: this solution or layer provides an ion for the electrode reaction and ensures ionic conductivity toward the electrolyte.
3. No contamination of the electrolyte from leakage of the inner filling: if the solution in a reference electrode contains ions interfering with the electrochemical reaction in the electrolyte, a bridge-electrolyte may be needed.

4. Low impedance: the inner resistance depends on the exchange current density. Additionally, high conductivity of the inner filling and the connection to the electrolyte is essential. High impedance results in noise issues but may also cause instability of the potentiostat.
5. Generally, simple usage (i. e., no need for gas supply) is preferred.
6. In applications beyond the lab, easy integration and suitability for miniaturization can be essential.

Requirement 1 is fundamental in all applications. Some use cases allow skipping the inner filling (requirements 2 and 3) and relying on ions from the electrolyte. One example is applying an Ag/AgCl electrode in the life science field by depending on the Cl^- ions in physiological media. Also requirements 5 and 6 strongly depend on the application.

4.2.1 Hydrogen reference electrodes

The hydrogen electrode relys on the equilibrium of hydrogen gas and protons in solution at an inert electrode, mainly platinum, because of its excellent catalytic properties for the hydrogen reaction:

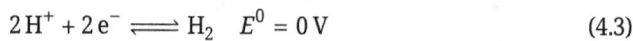

$$2\,H^+ + 2\,e^- \rightleftharpoons H_2 \quad E^0 = 0\,V \tag{4.3}$$

The equilibrium potential E depends on the proton activity a_{H^+} (the pH value) and the partial pressure p_{H_2} of the hydrogen gas:

$$E = \frac{RT}{F} \cdot \ln \frac{a_{H^+}}{\sqrt{p_{H_2}}} \tag{4.4}$$

This equation assumes the activity in $mol\,l^{-1}$ and the partial pressure in standard atmosphere (atm).[1]

Figure 4.2(A) shows the schematic arrangement. A high surface area of the platinum electrode is required to ensure enough contact between the gas and the solid phase, typically achieved by a platinized platinum electrode (platinum with a coating of roughly electrodeposited platinum). The gas and electrolyte must not contain oxygen, carbon dioxide, or carbon monoxide to avoid side reactions influencing the potential and, especially by the latter, electrode poisoning. The partial pressure of the gas also depends on the delivery depth, which needs attention in the electrode design. Several practical realizations are in use. Figure 4.2(B) shows a classical setup with a gas-filled electrode body.

---

[1] More strictly, the activity should be written as a_{H^+}/a_0 with $a_0 = 1\,mol\,l^{-1}$ and the partial pressure as p_{H_2}/p_N with $p_N = 101.325\,kPa$ to comply with the SI.

A

H$_2$

Platinized platinum

H$_2$ gas outlet

HCl

B

Wire

Salt bridge

H$_2$

HCl

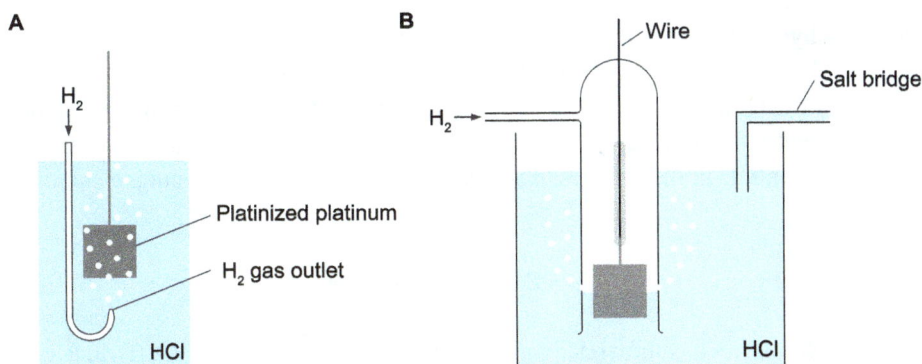

Figure 4.2: Principle of the hydrogen reference electrode: hydrogen gas bubbles at a platinized platinum electrode in an acidic solution (A). Practical realization of a hydrogen reference electrode half cell (B).

Poisoning of an electrode means the deactivation of an electrode by adsorption of substances resulting in lower catalytic properties or even a reduced accessible surface area. Apart from carbon monoxide, mainly sulfur-containing molecules are critical with noble metal electrodes. Electrode poisoning is especially crucial in life science, with electrolytes containing biological substances [24], i. e., amino acids or proteins. Besides avoiding those substances, electrode activation by chronoamperometric protocols is possible if the noble metal is the working electrode (e. g., as described in Sections 5.2.3 and 10.3.5).

Standard hydrogen electrode, normal hydrogen electrode

The electrochemical standard potentials all refer to the *standard hydrogen electrode* (SHE). The SHE is a hydrogen reference electrode with unit activity of protons (that is pH 0) and standard pressure for the hydrogen gas.[2]

The term *normal hydrogen electrode* (NHE) is ambiguous. Initially, it was a hydrogen reference electrode with a proton concentration of $1\,mol\,l^{-1}$. However, several textbooks describe the NHE as a synonym for the SHE. The difference between the two definitions is around 8 mV at room temperature.

■□□ Task 4.1: Calculate the potential difference between a normal hydrogen electrode (NHE) defined by the concentration and defined by the activity. Consider an aqueous solution of HCl as the acid and room temperature (25 °C). Take into account that the pH of a $1\,mol\,l^{-1}$ HCl solution is 0.13, and remember the definition of the pH scale (pH $= -\lg a_{H^+}$).

---

2 In contrast to most research areas, since 1982, the IUPAC recommends 100 kPa as standard pressure instead of $p_N = 101.325\,kPa$. However, at room temperature, the difference matters less than 0.2 mV for the electrode potential.

Reversible hydrogen electrode

The *reversible hydrogen electrode* (RHE) is a hydrogen reference electrode with standard pressure of the hydrogen gas but at any pH. In Section 2.8.1, you came to know about the RHE scale as a theoretical concept. In addition, the RHE is also an actual reference electrode. Potentials in RHE relates to SHE depending on the pH according to equation (2.72):

$$E_{RHE} = E_{SHE} + 2.303\frac{RT}{F} \cdot pH \tag{4.5}$$

Figure 4.3 shows a commercially available realization of an RHE. A cartridge generates hydrogen gas, which accumulates in the electrode's sleeve and reacts at a gas diffusion electrode made of platinum and palladium. For applications in which chloride ions from an Ag/AgCl electrode are not acceptable, such an electrode is the preferred choice. The glass-free setup also allows application in HF solutions.

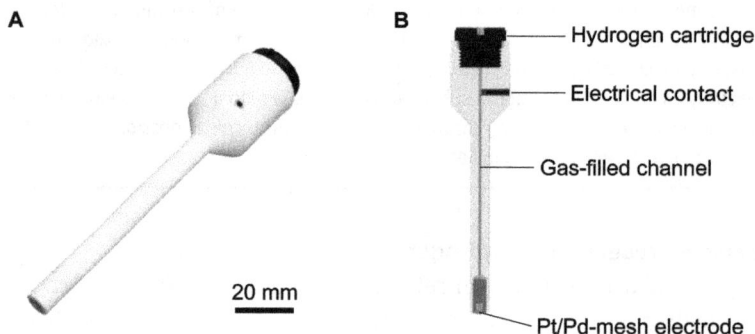

Figure 4.3: Practical realization of a reversible hydrogen electrode: the HydroFlex electrode contains a cartridge releasing hydrogen gas and a gas diffusion electrode made of palladium/platinum. Photograph and technical drawing reproduced with permission of Gaskatel GmbH, Germany.

! Because of the ambiguity of the NHE, I recommend using the SHE as the default scale when talking about the electrochemical potentials in general. For pH > 0, the RHE is the best choice to present data clearly and easily to understand.

4.2.2 Silver/silver chloride electrode

Class 2 electrodes (Section 4.1.1) are a good choice as reference electrode. Among them, it is the Ag/AgCl electrode, which is most widely applied. The equilibrium process involves the metallic silver, its salt, and chloride ions from the inner filling:

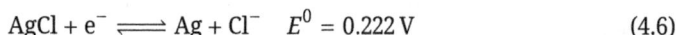

$$AgCl + e^- \rightleftharpoons Ag + Cl^- \quad E^0 = 0.222\,V \tag{4.6}$$

Figure 4.4A illustrates a possible setup with a silver wire coated by silver chloride. The inner filling is an aqueous solution containing chloride ions, e. g., 3 M KCl. For simplicity, one can assume chloride ions from the solution penetrate the AgCl and react with the metallic silver to form the salt or reverse. That means the equilibrium reaction happens at the boundary between Ag and AgCl, translating between electron conduction in the wire and ionic conduction in the aqueous solution (Figure 4.4(B)). A diaphragm links the inner filling and the electrolyte of the electrochemical cell, enabling ionic conduction while preventing a notable exchange of substance between the two compartments.

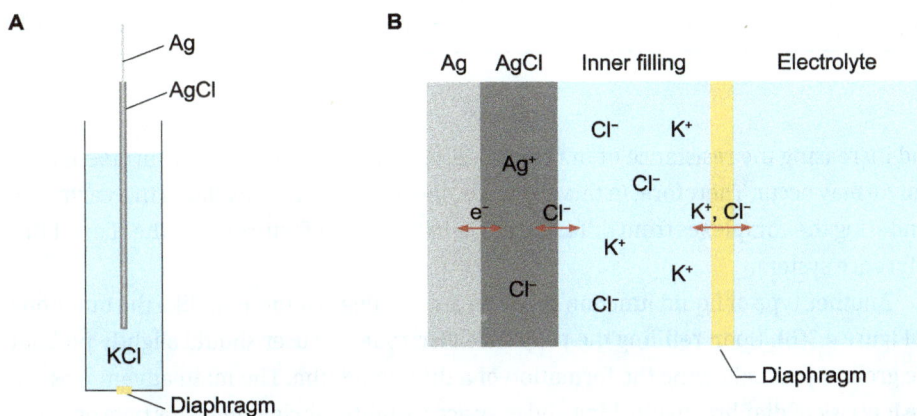

Figure 4.4: Ag/AgCl reference electrode: schematic setup with inner filling and diaphragm (A) and illustration of the different phases (B).

The equilibrium potential depends on the activity of the chloride ion in the inner filling according to the Nernst equation with unit activity for all solids:[3]

$$E = E^0 - \frac{RT}{F} \cdot \ln a_{Cl^-} \tag{4.7}$$

Table 4.1 shows the equilibrium potential for various KCl concentrations at 25 °C.

Most commercially available macroscopic Ag/AgCl electrodes have a design comparable to the illustration in Figure 4.4(A). An example is the electrode in Figure 4.5(A). The glass tube contains 3 M KCl and connects to the measurement electrolyte by a ceramic pin diaphragm at the bottom tip. Often 3 M KCl or a saturated KCl solution is the inner filling reducing liquid junction potentials (Section 4.2.5). An inherent issue of the high concentration KCl solution is the formation of complexes from the AgCl solution. Over time, the complexes accumulate and precipitate close to the diaphragm as it is typically the location with the lowest ion concentration. Apart from blocking the diaphragm

---

3 More strictly, the operand of the logarithm should be written as a_{Cl^-}/a_0 with $a_0 = 1 \, mol \, l^{-1}$.

Table 4.1: Potentials versus SHE for various reference electrodes at 25 °C. Values are calculated based on activity data from Hamer and Wu [25].

Electrode	Potential
SHE	0 V
Ag/AgCl (0.1 M KCl)	0.288 V
Ag/AgCl (3 M KCl)	0.206 V
Ag/AgCl (3.5 M KCl)	0.201 V
Ag/AgCl (KCl saturated)	0.196 V
Calomel electrode (0.1 M KCl)	0.333 V
Normal calomel electrode (NCE)	0.280 V
Saturated calomel electrode (SCE)	0.242 V

and increasing the resistance of the electrode, contamination of the measurement electrolyte may occur. Therefore, in this electrode, the Ag/AgCl is encapsulated in a cartridge, hindering the complexes from diffusing out, what the manufacturer calls the "Long Life" reference system.

Another type of liquid junction relies on a separable ground-joint like the outer one in Figure 4.5(B). Upon refilling the reference electrode, the user should slightly pull out the ground-joint, allowing the formation of a thin liquid film. The main advantages are the low risk of diaphragm blocking and easy accessibility for cleaning. This type of liquid junction is also beneficial when used in stirred electrolytes.

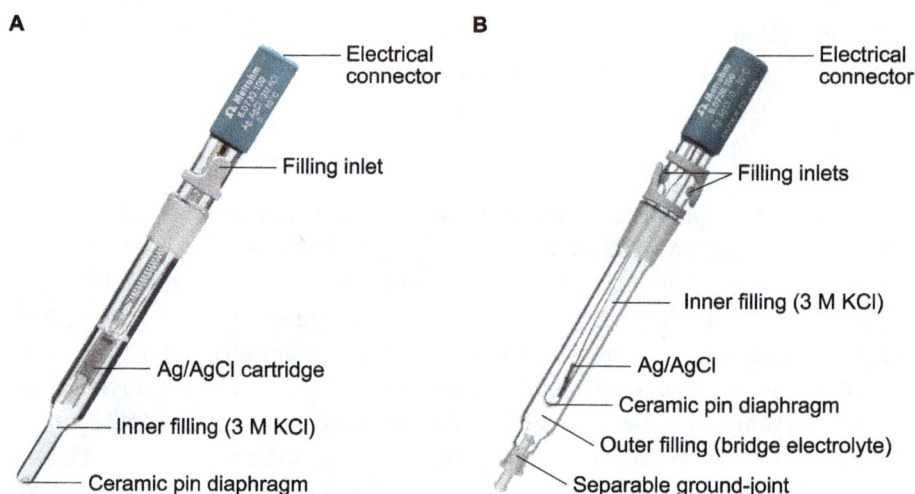

Figure 4.5: Ag/AgCl reference electrodes: single-chamber electrode with the "Long Life" reference system (A), and a two-chamber electrode comprising the bridge electrolyte (B). Reproduced with permission from Metrohm Autolab, The Netherlands.

Miniaturized Ag/AgCl electrodes often consist of a silver layer that is partially electrochemically converted to AgCl (see the infobox in Section 9.2.1). Typically, such electrodes come without a separate inner filling and rely on the chloride ions in the measurement electrolyte, e. g., at physiological levels in biomedical applications. Preferably such Ag/AgCl electrodes are covered with a hydrogel membrane ensuring stagnant fluidic conditions at the electrode/electrolyte interface and reducing the risk of contamination of the electrolyte by silver ions from the electrode.

Temperature dependency

Several factors defining the electrode potential are temperature dependent:

$$E(T) = E^0(T) - \frac{RT}{F} \cdot \ln a_{Cl^-}(T) \tag{4.8}$$

Typically, the temperature dependency of the chloride ion activity is negligible compared to $E^0(T)$ and the T in front of the logarithm. Only with saturated inner fillings, temperature variation matters for the activity because of the strong temperature dependency of the solubility.

Bates and Bower proposed a generally accepted polynomial for the temperature dependency of E^0 in the range $0\,°C$ to $90\,°C$ [26]:

$$E^0(\vartheta) = A + B \cdot \frac{\vartheta}{°C} + C \cdot \left(\frac{\vartheta}{°C}\right)^2 + D \cdot \left(\frac{\vartheta}{°C}\right)^3$$

$$A = 0.23659\,V, \quad B = -4.8564 \times 10^{-4}\,V \tag{4.9}$$

$$C = -3.4205 \times 10^{-6}\,V, \quad D = 5.869 \times 10^{-9}\,V$$

The polynomial uses the temperature $\vartheta = T - 273.15\,K$ in $°C$.

Table 4.2 shows equilibrium potentials around room temperature and $37\,°C$, which is relevant in biomedical applications. The values were calculated assuming negligible temperature influence on the activity.

Table 4.2: Potential of Ag/AgCl reference electrodes versus SHE for various temperatures. Values are calculated based on activity data from Hamer and Wu [25].

[KCl]	20 °C	25 °C	37 °C
0.1 M	0.289 V	0.288 V	0.282 V
1 M	0.237 V	0.235 V	0.227 V
3 M	0.209 V	0.206 V	0.197 V
3.5 M	0.204 V	0.201 V	0.192 V

4.2.3 Silver/silver bromide electrode

While the Ag/AgCl is a good match in life science applications, other systems may demand a chloride-free solution. When bromide is acceptable or even already part of the system, the Ag/AgBr electrode, another class 2 electrode, is a good choice. The equilibrium process is similar to the Ag/AgCl chloride but at lower absolute potential:

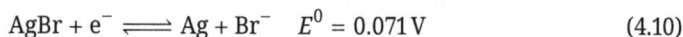

$$AgBr + e^- \rightleftharpoons Ag + Br^- \quad E^0 = 0.071\,V \tag{4.10}$$

The potential depends in the same way on the bromide ion as the Ag/AgCl electrode's potential on the chloride ion (equation (4.7)). Figure 10.18 shows a system with a miniaturized Ag/AgBr electrode for sensor application in microreactors.

4.2.4 Calomel electrode

The calomel electrode is another class 2 electrode, long the main alternative to the hydrogen reference. Its equilibrium reaction is a two-electron process:

$$Hg_2Cl_2 + 2\,e^- \rightleftharpoons 2\,Hg + 2\,Cl^- \quad E^0 = 0.268\,V \tag{4.11}$$

The potential depends on the chloride ion activity comparable to the Ag/AgCl:

$$E = E^0 - \frac{RT}{F} \cdot \ln a_{Cl^-} \tag{4.12}$$

Because of the mercury, today's applications should avoid the calomel electrode whenever possible. Still, the knowledge of the different calomel electrode's potential scales is essential to link to older electrochemical literature. The saturated calomel electrode (SCE) comprises an inner filling with a saturated KCl solution, while the normal calomel electrode (NCE) relays on a 1 M KCl solution. Table 4.1 includes the potentials for different calomel electrodes.

4.2.5 Diffusion potential, liquid junction potentials

At a liquid junction between two electrolytes with different concentrations or ion compositions, diffusion occurs. Caused by different mobilities of the cations and ions, a potential difference between the two electrolytes forms, called the *diffusion potential* or, more generally, the *liquid junction potential*.

Figure 4.6 illustrates the situation at a liquid junction between two electrolytes (phases α, β). The higher concentration in the left electrolyte (α) causes ions to diffuse through the permeable liquid junction to the right side. Because of the different ion mobilities (in this example, the larger cations are slower), charge separation occurs,

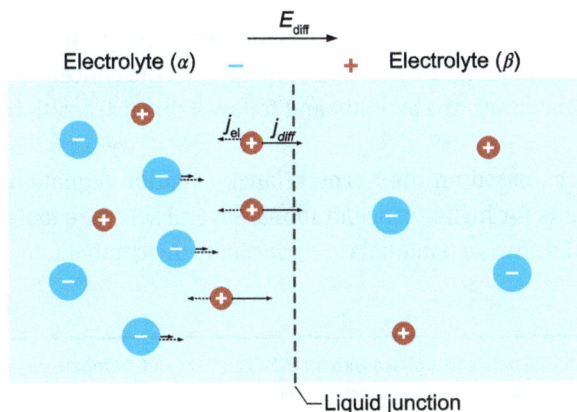

$$E_{diff}$$

Electrolyte (α) — + Electrolyte (β)

j_{el} j_{diff}

Liquid junction

Figure 4.6: Illustration on the formation of the diffusion potential: the concentration difference causes a diffusive species flux (j_{diff}). An electrical field forms because of different ion mobilities, which causes a counteracting species flux (j_{el}). Overall a diffusion potential E_{diff} establishes.

causing an electrical field. This field acts against the charge separation so that a dynamic equilibrium between diffusion and migration sets in. The resulting potential difference between the two sides of the liquid junction is the diffusion potential.

In the case of the same simple electrolyte (e. g., KCl or HCl solutions) in both phases (α, β), the diffusion potential at the liquid junction is

$$E_{diff} = \frac{\mu_+ - \mu_-}{\mu_+ + \mu_-} \cdot \frac{RT}{F} \ln \frac{a^\alpha}{a^\beta}. \tag{4.13}$$

The potential grows with the difference between the electrical mobility μ_+ of the cation and μ_- of the anion. Table 4.3 list mobility values for commonly used ions.

Table 4.3: Ion mobilities for infinite dilution at 25 °C.

Cation	μ_+ in m² V⁻¹ s⁻¹	Anion	μ_- in m² V⁻¹ s⁻¹
H^+	362.4×10^{-9}	OH^-	205.2×10^{-9}
K^+	76.2×10^{-9}	Br^-	80.9×10^{-9}
NH_4^+	76.2×10^{-9}	Cl^-	79.1×10^{-9}
Na^+	51.9×10^{-9}	NO_3^-	74.0×10^{-9}
Li^+	40.1×10^{-9}	Acetate⁻	42.4×10^{-9}

For more complex electrolytes and different ions in each phase, the Henderson equation provides an approximation for the diffusion potential:

$$E_{diff} = \frac{\sum_i \frac{|z_i|\mu_i}{z_i}(c_i^\beta - c_i^\alpha)}{\sum_i |z_i|\mu_i(c_i^\beta - c_i^\alpha)} \cdot \frac{RT}{F} \ln \frac{\sum_i |z_i|\mu_i c_i^\alpha}{\sum_i |z_i|\mu_i c_i^\beta} \tag{4.14}$$

The equation considers the valency z_i, mobility μ_i, and concentration c_i for each ion i in the phase α or β. The Henderson equation relies on two assumptions: the concentrations are low (i. e., the concentration equals the activity) and follow a linear transition between the two phases.

Other calculation methods, e. g., based on the Nernst–Planck equation (equation (2.47)), provide more accurate results for higher concentrations. Several software tools exist, simplifying the calculation of diffusion potentials and especially the cumbersome lookup of parameters.[4]

---

i ■□□ Task 4.2: Calculate the diffusion potential between a phase with 0.1 M HCl and 1 M HCl. Compare your results to the situation with 0.1 KCl and 1 M KCl. You should consider the activity coefficients 0.832 (0.1 M HCl), 0.741 (1 M HCl), 0.785 (0.1 M KCl), and 0.602 (1 M KCl). Approximate the ion mobilities by the values for infinite dilution.

Depending on the actual system, apart from diffusion, other effects may contribute to the liquid junction potentials. For example, in stirred solutions, electrokinetic phenomena can play a role, or at the interface between two immiscible electrolyte solutions (ITIES), a potential difference occurs depending on the actual charge transfer mechanism.

Minimizing liquid junction potentials

With a constant composition of the two phases' electrolytes, calibration or calculation allows for compensation of the diffusion potential. However, with varying or unknown analyte composition, minimizing diffusion potentials is the preferred approach.

There are mainly two strategies:

1. The best option is choosing electrolytes with similar electrical mobilities of the anion and cation. If not possible, a salt-bridge with a filling of a high concentration solution of a simple electrolyte with similar mobilities lowers the liquid junction potentials. Consult Table 4.3 for the ion mobilities. Good choices for the filling are solutions of KCl, KNO_3, or lithium acetate in the molar range. The actual selection also needs to consider, e. g., the need for a Cl^--free electrolyte, as is often the case in corrosion research.

2. A small but steady flow across a diaphragm is the other option to avoid liquid junction potentials. The manufacturer of the electrode in Figure 4.5A recommends keeping the filling inlets during measurements open to allow a constant outflow of the inner filling due to gravitation. The flow rate is low, specified up to $25\ \mu L\ h^{-1}$ depending on the diaphragm condition. With a separable ground-joint like the outer one in Figure 4.5(B), the flow rate can reach up to $100\ \mu L\ h^{-1}$.

---

4 For example, *LJPcalc* is an open-source tool to calculate diffusion potentials for common situations in electrophysiology using the Nernst–Planck equation: https://swharden.com/software/LJPcalc

4.3 Working and counter electrodes

The selection of the working electrode strongly depends on the application. The electrode material can participate in the redox process or only catalyze the reaction. In material science, it is the material to be investigated, which is the working electrode.

4.3.1 Metal electrodes

Metal is the classic electrode material since the beginning of electrochemistry. The metal can be part of the electrode reaction, be involved in surface reactions, or contribute through catalysis of the electrode reaction. Copper, zink, nickel, or silver electrodes are, in most cases, class 1 electrodes (Section 4.1.1), being consumed by the electrochemical reaction.

Platin group metal electrodes like gold, platinum, iridium, and palladium are involved in surface reactions with water but are essentially inert beyond the surface layers. Among them, platinum is one of the most exciting materials for research and application (more detailed discussion in Section 8.4). Palladium has a unique role because of its capability to absorb large amounts of hydrogen forming palladium hydride. All platinum group metals are excellent catalysts, especially for reactions involving oxygen, hydrogen, water, and hydrogen peroxide.

Table 4.4 summarizes some essential properties of the different metals used as electrodes. The equilibrium potential E^0 indicates the thermodynamic possibility for metal oxidation. However, the process does not necessarily need to occur at the given potential or does occur at all in a specific electrolyte.

Table 4.4: Properties of metal electrodes: atomic number (Z), density (ρ) at 25 °C, melting point (ϑ_{melt}), conductivity (σ) at 25 °C, its reaction equation, and the corresponding E^0.

Electrode	Z	ρ in g cm^{-3}	ϑ_m in °C	σ in S m^{-1}	Reaction	E^0 in V
Nickel	28	8.9	1455	14.0×10^6	$Ni^{2+} + 2\,e^- \rightleftharpoons Ni$	−0.26
Copper	29	9.0	1085	58.4×10^6	$Cu^{2+} + 2\,e^- \rightleftharpoons Cu$	0.34
Zinc	30	7.1	420	16.6×10^6	$Zn^{2+} + 2\,e^- \rightleftharpoons Zn$	−0.76
Palladium	46	12.0	1555	9.3×10^6	$Pd^{2+} + 2\,e^- \rightleftharpoons Pd$	0.95
Silver	47	10.5	962	61.8×10^6	$Ag^+ + e^- \rightleftharpoons Ag$	0.80
Iridium	77	22.5	2446	21.3×10^6	$Ir^{3+} + 3\,e^- \rightleftharpoons Ir$	1.16
Platinum	78	21.5	1768	9.3×10^6	$Pt^{2+} + 2\,e^- \rightleftharpoons Pt$	1.18
Gold	79	19.3	1064	44.3×10^6	$Au^+ + e^- \rightleftharpoons Au$	1.69
Mercury	80	13.5	−39	1.0×10^6	$Hg^{2+} + 2\,e^- \rightleftharpoons Hg$	0.85

For application as a working electrode, the metal electrodes are usually embedded in an electrically isolating material (glass, polymer), and only a disk with a defined diameter faces the electrolyte. Thus, polishing of both electrode material and isolation is possible. However, forming a gap between metal and isolator may occur and could cause unexpected behavior because of substances in trace amounts from previous experiments. Figure 4.7 displays typical macroscopic electrodes.

Figure 4.7: Different macroscopic disk electrode. The disk diameters are 2 and 3 mm, the outer material of the two left most electrodes is glass, the others are polymers. The electrode materials are Pt, glassy-carbon (GC), Au, Pt, GC, and Au (from left).

4.3.2 Mercury electrodes

Mercury electrodes are the only liquid metal electrode at room temperature and form an entirely different working electrode category. Historically, the mercury electrode was the preferred electrode for analytical purpose dating back to the introduction of the dropping mercury electrode in 1922 by Jaroslav Heyrovský. He coined the term *polarography*, which means voltammetry (Section 5.3) using a periodically renewed mercury electrode, which was for long synonymous with electroanalysis. Significant benefits of a mercury electrode are the simple possibility to obtain a new contamination-free electrode surface by the generation of another mercury drop, the smooth electrode surface, and low background current from the electrode material itself. The potential window extends largely below 0 V versus SHE but is limited on the oxidation side early by the dissolution of the electrode. The severe disadvantages are the toxicity of mercury (including the vapor) and environmental concerns.

Initially, gravitation was the driving force to generate new drops (Figure 4.8). Either by adjusting the reservoir's level or an additional valve, the experimenter could control the drop generation. Modern mercury electrode setups pressurize the mercury with a

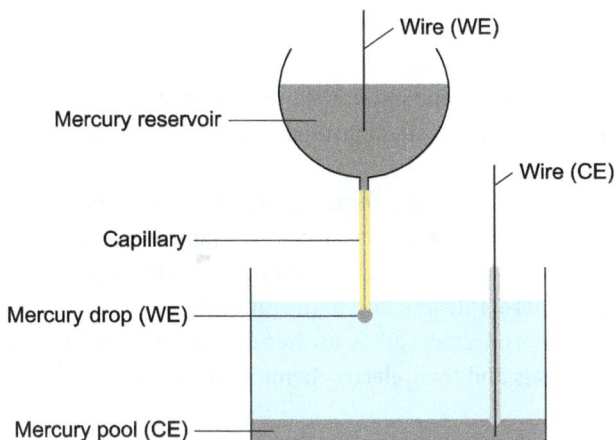

Figure 4.8: Dropping mercury electrode: periodically renewed drop electrode as working electrode (WE), optionally the excess mercury pool can be used as the counter electrode (CE).

gas to automatically control dosage and periodical renewal of the droplet. The presence of oxygen causes the mercury to oxidize and, therefore, requires an inert gas. The drops evolve at the end of a capillary with an inner diameter of several ten micrometers. The typical drop generation duration is up to 5 s with a maximum drop diameter of 1 mm.

Excess mercury from fallen drops accumulates at the bottom of the electrochemical cell. One possibility is to connect this mercury pool electrically and use it as the counter electrode as indicated in Figure 4.8.

Different operation modes are possible:

1. The classical method is the *dropping mercury electrode* (DME) with a continuous drop generation and falling once the drop reached its maximum diameter. During the measurement, the electrode area continuously changes, resulting in the characteristic curves with a fluctuating current band.
2. The *hanging mercury drop electrode* (HMDE) is the static variant in which, after the formation of the drop, the mercury flow stops resulting in a fixed diameter.
3. Both continuously renewed electrode surface and constant electrode area are features of the *static mercury drop electrode* (SMDE). Here, new drops periodically form, but during voltammetry, each voltage step uses one fresh drop. Synchronization of drop generation and measurement allows for a constant electrode area.

The *mercury film electrode* (MFE) is an alternative to the HMDE with a mercury layer deposited on a solid electrode, mainly glassy carbon [27]. Because of a film thickness of less than 100 µm, the total amount of mercury is low. At the same time, this approach sacrifices the benefit from the continuously renewed electrode surface. Also, the deposition of mercury on platinum is possible [28]. In this work, the focus was on the application at UMEs for scanning electrochemical microscopy (Section 6.2.1).

4.3.3 Carbon electrodes

Carbon is another important group of electrode materials. In contrast to (noble) metal electrodes, the absence of catalytic support for the decomposition of water provides a large water window.

The variety of carbon electrodes is enormous. Here, we focus on the different carbon allotropes most relevant for macroscopic electrodes. Table 4.5 summarizes the properties of the various electrodes, namely graphite, glassy carbon, and boron-doped diamond. Sharma provides a helpful general introduction to the different forms of carbon materials and their physicochemical properties [29]. A further discussion of the broad field of nano-featured carbon materials and their electrochemical application is in Section 9.4.1.

Table 4.5: Properties of carbon electrodes: density ρ and conductivity σ at 25 °C.

Electrode material	ρ in g cm^{-3}	σ in S m^{-1}
Glassy carbon (GC)	1.5	20×10^3
Graphite	2.2	$0.3 \times 10^3 \perp$ basal plane
		$200–400 \times 10^3 \parallel$ basal plane
Boron-doped diamond (BDD)	3.5	$0.2–50 \times 10^3$

Graphite

Graphite is a crystalline carbon structure with sp^2 hybridization (Figure 4.9). The material consists of layers with a two-dimensional hexagonal lattice (graphenes). Besides the naturally occurring α-graphite with an alternating stacking of the graphene layers, shear forces can lead to the β-graphite with a rhombohedral lattice due to different stacking. However, the overall properties of the two forms are similar.

The conductivity of graphite is highly anisotropic. The conductivity parallel to the basal plane of the lattice is three orders of magnitude high than perpendicular to it. Macroscopic graphite electrodes show an apparent isotropic conductivity because of their polycrystallinity.

> **i** The general availability of graphite as nowadays pencil lead made it an attractive material for teaching or demo experiments with public outreach. Also, in the scientific community, some interest in the so-called *pencil graphite electrode* came up, especially discussing it in sensor application for low-resource settings because of the ubiquitous availability of pencils at low cost. However, it is questionable whether the electrode material governs the sensor material and fabrication cost or availability issues in a low-resource setting. Annu et al. summarized in a recent review many publications around the pencil graphite electrode [30].

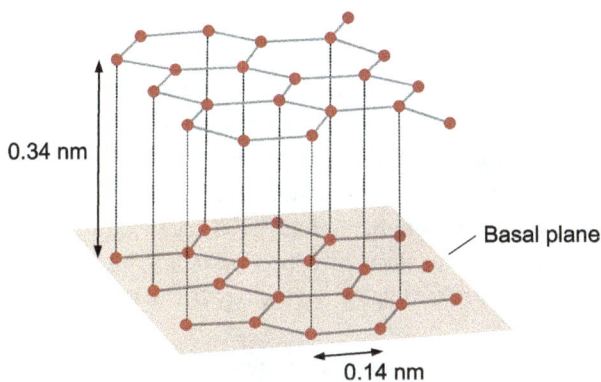

Figure 4.9: Graphite: a stack of layers with a two-dimensional hexagonal lattice (graphenes). In *a*-graphite, the third layer matches the position of the first layer. The bonds between the carbon atoms in the hexagonal lattice are covalent. Van der Waals forces hold the layers together.

Glassy carbon (glass-like carbon)

Glassy carbon (GC) is another carbon allotrope that is nongraphitizing and shows a glass-like appearance. The IUPAC recommends *glass-like carbon* instead of the widely used term glassy carbon as it has no structural similarity with silicate glasses. Like graphite, it is fully sp^2 hybridized. Besides its good conductivity, in between the two values for the directions in graphite, in combination with well-defined electrochemical properties and hardness, GC electrodes are often the preferred macroscopic electrodes when cost aspects do not dominate. Other features are GC's defacto impermeability for gases or liquids and its low density.

Diamond

Diamond, a complete sp^3 hybridized carbon crystal, is a perfect isolator, not suited as electrode material. It is a semiconductor with a wide bandgap (5.5 eV at room temperature). However, boron doping provides electrical conductivity to the diamond and enabling its usage as a *boron-doped diamond* (BDD) electrode. While a high dopant concentration can produce metallic conductivity, the BDD's electrochemical behavior deviates from real metal electrodes. An essential aspect is the termination of the carbon atoms at the surface by either hydrogen or oxygen. The surface termination depends on the fabrication method but can also be electrochemically modified afterward. Besides the large water window typical for carbon, the BDD electrode features lower background currents than GC and graphite electrodes. Macpherson provided a helpful guide for the application of BDD electrodes [31]. Deposition of BDD films on substrates such as quartz allows for optically transparent electrodes.

4.3.4 Metal oxide electrodes

Among the metal oxide electrodes, it is iridium oxide, which has the widest variety of possible applications. For example, it is a common choice as the indicator electrode in potentiometric pH microsensors. Accordingly, an iridium oxide layer can serve as a reference electrode if the pH is constant or know. In contrast, iridium oxide is a candidate for neurostimulation as well. Here, it is beneficial that the charge from the surface processes is large compared to noble metal electrodes (Section 10.3.3). Historically, iridium oxide was of interest as a candidate for display technology because of its electrochromism.

Iridium oxide electrodes are typically films coated on a metal electrode. The film thickness can range from several hundred nanometers to tens of micrometers. There are three common fabrications methods:
- Reactive sputtering in an argon/oxygen plasma with an iridium target results in a *sputtered iridium oxide film* (SIROF). The process conditions control the morphology and composition of the oxide [32]. The obtained layers are per se anhydrous.
- Repetitive potential cycling of an iridium electrode in an aqueous electrolyte is another possibility to form iridium oxide, called an *activated iridium oxide film* (AIROF). In contrast to the previous method, the layers are largely hydrated.
- Electrodeposition of iridium oxide is also possible, resulting in an *anodically electrodeposited iridium oxide film* (AEIROF). Most of the recipes in literature go back to the formulation of Yamanaka [33]. Section 9.2.3 discusses this process in more detail.

The different fabrication methods result in different film morphology and degree of hydration of the oxide film. The latter strongly influences the pH sensitivity when used in a potentiometric sensor (Section 10.1.1).

Other oxide electrodes, e. g., ruthenium oxide, are in use, too. Besides their application as a pH-sensitive electrode, different oxides got attention for potential application in supercapacitors or electrolysis.

Indium tin oxide (ITO) is an optically transparent material usually composed of 74 % indium, 18 % oxygen, and 8 % tin. Depending on the thickness, it has a colorless or yellowish appearance. The transparency in combination with a conductivity around 1×10^6 S m^{-1} makes ITO a valuable material for displays, solar cells, or shielding. Besides, ITO electrodes are an exciting material for spectroelectrochemistry (Section 6.3.4) and play a role in electroanalysis [34]. However, the possible pH range for ITO electrodes is limited to neutral and alkaline solutions.

4.3.5 Counter electrodes

All working electrode materials may serve as counter electrodes, too. It strongly depends on the application if analytical optimum or cost dominates the selection. Several questions need to be addressed:

- Is there any issue with the contamination from the CE material?
- Can the CE material itself interact with the compounds in the electrolyte?
- Does the size or weight of the CE play a role? If yes, a material with a high exchange current density for the CE processes and geometries with high accessibility are the best choices.
- What are the technological options, especially when thinking about microsystems?
- Do material or fabrication costs play a role?

Metal wires or sticks, with or without a holder, meshes or foils, are a suitable electrode form to be used as the counter electrode. For noble metal analysis, the counter electrode is preferably the same material as the working electrode to avoid contamination. In the case of redox active species in the electrolyte, noble metal electrodes may catalyze a decomposition reaction and, therefore, carbon is the better choice. If the counter electrode's possible area is limited, depending on the current to pass through, a higher (or lower) CE potential occurs. Especially with integrated potentiostats, the maximum compliance voltage may limit the cell current. Here, noble metal electrodes are beneficial over carbon materials. If cost dominates the decision, stainless steel plates or grids are a good compromise.

4.3.6 Ultramicroelectrodes

In electrochemistry, the term microelectrode referred to small electrodes causing a hemispherical diffusion profile ("microelectrode effect") even when the characteristic length of the electrode was in the millimeter range. To clearly distinguish the term, ultra-microelectrode (UME) was coined addressing electrodes with the size in the micrometer scale (often defined by at least one dimension small than 25 µm).

Initially, UME described tiny wires embedded in a glass cladding with a polished tip so that a disk electrode with, e. g., 10 µm diameter is formed. Those electrodes show massive faradaic currents due to the spherical diffusion profile but low capacitive contributions and small ohmic losses.

The characteristic quantity to describe the relationship between the sheath thickness and the electrode diameter is the RG value:

$$RG = \frac{r_T}{a} \tag{4.15}$$

r_T is the outer radius of the tip, a the radius of the electrode. A low RG value indicates less disturbance of the diffusion profile by the sheath. Typical state-of-art electrodes have RG values between 3 and 10.

Figure 4.10 shows commercially available UMEs. Similarly, electrodes with more than one metal core are possible. For example, in what is called a heptode, seven UMEs

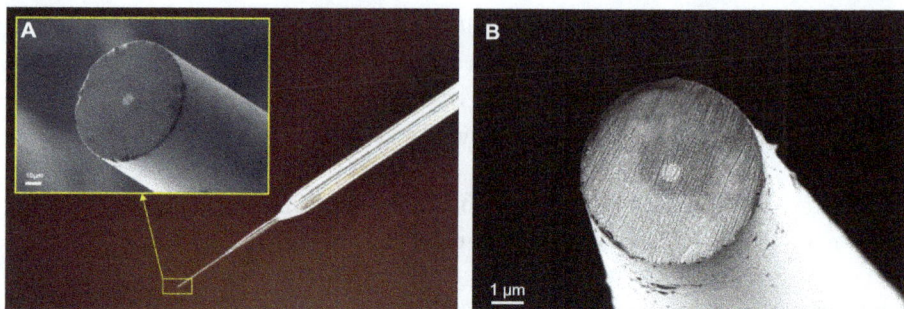

Figure 4.10: Commercial ultramicroelectrodes (UMEs): a 10 µm diameter platinum wire in a quartz glass fiber (A) and an electrode with a diameter around 0.5 µm (B). Reproduced with permission of Thomas RECORDING GmbH, Germany, www.ThomasRECORDING.com.

are embedded in the same fiber for advanced applications in a scanning electrochemical microscope (SECM, see Section 6.2.1) [35, 36].

Several researchers fabricate their application-specific UMEs by themselves. A process to manufacture UMEs by glass-pulling was described, e. g., by Danis et al. [37]. 1993, Heinze summarized the features of different UME geometries in [38] and presented what this book addresses in the next section as microfabricated electrodes.

4.3.7 Microfabricated electrodes

The further miniaturization of the UMEs based on wires in glass comes along with very low currents. Classical microtechnology offers an alternative in the form of electrode arrays with many small electrodes (more precisely: electrode openings) in parallel.

Figure 4.11(A) shows such an arrangement in top view. Electrode diameters could be in the range of a few micrometers or even below. Depending on the desired current, thousands of such electrodes form the array. Practically, a metal thin-film is the electrode material with an isolation layer on top. As shown in Figure 4.11(B) as a cross-section, each electrode is a round opening in the isolation layer. Therefore, all electrodes are connected in parallel. A step response causes an evolving hemispherical diffusion profile like from a single disk electrode with the same diameter but with a much higher current. However, for longer on-time, the individual hemispherical diffusion profiles will overlap depending on the distance between the electrodes, and the benefit of the hemisphere gets lost. But even in this situation, the actual electrode area is lower than an all planar electrode, which also translates in lower double layer capacity and background currents.

Another typical microfabricated electrode design is the interdigitated electrode array (IDA), as shown in Figure 4.11(C). It consists of two comb-like structures, which allow two separate electrical connections. Section 10.1.7 discusses electrochemical methods

A

C

B

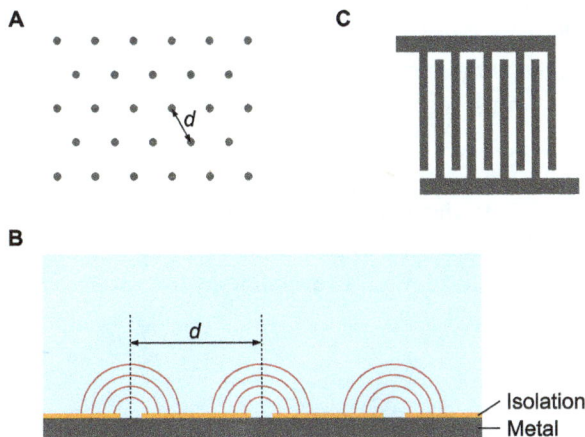

Figure 4.11: Microfabricated electrodes: Top-view of a microelectrode array (A) and corresponding cross-section (B) with typical diffusion profile. An interdigitated electrode array (IDA) consists of two pairs of fingers (C).

benefiting from this arrangement. Also, for electrochemical impedance measurement, those structures are helpful (Chapter 7).

4.3.8 Gas-diffusion electrodes

The key element of the gas-diffusion electrode is the porous electrode material from one side partly filled with the liquid electrolyte and from the other side filled with gas (Figure 4.12). The primary benefits are the high surface area for the dissolution of the gas and the fast gas exchange in the gaseous phase. At the electrode, the redox reaction involving the dissolved gas can take place.

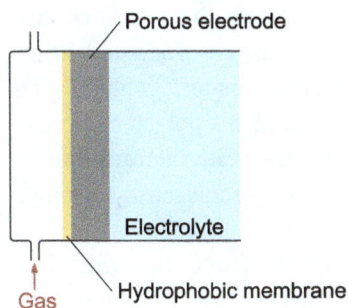

Figure 4.12: Gas-diffusion electrode with a porous electrode material separating the liquid electrolyte from the gas phase, the hydrophobic membrane is optional.

The gas pressure and the capillary pressure of the electrolyte need to be balanced. The introduction of a hydrophobic membrane on the porous catalyst layer allows for higher gas pressure and, at the same time, more flexibility in selecting the electrode material.

Typical applications of a gas-diffusion electrode are fuel cells, electrolysis cells, or electrochemical CO_2 conversion. Also, batteries like the zinc-air battery (the classical cells for hearing aids) or the more recent lithium-air battery use such a setup. The platinum/palladium mesh of the practical realization of a reversible hydrogen electrode (Figure 4.3) is a gas-diffusion electrode as well.

4.3.9 Electrode polishing

The surface of the macroscopic metal and carbon electrodes requires polishing before usage resulting in both removals of adsorbates and a smooth surface. During the manufacturing process, the electrodes are polished, often providing excellent surface quality. However, once an electrode was in use, the surface roughness might have increased, and potential adsorbates may affect measurement results. Latest then, the electrodes need polishing by the operator. Alternatively, scientists polish each electrode immediately before the measurement to ensure reproducible starting conditions. Inspection of the electrode surface by light microscopy and determining the surface roughness by electrochemical methods (Section 8.4.2) is essential to judge about necessity and progress of polishing. Especially noble metal electrodes like platinum or gold require some potential cycling after the polishing process to reach a stable surface constitution.

Polishing typically starts with a coarse grain (e. g., 5 or 1 µm) followed by one or more steps to a fine grain (e. g., 200 or 50 nm). Without visible scratches, for adsorbates removal, it is sufficient to work with fine grain only. A polishing slurry can be alumina (Al_2O_3) or diamond powder with a defined grain size in water. Alternatively, pre-made pastes can be applied. The electrode is moved manually over a polishing pad with the slurry or paste. It is essential to ensure that the polishing acts planar and without any predominant direction (patterns of eights are a better choice over circular movement). In between different grain sizes and at the end, the electrode needs careful rinsing. The operator must use each pad with one electrode material only to avoid cross-contamination. Especially with disk electrodes embedded in a polymeric or glass cladding, the risk is that contaminations can accumulate at the material transition not accessible to or even worsened by polishing. Ideally, for each type of electrolyte/analyte combination, a dedicated electrode should be used.

Also, electrode polishing machines are available, mainly used for UMEs. While those apparatuses ensure reproducible results independent of the operator, well-trained staff can manually achieve equally good or even better results. Thin-film microelectrodes with a typical film thickness of 100 to 300 nm cannot be polished.

4.4 Electrochemical cells – labware

Practical realization of an electrochemical cell with liquid electrolyte can be any vessel containing the electrolyte. However, typical labware consists of a setup as shown in Figure 4.13, often with only a subset of the features. For example, the electrolyte vessel can be double-walled to allow liquid to pass through the jacket for thermostatic control. A lid can be both holder for the electrodes and ensure limited gas exchange with the environment. Latter plays an important role when an oxygen-free electrolyte is required. Inert gas bubbling through the solution helps drive out the dissolved oxygen; a diffusor can reduce the time needed. However, the continuous gas flow causes convection in the electrolyte, disturbing mass-transport limited reactions. Therefore, the inert gas supply can be switched to a second inlet flooding the space above the liquid level to prevent fresh oxygen from entering. Argon is an excellent choice as an inert gas because of its higher density than air. When a lab provides nitrogen gas from a bigger tank, nitrogen is favored over argon because of the lower cost and convenient use.

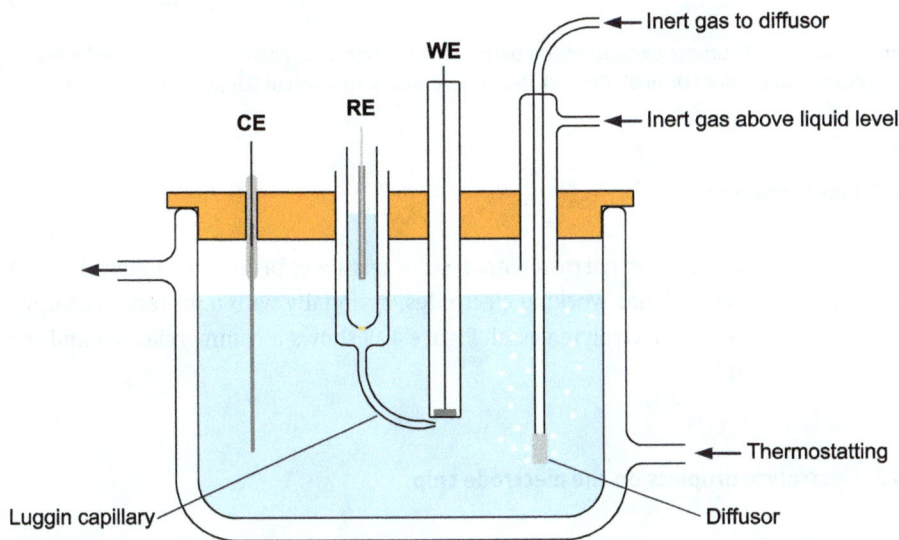

Figure 4.13: Typical electrochemical cell with many features: the jacket can control the temperature of the electrolyte vessel; a Luggin capillary allows direct access to the potential next to the working electrode; a two-way inert gas inlet ensures oxygen-free electrolyte.

The drawing in Figure 4.13 also contains a *Luggin capillary* to access the potential very close to the working electrode surface. Especially with high currents, low electrolyte conductivity, or large working electrodes, the introduction of a Luggin capillary can reduce the potential drop (Section 3.2.4) tremendously.

Many different commercial cells are available. Figure 4.14 shows two typical realizations. The cell in (A) provides all features discussed above, while the cell in (B) focuses on small electrolytes volumes. In contrast, according to ASTM[5] standards, corrosion analysis requires a larger cell, holding an electrolyte volume of one liter.

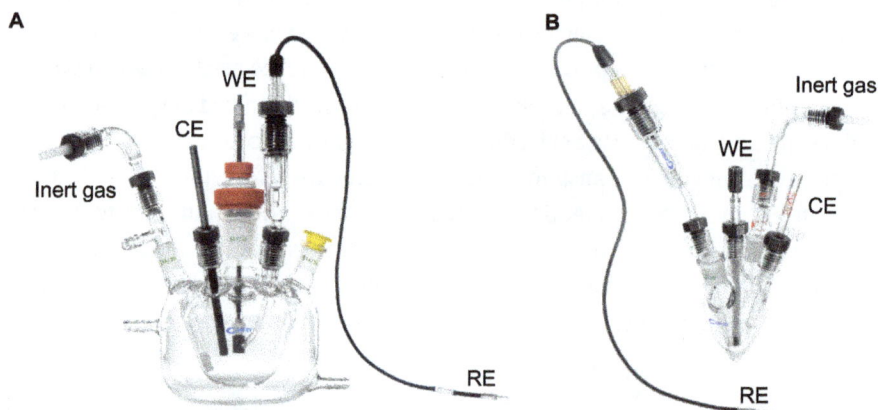

Figure 4.14: General-purpose electrochemical cell (A) and cell dedicated to handling small electrolyte volumes (down to ml), called "Dr. Bob's Cell" (B). Reproduced with permission of Gamry Instruments, USA.

4.4.1 Flat sample cell

Another type of cells used in corrosion analysis or testing of protecting materials is the flat sample cell. Here, planar working electrodes, optionally with a protective coating, form the bottom of the electrolyte vessel. Figure 4.15 shows a commercial cell and the application principle.

4.4.2 Electrolyte droplets on the electrode chip

Figure 4.16(A) shows a ceramic chip with screen-printed electrodes. The chips feature a three-electrode setup with the possibility to place an electrolyte droplet directly on the electrodes, as indicated in Figure 4.16(B). When experimental duration is short enough that evaporation does not play a role, electroanalysis in small volumes (less than 100 µl) is feasible. While such systems are primarily in use for student labs, also scientific ap-

---

5 ASTM International, initially the American Society for Testing and Materials, is now a global standardization organization. Standards from Section G deal with corrosion, deterioration, and degradation of materials, among them several describing in detail the equipment and methods for corrosion analysis.

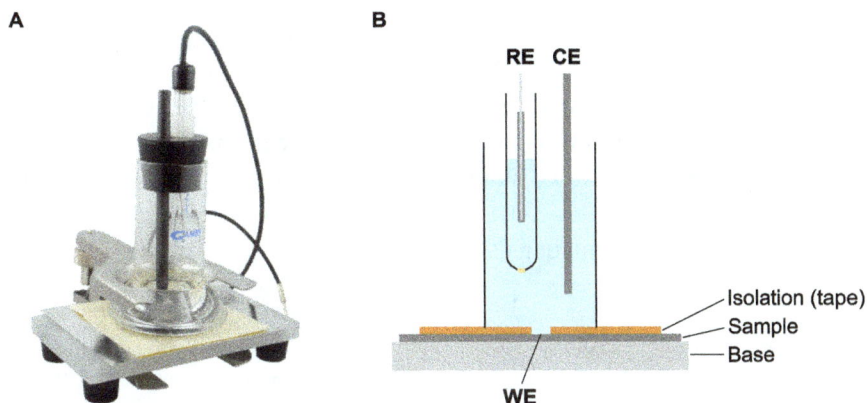

Figure 4.15: Flat sample cells: commercial cell, called "Paint Test Cell," (A) and schematic illustration (B). A: reproduced with permission of Gamry Instruments, USA.

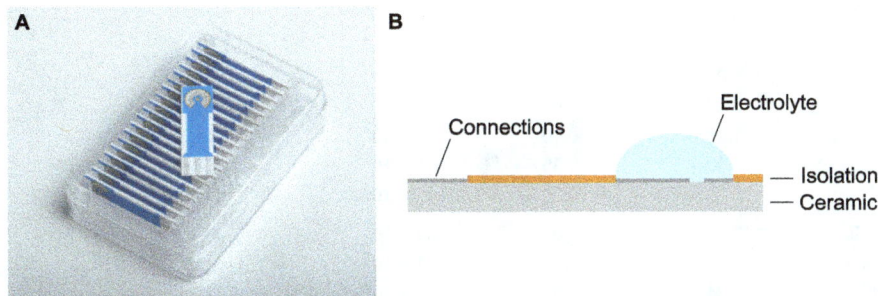

Figure 4.16: Electrochemistry in an electrolyte droplet: commercially available electrode chips (A), and illustration of the chips' operation (B), the electrolyte volume can be less than 100 µl.

plications are possible. In addition to the shown chip, a similar approach is available as an array compatible with the format of standard microtiter plates.

4.5 Electrolytes

An electrolyte is an ion conductive phase, either liquid or solid. The ions can stem from the liquid or solid itself. When there are intrinsically not enough ions, dissolution of salt or doping improves the conductivity.

4.5.1 Aqueous electrolytes

Most basic and typically seen as standard are aqueous electrolytes. Traditionally, many electrochemical experiments used acidic solutions, preferably from sulfuric acid, or if the presence of the chloride ions does not harm, from hydrochloric acid. Still, the one-normal sulfuric acid solution (0.5 M H_2SO_4 in water) is the first choice in many electrochemical fields. Alternatively, classical electrochemistry considered alkaline solutions (e. g., 1 M KOH or 1 M NaOH in water) to evaluate the influence of pH or use electrodes that are not stable at low pH. Table 4.6 summarizes common aqueous electrolytes along with their most relevant properties. $\Delta RHE = 2.303\frac{RT}{F} \cdot pH$ is the offset needed to relate SHE and RHE at the given pH (equation (2.72)).

Table 4.6: Properties of common aqueous electrolytes at 25 °C. Here, I denotes the ion strength and σ the conductivity (specific conductance). The data is calculated with the software *aqion PRO*, version 7.4.2, Harald Kalka, Germany. ΔRHE is the voltage added to the potential versus SHE to obtain the potential versus RHE for the electrolyte's pH. See Table 4.7 for the PBS recipe.

Electrolyte	pH	I in mol l^{-1}	σ in μS cm^{-1}	ΔRHE in mV
0.5 M H_2SO_4	0.39	0.58	173×10^3	23
0.25 M H_2SO_4	0.66	0.31	96×10^3	39
0.1 M H_2SO_4	1.01	0.14	42×10^3	60
1 M HCl	0.13	1.00	306×10^3	8
0.1 M $HClO_4$	1.08	0.10	38×10^3	64
PBS (recipe II)	7.40	0.35	21×10^3	438
1 M KOH	13.74	0.90	149×10^3	812
0.5 M K_2SO_4	7.32	1.07	59×10^3	433
0.5 M Na_2SO_4	7.33	1.10	52×10^3	433
1 M NaCl	6.94	1.00	81×10^3	410
0.1 M NaCl	6.98	0.10	11×10^3	413
1 M KCl	6.93	1.00	90×10^3	410
0.1 M KCl	6.98	0.10	13×10^3	413
0.1 M KNO_3	6.98	0.10	12×10^3	413

The neutral pH range most relevant for life science applications was not considered much and is still not the preferred range in classical electrochemistry. At neutral pH, electrolytes can be simple, unbuffered salt solutions (NaCl, KCl, Na_2SO_4, or K_2SO_4 in water) but with the risk of unclear pH when the redox reaction of interest involves protons. Changes of the pH directly at the electrode are critical, as the potential versus SHE at which a reaction occurs depends on the pH, that is, what the RHE scale accounts for (Section 2.8.1).

Therefore, often buffered neutral pH electrolytes are preferred. Among them, phosphate-buffered saline (PBS) solutions are the most common. PBS is a phosphate buffer adjusted to imitate the physiological ion strength. In many biological fields, solutions like the recipe I in Table 4.7 are an unofficial standard. Some recipes also contain $CaCl_2$ or $MgCl_2$ in the mM range. For electrochemical measurement, recipe II is a better

Table 4.7: Composition of phosphate-buffered saline (PBS) solutions with pH 7.4. Recipe I is commonly used in biology, recipe II is the preferred choice for electrochemistry.

PBS (recipe I)		PBS (recipe II)	
10 mM	Na_2HPO_4	85 mM	Na_2HPO_4
1.8 mM	KH_2PO_4	15 mM	NaH_2PO_4
137 mM	NaCl	100 mM	NaCl
2.7 mM	KCl		
adjust to pH 7.4 with HCl		no adjustment needed	

choice. Besides the higher buffer capacity and ion strength (0.35 vs. 0.17 mol l^{-1}), for this recipe, there is no need to adjust the pH manually.

Premade PBS solutions or ready-to-use tablets for biological application are often inappropriate for electrochemical experiments and, e. g., show unwanted peaks when recording a platinum cyclic voltammogram. It is recommended to prepare your solution based on good purity chemicals. While selecting the appropriate chemicals, the focus should be primarily on trace metals. Water should always be demineralized or distilled. Depending on your facilities, buying bottled water (preferably "ultrapure water") may be required. Never even think of preparing analytical solutions with tap water.

4.5.2 Nonaqueous liquid electrolytes

In addition to the aqueous electrolytes discussed above, nonaqueous systems play an essential role in many applications. Motivations are the absence of water or protons, a larger possible temperature range, and generally a larger potential window. The electrolytes comprise a solvent and a conducting salt. Optionally practical formulations contain additives like wetting agents, corrosion inhibitors, or flame retardants. Besides analytical applications, nonaqueous electrolytes have their place in electrodeposition and energy storage, e. g., in lithium-ion batteries or electrochemical double-layer capacitors.

Table 4.8 shows some basic properties of common solvents. For example, in electroanalysis, acetonitrile (AN) or methanol are general-purpose solvents. Propylene carbonate (PC) features a considerable temperature range. In addition, its higher viscosity helps to eliminate the influence of thermal convection. Tetrahydrofuran (THF) shows low permittivity, which comes along with the disadvantage of lower salt solubility.

Common cations are tetrabutylammonium (TBA) or lithium ions. Among the wide variety of anions, typically tetrafluoroborate, hexafluorophosphate, or perchlorate ions are common. For the application in lithium-ion cells $LiPF_6$, $LiBF_4$, or $LiClO_4$ are widely used salts. Table 4.9 lists the conductivity for some electrolyte formulations. The values are nearly as high as the conductivity of aqueous electrolytes (Table 4.6).

Table 4.8: Properties of common nonaqueous solvents and water for comparison. ϵ_r is the relative permittivity and μ the dynamic viscosity at 25 °C. The temperature range is given by the melting point ϑ_m and the boiling point ϑ_b.

Solvent	ϵ_r	μ in mPa s	ϑ_m in °C	ϑ_b in °C
Acetonitrile (CH_3CN, AN)	35.7	0.37	−44	82
Methanol (CH_3OH)	32.6	0.54	−98	65
Propylene carbonate (PC)	64.9	2.50	−49	242
Tetrahydrofuran (THF)	7.4	0.46	−109	65
Water	*78.4*	*0.89*	*0*	*100*

Table 4.9: Conductivity for some common electrolyte formulations with acetonitrile (AN) and propylene carbonate (PC) at 25 °C Data from [39].

Electrolyte	σ in µS cm^{-1}	Electrolyte	σ in µS cm^{-1}
1 M $LiBF_4$ in AN	18×10^3	1 M $LiBF_4$ in PC	3.4×10^3
1 M $LiPF_6$ in AN	50×10^3	1 M $LiPF_6$ in PC	5.8×10^3
1 M $LiClO_4$ in AN	32×10^3	1 M $LiClO_4$ in PC	5.6×10^3
1 M $TBABF_4$ in AN	32×10^3	1 M $TBABF_4$ in PC	7.4×10^3
1 M $TBAPF_6$ in AN	31×10^3	1 M $TBAPF_6$ in PC	6.1×10^3
1 M $TBAClO_4$ in AN	27×10^3	1 M $TBAClO_4$ in PC	6.0×10^3

For many applications, even a trace amount of water in the electrolyte is critical. Therefore, the water content of the solvents needs to be checked, e. g., by Karl Fischer titration. Preparation of the solutions may require cleaning the solvent with molecular sieves and heat treatment of the conducting salt. The electrolyte storage and the measurement requires pure inert gas.

In contrast to the many options for the reference electrode in aqueous electrolytes, good possibilities for nonaqueous systems are lacking. For example, with acetonitrile, sometimes an $Ag/AgNO_3$ electrode is used but with stability limitations. As a result, most experiments rely on a silver wire as a pseudo-reference electrode and calibrate the actual setup each time with a known redox couple. First, the silver wire needs to be freshly polished to obtain an oxide-free surface. The established potential then depends on the formation of new Ag_2O. Ferrocene (Section 4.6.2) is a typical redox couple used to relate the potential scale.

---

i An ionic liquid (IL) is a salt that is liquid below 100 °C. ILs are exciting candidates as an electrolyte because of their inherent conductivity and often large potential window. Possible applications are in energy storage and catalysis, but also gas sensors can benefit from IL as the electrolyte. Especially room-temperature ionic liquids (RTILs), ILs being liquid at room temperature, are of great interest.

4.5.3 Solid electrolytes

While classical electroanalysis focuses on liquid electrolytes, for a long time, gas sensors use solid electrolytes. More recently, the field of solid electrolytes had greater attention because of the demand for energy storage applications, e. g., in all-solid-state lithium-ion batteries. Historically, Ag_2S and PbF_2 were the first solid-state materials that scientists found to be ion-conductive. Later metal oxides and polymeric materials came into the arena.

ZrO_2 with Y_2O_3 added for stabilization ("yttria-stabilized zirconia") is a widely used ceramic electrolyte for gas sensors showing good conductivity for oxygen ions at higher temperatures ($300\,°C$). Its widespread application comes from the application in oxygen sensors to measure potentiometrically the oxygen content in the exhaust gas of cars (called lambda sensor because of the parameter λ describing the air-fuel ratio). Also, in fuel cells, those ceramics, conductive for the oxygen ion, are in use.

Polymer electrolytes gain a lot of interest for battery application. However, dry polymer electrolytes still suffer from too low conductivity at room temperature. Therefore, currently, mainly gel polymer electrolytes are used. Bocharova et al. provide an excellent introduction to, and overview on this field [40].

The lab practices with solid electrolytes strongly deviate from the before-presented liquid electrolytes. In addition, each material requires specific fabrication capabilities in contrast to macroscopic electrodes' "plug-and-play" possibilities in liquid solutions. However, all the electrochemical methods and the electrochemical theory apply to solids as well.

4.6 Redox couples

This section provides some practical realizations of the general redox couple

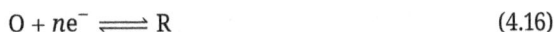

$$O + ne^- \rightleftharpoons R \tag{4.16}$$

introduced in the theory chapter. The discussed systems with a focus on aqueous electrolytes are summarized in Table 4.10.

Table 4.10: Commonly used redox couples for aqueous electrolytes.

Redox couple	E^0 in V	Mechanism	Example solution
$[Ru(NH_3)_6]^{3+}/[Ru(NH_3)_6]^{2+}$	0.10	Outer-sphere	1 mM $Ru(NH_3)_6Cl_3$ in 0.1 M KCl
FcMeOH/FcMeOH$^-$	0.40	Outer-sphere	0.5 mM FcMeOH in 0.1 M KCl
$[Fe(CN)_6]^{3-}/[Fe(CN)_6]^{4-}$	0.36	Outer-sphere	1 mM $K_3Fe(CN)_6$ in 0.1 M KCl
Fe^{3+}/Fe^{2+}	0.77	Inner-sphere	10 mM $Fe(SO_4)_3$ in 0.5 M H_2SO_4

4.6.1 Ruthenium hexamine

Ruthenium hexamine is a good choice as a general-purpose, positively charged redox couple for aqueous electrolytes. Its one-electron redox process is a good representation of the fictive species O and R:

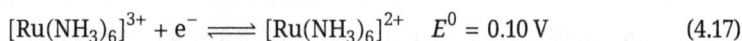

$$[Ru(NH_3)_6]^{3+} + e^- \rightleftharpoons [Ru(NH_3)_6]^{2+} \quad E^0 = 0.10 \text{ V} \tag{4.17}$$

The redox couple comes in the form of aqueous solutions from ruthenium hexamine (III) chloride ($Ru(NH_3)_6Cl_3$) or ruthenium hexamine (II) chloride ($Ru(NH_3)_6Cl_2$), with the first one the more common choice. Typical solutions are 1 mM to 10 mM $Ru(NH_3)_6Cl_3$ in 0.1 M KCl, in 0.1 M KNO_3, or in PBS.

4.6.2 Ferrocene

Ferrocene (Fc) is an organometallic compound with a central iron atom (Figure 4.17). The uncharged ferrocene forms a redox couple with the ferrocenium cation $[Fc]^+$ by change of the iron's valency:

$$[Fc]^+ + e^- \rightleftharpoons Fc \quad E^0 = 0.400 \text{ V} \tag{4.18}$$

Ferrocene is the defacto standard in nonaqueous electrolytes. A typical formulation is 1 mM ferrocene in acetonitrile with 0.1 M $TBAClO_4$ as conduction salt.

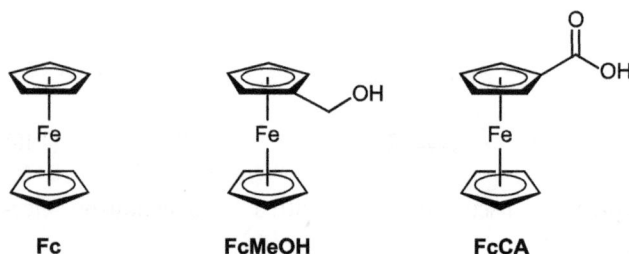

Figure 4.17: Ferrocene (Fc) and two water-soluble derivates: ferrocene methanol (FcMeOH) and ferrocene carboxylic acid (FcCA).

While ferrocene itself is not suitable for aqueous electrolytes, ferrocene methanol (FcMeOH) is a water-soluble form of ferrocene. Still, solubility is weak; increasing the temperature or sonification is beneficial. Predissolving in other solvents such as methanol would help too but may cause unwanted redox reactions, e. g., at noble metal electrodes. Another water-soluble derivate is ferrocene carboxylic acid (FcCA).

4.6.3 Ferrocyanide/ferricyanide

The ferrocyanide-ferricyanide redox couple ("hexacyanoferrate") is a very famous system despite its inferior properties. Potassium ferricyanide ($K_3[Fe(CN)_6]$) is also known as *Prussian red* and the salt of the reduced form potassium ferrocyanide ($K_4[Fe(CN)_6]$) as *Prussian yellow* because of the salts' color. The latter's ion can react with iron ions to the insoluble ferric ferrocyanide ($Fe_4[Fe(CN)_6]_3$), better known as *Prussian blue* (Section 10.1.3).

The redox-active substances of the ferrocyanide-ferricyanide redox couple are negatively charged and linked by a one-electron process:

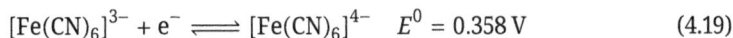

$$[Fe(CN)_6]^{3-} + e^- \rightleftharpoons [Fe(CN)_6]^{4-} \quad E^0 = 0.358 \text{ V} \quad (4.19)$$

In general, the solutions are not too harmful. Only in contact with strong acids, the severely toxic gas HCN can evolve. Possible precipitation and sensitivity to the status of the electrode surface are disadvantages of this system, suggesting using one of those mentioned above.

4.6.4 Inner-sphere versus outer-sphere electron transfer

All so far discussed redox couples have an *outer-sphere electron transfer*. In the outer-sphere mechanism, the metal atom remains linked to the ligand during the reaction. Therefore, the electroactive species remains in the region of the outer Helmholtz-plane. When the metal atom is without ligand (e. g., Fe^{3+}/Fe^{2+}), the reactant can link to the electrode during the transition state and an *inner-sphere electron transfer* takes place.

$$Fe^{3+} + e^- \rightleftharpoons Fe^{2+} \quad E^0 = 0.771 \text{ V} \quad (4.20)$$

The reaction takes place at the inner Helmholtz-plane. Figure 4.18 illustrates this simplified explanation. The *Marcus theory* treats the outer-sphere and inner-sphere mechanisms more comprehensively and discusses the consequences for the reaction rate. The inner-sphere electron transfer strongly depends on the electrode material, while the outer-sphere electron transfer is ideally independent of the electrode's catalytic activity.

Practical realizations of electrolytes with this inner-sphere redox couple are $FeCl_3$ in HCl or $Fe(SO_4)_3$ in H_2SO_4 solutions. Tanimoto et al. described a student lab experiment investigating the difference between an inner- and outer-sphere electrode reactions by cyclic voltammetry [41].

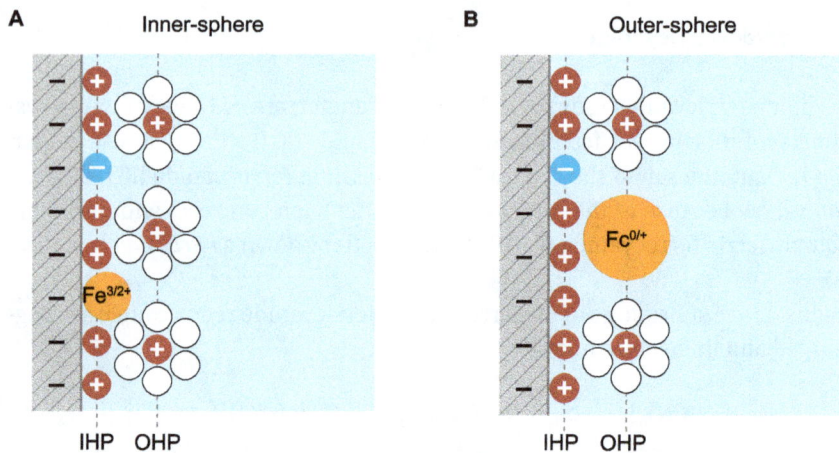

Figure 4.18: Illustration of the Fe^{2+}/Fe^{3+} redox couple as an example for an inner-sphere electrode reaction (A) and the ferrocene system ($Fc/[Fc]^+$), which is a typical outer-sphere redox couple (B).

Part II: **Methods**

"I have far more confidence in the one man who works mentally and bodily at a matter than in the six who merely talk about it—and I therefore hope and am fully persuaded that you are working."

Michael Faraday in a letter to John Tyndall (1851)

5 Classical methods

This chapter describes the classical electrochemical methods dealing with an electro-chemical cell and electrical instrumentation. Any method requiring further setups, such as for hydrodynamic control are considered as combined methods described in the next chapter. The electrochemical impedance spectroscopy can be seen also as part of the classical methods but is treated in another, separated chapter because of the completely different approach and different nomenclature.

In principle, all classical methods can be applied in a macroscopic electrochemical cells as well as in microsystems. In case there are specific aspects to be considered at micro- and nanoscale, it is mentioned along with the method. The methods can be grouped into four categories:

- *Potentiometry* is the measurement of voltage without current flow. The electrode reactions are in thermodynamic equilibrium.
- In *amperometry*, the current is recorded as function of time. Typically, a constant potential or a potential step protocol with a sequence of constant phases is applied. The current flow causes the electrode reactions to be out of electrochemical equilibrium.
- In *voltammetry*, the current is recorded as function of potential. Therefore, typical excitation is a linear potential ramp or a cyclic waveform.
- In current-controlled techniques, most prominently *chronopotentiometry*, the potential is recorded as a function of time. Typically, a constant current or a current step protocol with a sequence of constant phases is applied.

5.1 Potentiometry

5.1.1 Open circuit potential

The *open circuit potential* (OCP) is the measured voltage between two electrodes in a two-electrode setup without current flow.[1] Typically, one electrode is a reference, while at the other electrode, the indicator electrode, the reaction of interest occurs (Figure 5.1). It is essential to use a high impedance voltmeter to closely approximate zero current flow (Section 3.4).

Different applications of potentiometry are possible:
- Potentiometric sensors, especially ion-selective sensors
- Calibration of reference electrodes
- Measurement of corrosion potential

---

[1] Potentiometry should not be confused with chronopotentiometry, a term sometimes used for a current-controlled technique and therefore a non-equilibrium method.

https://doi.org/10.1515/9783111488844-005

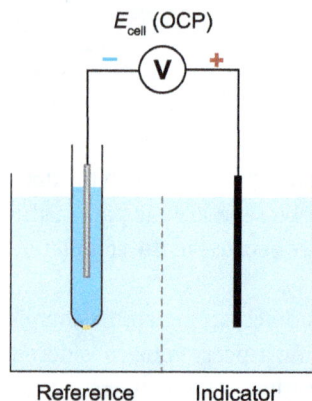

E_{cell} (OCP)

Reference   Indicator

Figure 5.1: Open circuit potential (OCP) measurement: the potential of an indicator electrode (e. g., a metal oxide or pH glass electrode) is compared to a reference electrode.

Each domain comes with a slightly different interpretation of the electrochemical theory and its own set of equations. Especially the field of corrosion uses a different concept to interpret the data. In the technical application of ion-selective electrodes, unique formulations of the equations and sensor parameters, e. g., the limit of detection, are common. I invite you to consider the sections in this textbook not precisely matching your field, too. For example, when developing potentiometric sensors, a sound understanding of the mixed-potential interpretation like for a corrosion process can be beneficial.

> **i** Measuring the potential without current flow seems a simple method, but in practice, the data interpretation can be challenging, especially with more than one equilibrium reaction involved. Ideal systems allow description by the Nernst equation or as a Donnan potential (Section 5.1.2). In the general case, measured values are mixed potential as discussed in the context of corrosion (Section 8.2.1). Diffusion potentials (Section 4.2.5) can contribute or be the sole source of the measured difference.

5.1.2 Donnan potential, membrane potential

The observed open circuit potential can have different causes, often even a combination of them. For example, faradaic processes in equilibrium as the potential forming mechanism follow the Nernst equation (equation (2.10)). On the other hand, in case diffusion of species with different ion mobility is the cause, the obtained difference is a diffusion potential (Section 4.2.5). Finally, a perm-selective membrane can cause the so-called *membrane potential* or *Donnan potential*.

Figure 5.2 illustrates the situation with a perm-selective membrane separating two electrolyte phases, α and β. Commonly, the selectivity depends on charge or size, but other selection principles are possible, too. The illustration considers a membrane al-

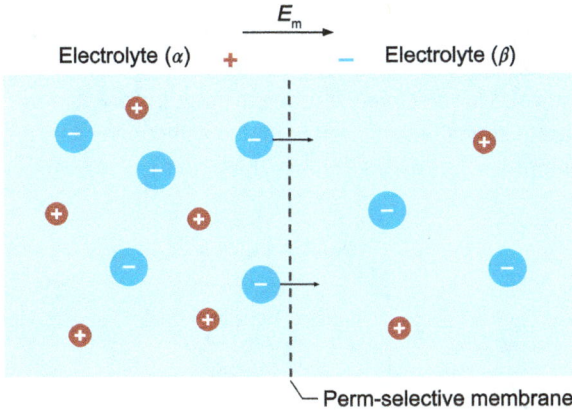

Figure 5.2: A perm-selective membrane separating two electrolyte phases, α and β, causes a membrane potential. Here, the membrane is selective for the anions.

lowing only the anions to pass. Accordingly, charge separation occurs, and a Donnan potential for the ion i establishes

$$E_\mathrm{m} = \frac{RT}{z_i F} \cdot \ln \frac{a_i^\alpha}{a_i^\beta} \tag{5.1}$$

z_i is the ions' valency, and a_i^α and a_i^β the activities on the two sides of the membrane. The Donnan potential is a common way to describe the membrane potential in ion-selective electrodes.

Biopotentials: the Donnan equation (equation (5.1)) also allows to approximate potentials in biological cell membranes. However, the *Goldman equation* is needed in the general case considering different ions contributing to the membrane potential.

5.1.3 Ion-selective electrode

Ion-selective electrodes (ISE) use a polymer or liquid membrane as the ion-selective element. Their selection mechanism distinguishes the different membranes:
- Neutral carrier membrane
- Charged carrier membrane
- Ion-exchanger-based membrane

Today, most commonly neutral carrier membranes are in use, especially for cations. The ion carrier, called ionophore, selectively binds to the desired ion. Those molecules have in nature often the role of an ion carrier. However, in an ISE, the carrier mainly acts as a trap for the ions and does not move much depending on the exact principle. Therefore, the term ion carrier can be misleading.

Valinomycin is a natural neutral carrier, selectively binding to potassium. It was one of the first neutral carrier molecules technically exploited in an ion selective-electrode, in this context, called potassium ionophore I. Figure 5.3 shows the structure of a valinomycin molecule.

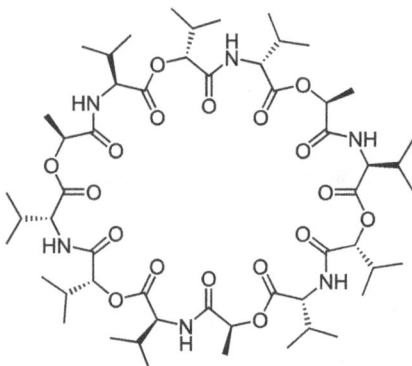

Figure 5.3: A valinomycin (potassium ionophore I) molecule can selectively incorporate a potassium ion into its polar inner.

The appropriate embedding of the ionophore into the membrane is detrimental to the quality of the ISE. Two variants exist; Figure 5.4 illustrates typical electrodes with a polymer membrane (A) incorporating the selective molecule or a liquid membrane (B). The electrodes have an inner electrolyte and an electrode class 2 (typically Ag/AgCl) to connect to the potential measurement. The schematic presentation of the different layers in Figure 5.4(C) illustrates the working principle. On both sides of the membrane, an exchange of the analyte ion occurs, with the analyte ions (here: K^+) and counter ions (represented by R^-) being present in the bulk of the membrane. The inner filling in the ISE contains an aqueous solution comprising the analyte ion and the appropriate counter ion to match the class 2 electrode. This example is KCl; for an ammonium-selective electrode, it would be NH_4Cl, etc. The translation between ionic and electron conduction occurs at the Ag/AgCl.

Figure 5.5 shows a complete setup of a potentiometric measurement with an ISE. It is noteworthy that the voltmeter should have a high internal impedance (Section 3.4). Typically, ISEs are used to measure ions in a concentration range over decades. Therefore, it is essential to use a reference electrolyte with an appropriate bridge electrolyte to avoid contamination of the analyte by the reference's internal filling and minimize diffusion potentials.

A sketch of the potential trace along this setup is in Figure 5.6. Interestingly, in the discussion of the trace, all three types of potential forming mechanisms possibly occur-

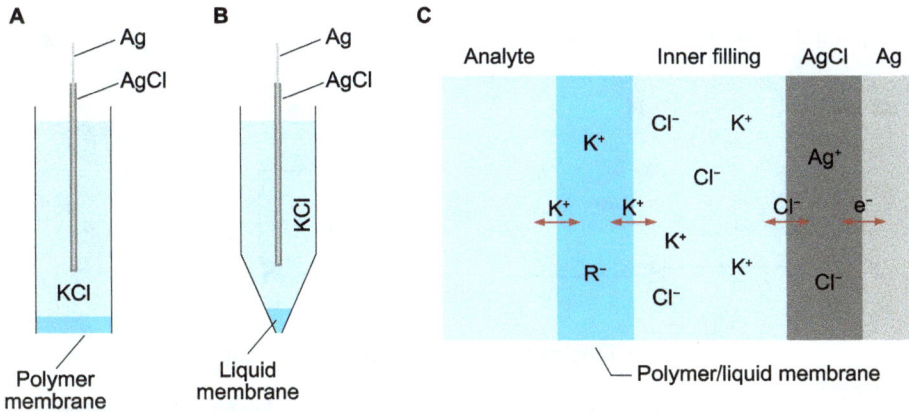

Figure 5.4: Ion-selective electrode (ISE) with polymer (A) or liquid membrane (B). The working principle (C) illustrates the situation for a potassium-selective ISE.

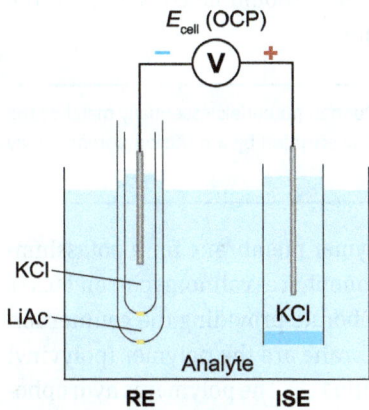

Figure 5.5: Measurement setup with a potassium-selective ISE. The reference electrode incorporates a salt-bridge filled with lithium acetate (LiAc).

ring in potentiometry contribute. The step from the Ag/AgCl to the inner filling in the reference electrode depends on the faradaic equilibrium reaction of the class 2 electrode and follows the Nernst equation. Because of the zero-current condition, no voltage drops in the liquid phases (inner fillings, salt-bridge, analyte). Across the diaphragms, diffusion potentials occur. Especially the potential drop at the diaphragm separating the salt-bridge from the analyte should be small as the analyte concentration may change but should not affect the reference electrode side. Following the discussion in Section 4.2.5, lithium acetate is a good choice in a potassium sensor (high concentrated KCl could contaminate the analyte). On both sides, Donnan potentials occur at the ISE membrane, while no drop occurs within the membrane. Finally, the Ag/AgCl in the ISE completes

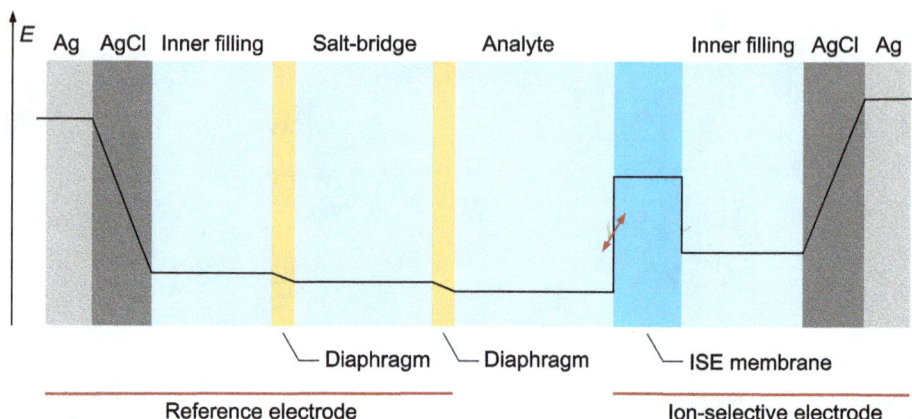

Figure 5.6: Sketch of the potential trace along the setup in Figure 5.5.

the setup. All potentials along the trace through the setup should be constant, but the Donnan potential at the outer side of the ISE membrane.

> **i** The interpretation of the potential at the ISE membrane as two Donnan potentials essentially matches the actual situation with macroscopic membranes. For microsensors, description by a diffusion potential may better suit some configurations.

Table 5.1 shows a typical sample composition of a polymer membrane for a potassium-selective electrode. The membrane contains only 1 % ionophore (valinomycin) and 0.5 % of the organic salt potassium tetrakis(4-chlorophenyl)borate providing the counter ion (R^- in Figure 5.4). The main components in the membrane are the polymer (polyvinyl chloride) and the plasticizer (dioctyl sebacate) in the ratio 1 : 3. The polymer is hydrophobic, preventing water uptake (besides vapor). The high plasticizer content is essential to provide good mobility for the organic salt, ensuring sufficient conductivity.

Table 5.1: Composition of a polymer membrane for a potassium-selective ISE. All fractions relate to weight.

1.0 %	Valinomycin
0.5 %	Potassium tetrakis(4-chlorophenyl)borate
65.5 %	Dioctyl sebacate
33.0 %	Polyvinyl chloride

In microsensors, the integration of a liquid inner filling is problematic and generally avoided. Instead, a solid-state contact replaces the inner filling and the class 2 electrode. The solid-state contact typically includes a transducer layer which translates between ion conduction in the membrane and electron conduction at the connection. Section 10.1.1 discusses ISE microsensors in more detail.

Nikolsky–Eisenman equation

The most critical aspect in the application of ion-selective electrodes is selectivity. The Nikolsky–Eisenman equation describes the measured potential E as a combination of the influence of the interfering ions j to the primary ion i:

$$E = E_0 + \frac{RT}{z_i F} \cdot \ln\left(a_i + \sum_j K_{ij} \cdot a_j^{z_i/z_j} \right) \tag{5.2}$$

E_0 is the zero-point of the calibration summing up all unknown but constant contributions to the potential in the setup, as illustrated in Figure 5.6. K_{ij} is the selectivity coefficient, describing how much the interfering ion j influences the measurement of the primary ion i. In some cases, K_{ij} values even above 1 are possible, which might be acceptable when the interfering ion is in a much lower concentration than the primary ion. Often, the selectivity coefficients are provided in the form $\lg K_{ij}$. For a potassium ISE with sodium as the interfering ion a typical value is $\lg K_{K,\,Na} = -4$.

The Nikolsky–Eisenman equation assumes the slope as in the Nernst equation. Often practical ISE show slightly lower slopes, e. g., 50 mV/dec at room temperature for $z_i = 1$. Some ISE shows a larger slope than predicted by $RT/z_i F$, and the behavior is called a *super-Nernstian* response.

Selectivity: separate solutions method

Measuring the potential for the primary ion and interfering ions in separate solutions allows determination of the selectivity, assuming that superposition of the individual responses is possible. This method leads to the selectivity coefficient according to the Nikolsky–Eisenman equation:

$$K_{ij} = \frac{a_i}{a_j^{z_i/z_j}} \cdot 10^{\frac{z_j F}{2.303 RT}(E_j - E_i)} \tag{5.3}$$

Figure 5.7 shows calibration curves illustrating this method. Both the primary and interfering ions need to be within the linear range at the activity of interest.

In case of a slope deviating from Nernstian behavior, the equation needs to be adjusted accordingly. Often, reported K_{ij} values assume $a_i = a_j$. With a slope of nominal $\pm\frac{F}{2.303 RT}$, the equation simplifies for simple ions ($z_i = z_j = \pm1$):

$$K_{ij} = 10^{\frac{E_j - E_i}{\text{slope}}} \tag{5.4}$$

Selectivity: fixed interference method

Another method to determine selectivity is the measurement of solutions with the primary ion in different concentrations, each with the same concentration of the interfering ion added. Figure 5.8 shows a typical calibration curve and its evaluation.

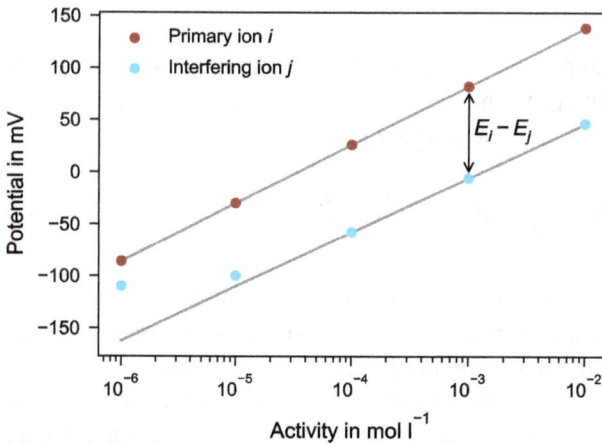

Figure 5.7: Separate solutions method to determine the selectivity coefficient: the calibration curves for the primary ion *i* and interfering ion *j* allow calculation of K_{ij}. Often the same activity for *i* and *j* is considered.

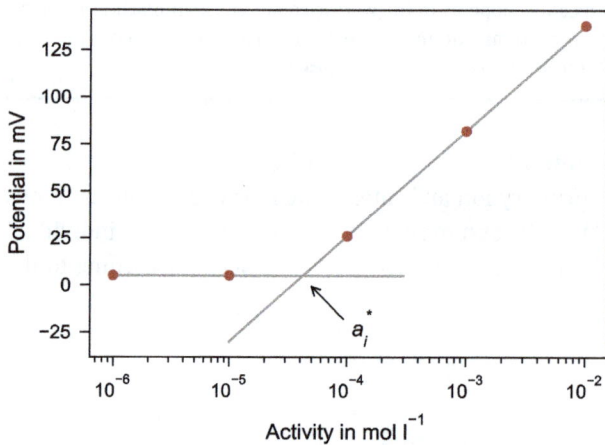

Figure 5.8: Fixed interference method to determine the selectivity coefficient: each measurement solution for the primary ion *i* contains a constant addition of *j*. The intersection point of the two lines a_i^* allows calculation of K_{ij}.

From the activity a_i^* at the intersection point, the selectivity coefficient can be calculated:

$$K_{ij} = \frac{a_i^*}{a_j^{z_i/z_j}} \tag{5.5}$$

For example, a dilution series of KCl solutions between 1 and 1×10^{-6} M and each containing 0.01 M NaCl would allow the determination of $K_{\mathrm{K,Na}}$. But again please pay attention to consider the activity instead of the concentration during the calculations. While reporting K_{ij} values, it is essential to mention the exact procedure as the results may depend on the method and the concentrations.

Limit of detection

Besides sensitivity and selectivity, typical performance parameters of ISE are the linear range and the lower/upper detection limit. The IUPAC definition for the limit of detection (LOD) assumes ideal Nernstian behavior:

$$\Delta E = \pm \frac{RT}{z_i F} \cdot \ln \frac{a_{\text{measured}}}{a_{\text{target}}} = \pm \frac{RT}{z_i F} \cdot \ln 2 \tag{5.6}$$

This definition of the detection limit for ISE deviates from the conventional formulation of LOD for other sensors. Although better matching the logarithmic transfer function, its applicability with calibration curves deviating from ideal Nernstian behavior is disputable. At room temperature, the maximally allowed deviation from linearity is 18 mV ($z_i = 1$) or 9 mV ($z_i = 2$). Figure 5.9 illustrates the LOD definition according to IUPAC.

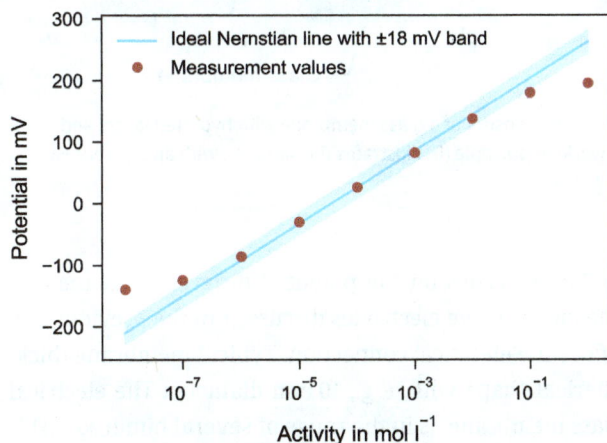

Figure 5.9: Calibration curve of an ISE with $z = 1$, the line represents the ideal Nernst slope, which the IUPAC considers in the LOD definition (blue band).

5.1.4 pH glass electrode

Another kind of an ion-selective electrode is the pH glass electrode. Instead of a polymeric membrane as describe in the previous section, a pH-sensitive glass is the key of the sensor mechanism. The glass membrane is an amorphous silicate matrix containing lithium ions. In contact with an aqueous solution, the topmost glass layer (around

100 nm) gets hydrated and forms a gel, which shows cation-exchange properties with high selectivity toward protons. The electrodes need soaking and storage in an aqueous solution to develop and maintain a stable, hydrated layer.

Figure 5.10(A) shows a typical setup of the pH glass electrode. The glass membrane forms two hydrated layers, inside the electrode with the inner filling and outside in the aqueous analyte. Inside is an aqueous HCl solution, or better, a buffer containing KCl at pH 7 to fulfill the criteria of stable pH and provide chloride ions to the Ag/AgCl electrode.

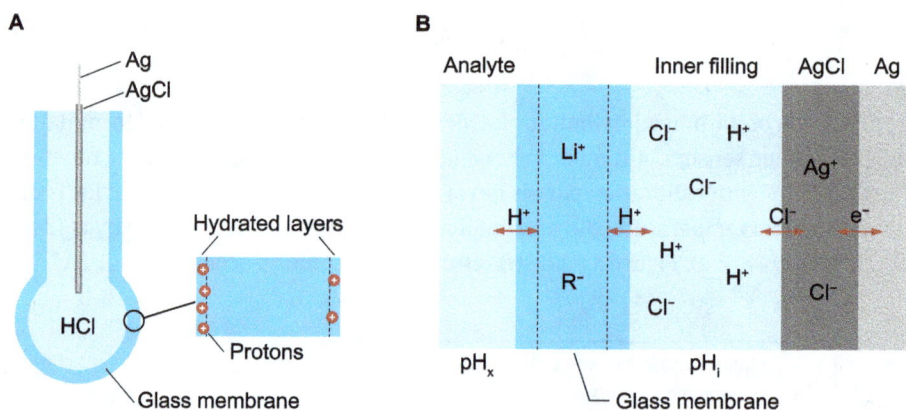

Figure 5.10: pH glass electrode: the electrode consists of a glass membrane with hydrated layers and an internal Ag/AgCl electrode (A). The working principle (B) illustrates the situation with an aqueous HCl solution as the inner filling.

The working principle (Figure 5.10(B)) relies on the potential differences at the two hydrated layers analogous to the ion-selective electrodes discussed in the previous section. An Ag/AgCl electrode provides the electrical connection. Typical membrane thickness is around 0.5 mm in a spherical shape with, e. g., 10 mm diameter. The electrical resistance, dominated by the glass membrane, is in the range of several hundred MΩ.

The potential difference across the glass membrane can be described as a Donnan potential (equation (5.1)) as a function of the proton activities:

$$E_m = \frac{RT}{F} \cdot \ln \frac{a_{H^+, i}}{a_{H^+, x}} \tag{5.7}$$

The index "i" refers to the inner filling, "x" to the analyte. With the definition of the pH scale by $pH = -\lg a_{H^+}$ (see the infobox below), the potential difference across the glass membrane is

$$E_m = \frac{2.303RT}{F} \cdot (pH_x - pH_i) \tag{5.8}$$

pH_i is the known pH of the inner filling, pH_x the value of interest. Another way to understand the potential difference is to assume diffusion potentials between the hydrated layers and the solutions.

Alkali metal ions can interfere at low proton concentrations (pH > 12), most critically sodium. This effect is called "alkali error" and can be reduced by a specific glass composition less sensitive to alkali metal ions. **!**

Analog to the setup in Figure 5.5, a reference electrode closes the circuit. Practical realizations are often combination electrodes having the pH and the reference electrode in the same rod (Figure 5.11).

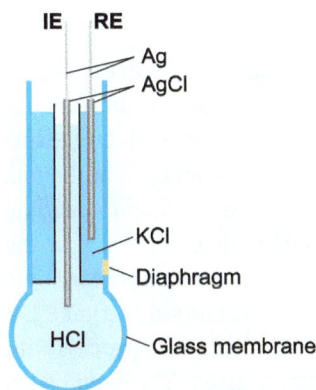

Figure 5.11: Combination electrode: the pH-sensitive indicator electrode (IE) and the reference electrode (RE) are combined in a single rod.

Miniaturization of the glass membrane is technologically challenging and not recommend for microsensor applications. Here, metal oxide electrodes (mainly iridium oxide) allow for miniaturized potentiometric pH sensors (Section 10.1.1).

Sørensen introduced the *pH value* as a scale to measure the acidity of an aqueous solution: **i**

$$pH = -\lg a_{H^+} \qquad (5.9)$$

The equation implicitly assumes the division of the activity by 1 mol l^{-1}.

As the dissociation constant of water $K_w = [H^+] \cdot [OH^-]$ is around 1×10^{-14} M^2, the mid-point (pH 7) refers to a neutral solution (equal amount of H$^+$ and OH$^-$ ions). For pH < 7, the protons prevail (acidic solutions), and for pH > 7, the OH$^-$ dominates (alkaline solutions).

The concept of acidity of Brønsted (acids are proton donors) extends beyond aqueous solutions. Therefore, Himmel et al. introduced the *unified pH scale* (pH$_{abs}$) based on the chemical potential $\mu_{abs}(H^+)$ of protons [42]:

$$pH_{abs} = -\frac{\mu_{abs}(H^+)}{RT \cdot \ln 10} \tag{5.10}$$

The authors defined the zero-point of this scale by $\mu_{abs}(H^+) = 0$ for standard pressure at 25 °C. In an aqueous system, the unified and conventional pH differ only by the zero-point, at 25 °C:

$$pH_{abs} = 193.5 + pH \tag{5.11}$$

Recently, an IUPAC Technical Report introduced the pH_{abs} [43], emphasizing its broad applicability in different solvents in liquid, gaseous, or solid states.

5.1.5 Active potentiometry

Active potentiometry, the combination of potentiometric measurement with active preconditioning is a variant of classical potentiometry in thermodynamic equilibrium. Especially with metal electrodes, the potentiometric behavior strongly depends on the surface status of the electrode, e. g., if a surface oxide prevails or bare metal is at the top. During the preconditioning, the electrode surface reaches a specific state. Depending on the electrode material, either the reaction is in thermodynamic equilibrium or the electrode reactions are in pseudo-equilibrium, appearing as constant on a shorter time scale. In the latter case, the measured potential is a mixed potential (see Section 8.1.4) consisting of a redox process with the analyte and at least one other redox process with opposite sign, e. g., a surface reaction of the electrode material itself. This method requires a potentiostat for the active preconditioning, which additionally needs the possibility to record open-circuit potentials (what most modern potentiostats offer).

Figure 5.12 shows an active potentiometry measurement of dissolved oxygen using a platinum working electrode [44]. The final potential of the active preconditioning steps ensures an oxide-free platinum surface. During the potentiometric measurement phase, a mixed potential consisting of the dissolved oxygen (O_2) reduction (analyte reaction) and the formation of PtO (the surface process of the electrode material) is measurable. In this pseudo-equilibrium, the transfer function is logarithmic, similar to the case of classical potentiometry.

For very long potentiometric measurement periods, the electrode surface approaches full coverage with PtO making it insensitive for further analyte reduction. The potentiometric measurement of oxygen at platinum is a representative example in which active preconditioning of the electrode surface enables reproducible results. Classical potentiometry in thermodynamic equilibrium would not work as the electrode surface would be fully covered with PtO once the electrode reaches equilibrium.

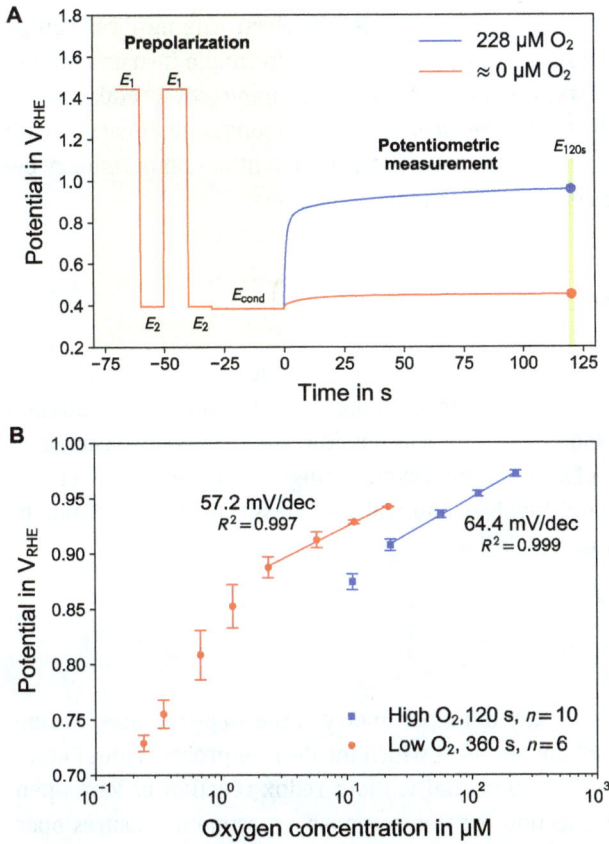

Figure 5.12: Active potentiometric measurement of dissolved oxygen: the surface of the platinum working electrode is preconditioned by appropriate prepolarization, followed by the potentiometric measurement phase (A). The resulting calibration curve (B) depends on the duration of the potentiometric measurement as the electrode is in pseudo-equilibrium [44]. Reproduced from Zimmermann et al., 2018, DOI 10.3390/s18082404, under CC BY 4.0.

Sensors based on active potentiometry, the fusion of potential-controlled and open-circuit phases, transfer the complexity from the sensor device to the operation and data evaluation. Advanced electrochemical protocols benefit from a combination of chronoamperometry and potentiometric steps so that quantitative and (semi)selective measurements are possible even with a bare platinum electrode that may be already available in an application (e. g., stimulation electrodes in neurotechnology, see Section 10.3.5). **i**

5.2 Amperometry

Amperometry is the most commonly applied technique with classical biosensors. The potential is constant during the whole measurement, and the current is, in the ideal case, proportional to the analyte concentration. Besides the electrochemical cell config-uration, the only degree of freedom is the potential leading to simple instrumentation and data interpretation. Therefore, this method seemed a simple compromise to the more complex electroanalysis by voltammetric methods for a long time. However, with the introduction of transient protocols involving different potentials, amperometry be-came equally mighty and complex as the voltammetry methods.

5.2.1 Single-potential amperometry

In single-potential amperometry, the electrode is at a constant potential at which the de-sired process occurs, and preferable, reaches mass transport limitation. For an oxidation reaction, the applied potential should be as high as possible, for a reduction reaction as low as possible without the risk of another process occurring, e. g., decomposition of the electrolyte. Without other reactions involved and reaching mass transport limitation by diffusion control, the transfer function is linear.

5.2.2 Step-response amperometry

The next level of increasing complexity in amperometry is the step-response. At time zero, the electrode is polarized to a potential at which the desired process runs. Before, the electrochemical cell can rest at a potential without redox reaction or kept open-circuit. Typically, the potential selection for step-response experiments ensures oper-ation under mass-transport limitation. In that case, the current response consists of a faradaic component following the Cottrell equation (equation (2.55)) and a capacitive component from charging the double layer.

Diffusion constant measurement

The step-response amperometry is a well-suited method for measurements of diffusion constants. The electrode is polarized from OCP or from a potential at which no reaction occurs to a potential at which the analyte gets oxidized or reduced in mass-transport limitation.

The resulting current (Figure 5.13(A)) follows the Cottrell equation with exponential decay of the capacitive contribution at the beginning. A simple evaluation is possible from the current versus reciprocal square root (Figure 5.13(B)). Pure mass-transport-limited current appears linear in this presentation. Relevant capacitive contribution directly after switching on (at the right end of the curve) shows up by the deviation from linear behavior. In the linear region, the slope allows calculation of the diffusion

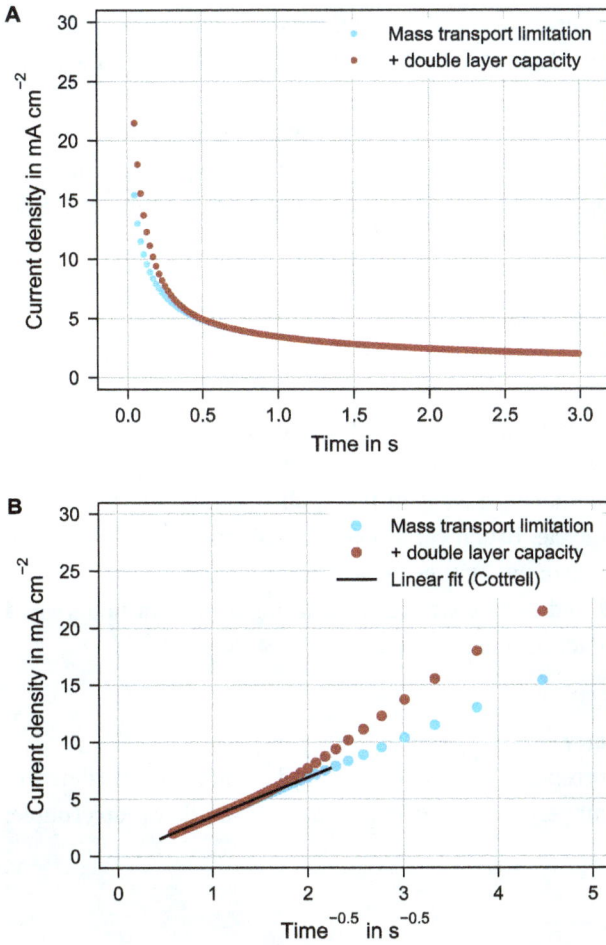

Figure 5.13: Step response for measurement of the diffusion constant: the capacitive current superimposes the Cottrell response at early time points (A). The current versus reciprocal square root representation allows evaluating the linear region after decaying of the double layer capacity current (B). Numbers are for an oxidizable species, $D = 1 \times 10^{-5}$ cm^2 s^{-1}, $z = 2$, $c = 1$ mM.

constant D:

$$D = \frac{\text{slope}^2 \cdot \pi}{(nFc)^2} \qquad (5.12)$$

c is both initial and bulk concentration of the analyte, F the Faraday constant, and n the number of electrons.

It is important to consider the characteristic diffusion length $\delta = \sqrt{Dt}$. At least 3δ should be within the region of interest, e. g., the membrane layer to be considered or the convection-free zone next to the electrode in free liquid. Depending on the expected diffusion constant value, this criterion leads to the maximal useful duration for evaluation. The onset of convection would appear in a plot like in Figure 5.13(B) as a deviation from linear behavior at the left end.

A considerable advantage of this method, especially with microsensor configurations, is the possibility to measure the diffusion parameters in situ, e. g., in a membrane stack resembling the final setup. Besides, other methods, e. g., with rotating disk electrodes (Section 6.1.1), offer more accuracy.

5.2.3 Chronoamperometry

Chronoamperometry (CA) is the more general method with a sequence of different potential steps. Often, a CA protocol, e. g., with three potential steps, repeat for a finite cycle number or until user interruption. Depending on the application, an off-time (open-circuit phase) can follow each cycle. Usually, the potentiostat samples many current readings during each step. For data evaluation, typical protocols need consideration of only some data, e. g., at the end of one step. Electrical engineers can understand chronoamperometry as a sequence of step-responses.

Double-step chronoamperometry

The logical continuation of step-response experiments as used to measure diffusion constant is the double-step chronoamperometry. Let us consider our familiar redox couple:

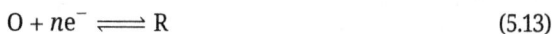

$$O + ne^- \rightleftharpoons R \tag{5.13}$$

A typical scenario could be the presence of only one type of species in the electrolyte (e. g., the reduced form R) and a double-step experiment with the oxidation followed by the reduction step.

After polarizing the cell, R gets oxides to O, and in the second step, the just-formed species is reduced back to R. Comparison of the charge allows, e. g., for evaluation of O's stability. The potential sequence is a potential E_{ox} higher than E^0 in the first step, followed by E_{red} lower than E^0 in the second (Figure 5.14(A)).

Assuming diffusion-limitation (E_{ox}, E_{red} are far enough from E^0) the current response (Figure 5.14(B)) follows the Cottrell equation:

$$i(t) = zFc\sqrt{\frac{D}{\pi t}} \tag{5.14}$$

The exact description of the reverse reaction in the second step is more complex. A common simplification is a formulation with a superimposed reduction process starting at

A

B

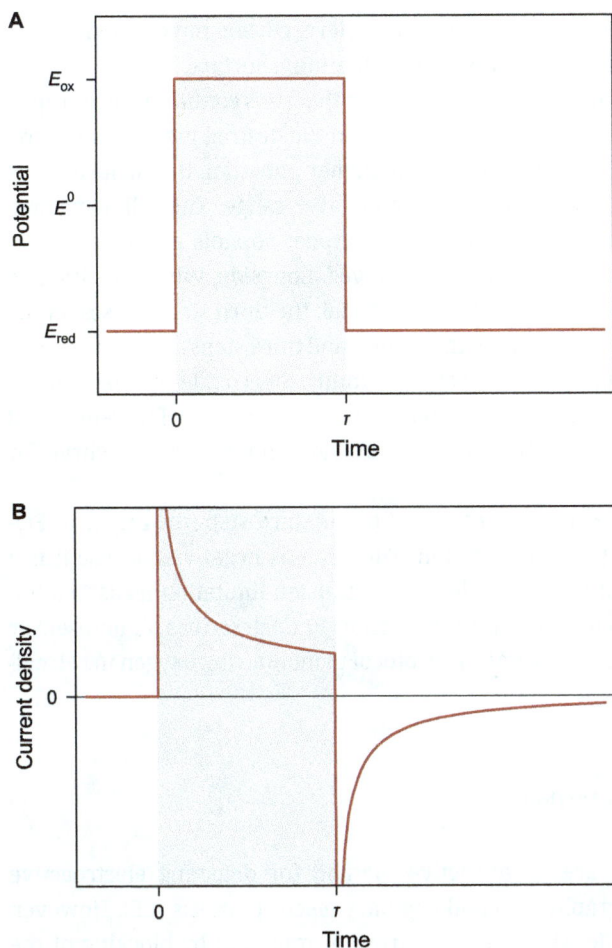

Figure 5.14: Double-step chronoamperometry with a redox couple: at the beginning, only the reduced form is available, which gets oxidized in the first step. In the second step, the reverse reaction occurs.

a time τ after the beginning of the first step:

$$i(t) = zFc\sqrt{\frac{D}{\pi}} \cdot \left(\frac{1}{\sqrt{t-\tau}} - \frac{1}{\sqrt{t}} \right) \tag{5.15}$$

Oxygen monitoring

A typical example in which both a redox-active substance in the electrolyte and reactions of the electrode material itself takes place is the measurement of dissolved oxygen (O_2) with a platinum electrode. As platinum is an excellent catalyst for oxygen, it is in an ideal electrolyte possible to measure the oxygen reduction current at a constant potential. However, in many applications, contaminants from the measurement medium may interact and block the platinum electrode over time. Here, chronoamperometry helps restore the electrode functionality by refreshing the platinum surface.

Figure 5.15 shows a typical oxygen sensor protocol[2] that runs continuously or repetitively but with some off-time in between depending on the desired measurement frequency. This protocol comprises three steps: At the higher potential, the surface of the platinum electrode gets oxidized to PtO in the monolayer range. The following step causes PtO reduction to present a new platinum electrode; possible adsorbates from contaminants in the measurement medium get removed alongside. While the first two steps deal with redox processes of the electrode material, the third step focuses on the dissolved oxygen. In reality, the processes in the second and third steps are not as strictly separated as indicated in the figure. As soon as a platinum site is oxide-free, oxygen reduction sets in. The main reason to use two separate steps is the faster PtO removal at the lower potential while benefiting from the step upwards for a more flat curve for taking the measurement value.

In a sensor application, the current at the end of the third step (the arrow in Figure 5.15(B)) measures the oxygen concentration. The value is negative because it is a reduction process; operation of the electrode under diffusion-limitation leads to a linear relationship with the analyte concentration. Section 10.1.8 describes a microsensor application using such a chronoamperometric protocol to monitoring oxygen in cell culture.

5.2.4 Pulsed-amperometric detection

Amperometric measurements are an attractive method for detecting electroactive species from liquid chromatography (LC) and capillary electrophoresis (CE). However, especially noble metal electrodes (Pt, Au) suffer from degradation by blocking of the surface. Here, *pulsed amperometric detection* (PAD) methods can support. Figure 5.16(A) shows a classical PAD protocol: after the measurement at the detection potential (E_{det}), the electrode surface is refreshed by polarizing to a potential E_{ox} at which the surface gets oxidized, followed by removal of the oxide in the third step (E_{red}). This sequence allows for a renewed, oxide-free electrode surface. Sampling the current with some delay after switching to the detection potential prevents a strong background from charging the double layer capacity.

For detection processes benefiting or depending on the metal electrode's surface oxide, a different cleaning step order provides an oxidized surface (Figure 5.16(B)). This approach is called reverse *pulsed amperometric detection* (RPAD). While a step directly

---

2 Experimental details: the working electrode was a platinum disk with 2 mm diameter in phosphate-buffered saline, pH 7.4. Oxygen was removed by nitrogen bubbling through before and nitrogen flushing above during the measurement of the "no oxygen" curve.

A

B

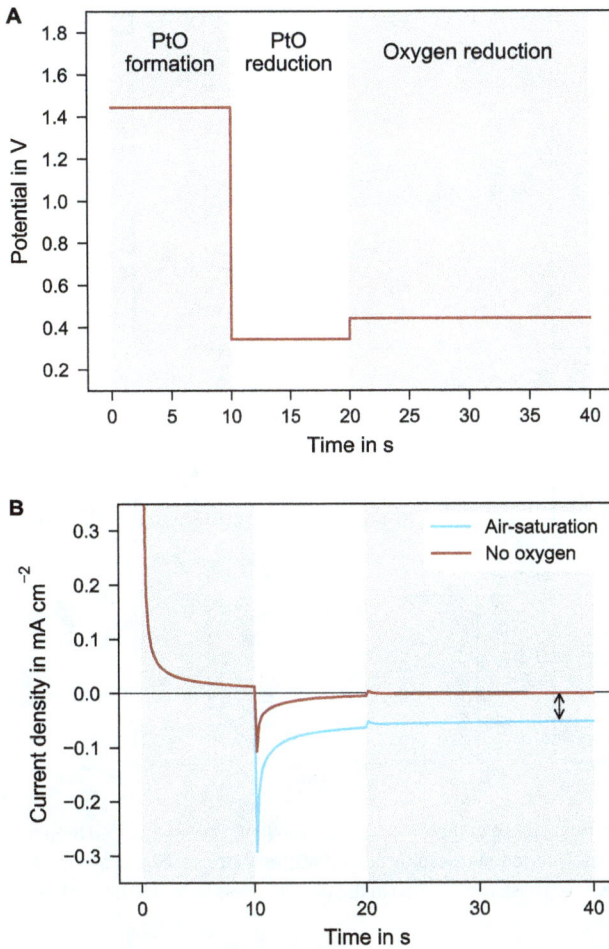

Figure 5.15: Chronoamperometric protocol (A) and typical current responses (B) for oxygen monitoring with a platinum electrode. At the end of the cycle, the current provides a measurement for the dissolved oxygen concentration (indicated by the arrow).

from the E_{ox} from the previous cycle to the detection potential produces a stronger background signal, and the oxide itself may see severe changes depending on the metal, the introduction of another step just before the detection can improve the performance (Figure 5.16(C)). This step prepares (activates) the electrode for the detection; therefore, the method's name *activated pulsed amperometric detection* (APAD).

The *integrated pulsed amperometric detection* (IPAD) goes even beyond the strict definition of amperometry (Figure 5.16(D)). For detection, the potential linearly ramps up and down between E_{min} and E_{max}. The output signal is the integrated current during this cycling. In the upward cycle, the signal comprises the charge from surface processes and the analyte. In the reverse scan, the charge from the surface processes is (partially)

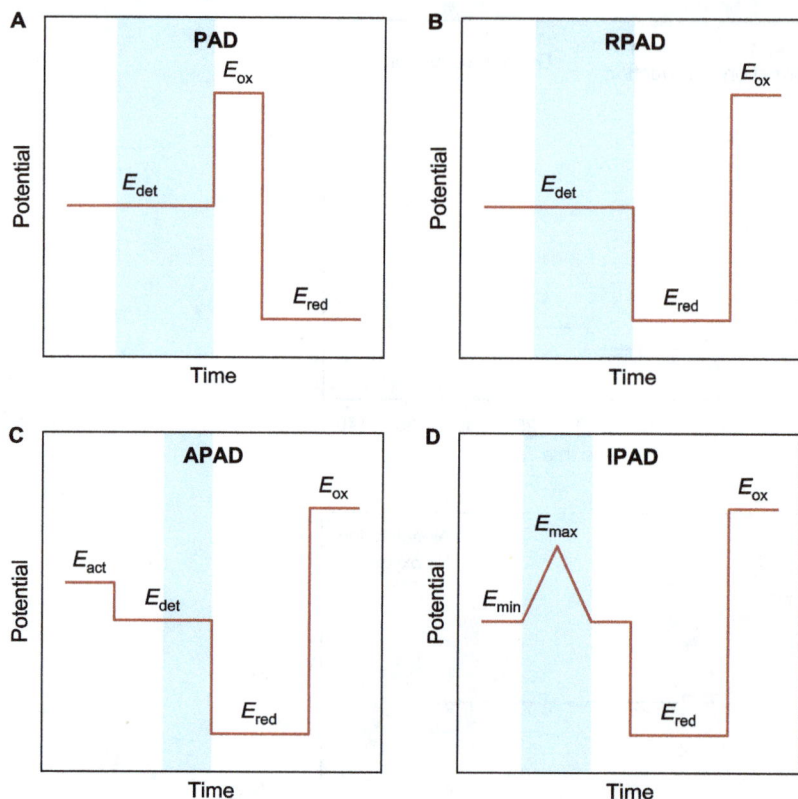

Figure 5.16: Different detection waveforms: pulsed amperometric detection (PAD), reverse pulsed amperometric detection (RPAD), activated pulsed amperometric detection (APAD), and integrated pulsed amperometric detection (IPAD). The current response is evaluated in the time intervals indicated by the blue regions.

recovered while the contribution from the analyte oxidation remains. IPAD helps reduce the background signal and is also more tolerant for minor changes in the potential scale (e. g., due to pH variation in the electrolyte).

Pulsed amperometric detection refers to fast protocols (the typical duration of a cycle is in the second range or below) and the application downstream a separation column. However, the term chronoamperometry would also fit but usually indicates slower pulsing or stand-alone sensor applications. A typical example of a protocol in between the classical PAD waveforms and generally slower chronoamperometry is glyphosate monitoring using a protocol with a total cycle time of 15 s [45].

Common to all, the pulsed amperometric detection method compared to amperometry with a constant potential is the chance of fewer electrode poisonings because of the shorter time at the detection potential and the possibility of electrode cleaning. In comparison to scanning methods (Section 5.3), the lower (capacitive) background currents allow for a lower limit of detection (LOD) and, maybe apart from IPAD, easier data acquisition and evaluation. For some analyte/electrode combinations, there are even more complex PAD protocols but all following a scheme related to the sequences shown in Figure 5.16. A good overview of the different methods and their application in the detection of organic molecules is provided in the review by Fedorowski et al. [46].

5.2.5 Chronocoulometry

Chronocoulometry closely relates to chronoamperometry but with the measured quantity charge instead of current. In theory, looking at charge instead of current is not a big deal but often allows for better data interpretation. Figure 5.17 shows the integral of the double-step chronoamperometry curve in Figure 5.14(B). Looking at the charge instead of current reveals quickly, e. g., if a reverse process completely reversed the action of the forward process.

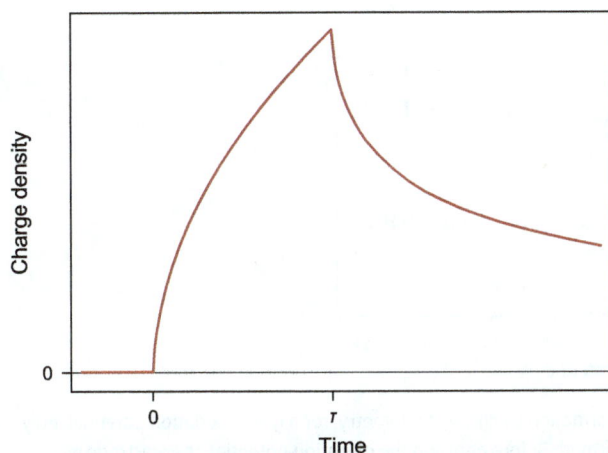

Figure 5.17: Chronocoulometry: integral of the double-step chronoamperometry curve in Figure 5.14(B). Within the second step, the electrode reduced not all O formed in the first step back to R.

Practically, with discrete current data, the obtained numerical integral may differ significantly from the accurate charge. A fraction or sometimes the majority of the charge change occurs immediately following the potential difference, much before the first current value is sampled. Increasing the data rate may milden the error but cannot avoid it. In contrast, the usage of an analog integrator (Figure B.5 in the Appendix) with the potentiostatic circuit provides accurate results.

> Chronocoulometry offers several benefits over chronoamperometry. Interpreting current spikes is difficult, while their effect on the charge is more intuitive. Additionally, the integration also smoothes the curves by eliminating the noise independent of the disturbing frequency band and possible interference with the sampling rate. However, commercial potentiostats rarely allow for true current integration or require a costly extension.

Platinum: surface oxidation

A typical application for chronocoulometry is the investigation of surface processes of the electrode material. Figure 5.18 shows the charge recorded for the formation of the surface oxide at a platinum electrode. Before the oxidation, a lower potential was applied to ensure an oxide-free electrode surface. At time zero, the integrator was discharged and the electrode polarized to the desired oxidation potential.[3]

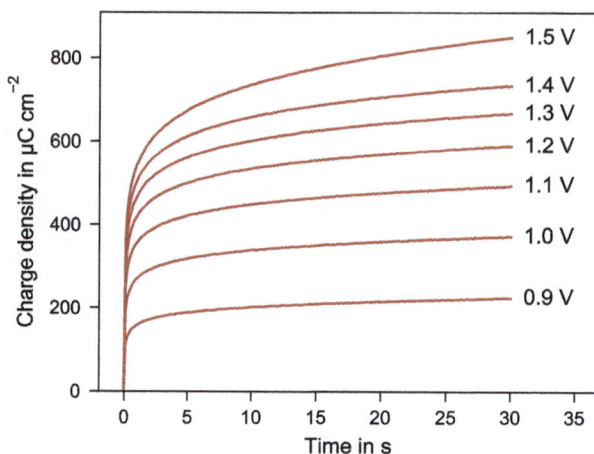

Figure 5.18: Investigation of the PtO formation by chronocoulometry: for a given oxidation potential, only a certain amount of surface oxide is formed. Before applying the oxidation potential, the electrode was kept at a lower potential to ensure an oxide-free surface. The curve labels are the oxidation potentials versus RHE.

Notable that for a given oxidation potential, only a certain amount of surface oxide is formed. For a smooth, polycrystalline platinum electrode, the formation of a PtO monolayer requires around $440\ \mu C\ cm^{-2}$ (see Section 8.4 for a more detailed discussion on the platinum surface oxide).

---

3 Experimental details: the platinum working electrode was cleaned by potential cycles close to the limits of the water window and prepolarized to 0.4 V versus RHE, the electrolyte was deoxygenated phosphate-buffered saline, pH 7.4.

5.3 Voltammetry

Voltammetry is the group of methods most common in electroanalysis. The defining property is the recording and data presentation of the current as a function of the potential. The most famous voltammetric method is cyclic voltammetry, with the potential ramped up and down linearly. Throughout this section, you will understand the powerful features of cyclic voltammetry for different fields and learn about some variants and advanced voltammetric techniques.

5.3.1 Sign conventions

In polarography, the voltammetry with mercury electrodes, the focus was on reduction processes. Consequently, the higher driving force for the reduction (more negative potentials) was on the x-axis to the right, and the y-axis was the reduction current. From a nowadays perspective, it is more logical to use the signs consistent with the derivation from theory and the currents with the polarity in agreement with measured values. The latter is also the recommendation of the IUPAC. Figure 5.19 illustrates the two possible conventions.

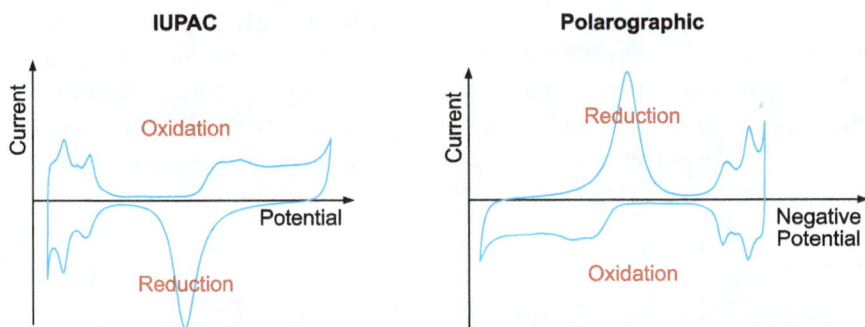

Figure 5.19: Sign conventions: the IUPAC recommendation is the more logical arrangement (A), the polarographic style (B) the more traditional one.

Unfortunately, older but often excellent publications and some textbooks mainly from the United States stick to the polarographic convention. Consequently, even in today's published data, you may find both variants and potentiostat manufactures provide in the software the option to switch between the two. Therefore, whenever you see a voltammetric plot, you need to figure out which convention it is, so whether it is "oxidation up" (IUPAC) or "reduction up" (polarographic). The curves in this book follow the IUPAC, and I strongly encourage you to do the same.

5.3.2 Cyclic voltammetry and linear scan voltammetry

Cyclic voltammetry (CV) is one of the most widely applied methods that can provide a plethora of information from a relatively simple experimental procedure. The applied potential in an electrochemical cell goes up and down, and the results are visualized as a *cyclic voltammogram* (CV) with the current (or current density) on the y-axis and the potential on the x-axis. Please note that the abbreviation CV refers to both the method and the diagram.

Figure 5.20 shows the applied waveform. The potential goes linearly between the two extremes (E_{min}, E_{max}) to define the energy domain of the expected processes. Often the turning points are set to the value where the decomposition of the electrolyte's solvent starts, i. e., the onset of the oxygen gas evolution limits E_{max} and the hydrogen gas E_{min} in an aqueous system. Another critical parameter is the scanrate v, the slope of the applied voltage ramp, per convention, always reported as a positive number. The scanrate sets the time domain of the experiment and thereby the visibility of different effects (electron transfer, mass-transport, capacitive). Accordingly, the scanrate directly links to electrochemical reversibility (Section 5.3.4). Finally, the waveform needs a start and end potential, which are often identical. It is a good choice for many experiments to use the open-circuit potential as E_{start} (some potentiostats can do this automatically). One cycle is when the potential reached again the start value after going into both extremes. Another parameter is the initial sweep direction.

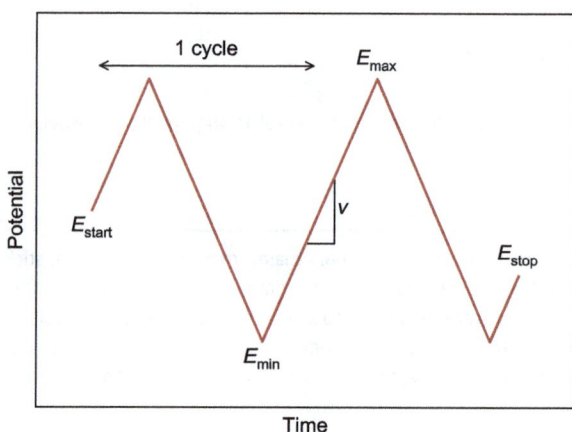

Figure 5.20: Applied waveform in cyclic voltammetry: voltage sweep between E_{min} and E_{max}, with a sweep rate (scanrate) v.

A subset of CV is linear scan voltammetry (LSV). Here, the applied potential increases or decreases linearly from a start value to an end value. Often LSV helps as a starting point in the discussion of more complex curves, which is especially true for CVs with redox-active substances in the electrolyte (Section 5.3.4). Also, LSV benefits from a combination with phases of constant polarization, e. g., in stripping voltammetry (Section 5.3.9).

Interpreting cyclic voltammograms can sometimes be counterintuitive as it seems that two different current values could exist depending on the scan direction at the same potential. It is essential to remind oneself that the *x*-axis is the applied potential rather than time what would be our general intuition. And at a given time point, of course, the history matters. **!**

The source of the redox-active component helps to differentiate the typical use cases of cyclic voltammetry:

1. Electrode material: when the surface processes of the working electrode material are the only or the dominant redox contributions, the CV allows for studies of the material properties. Here, CV provides a very sensitive method (sub-monolayer resolution) to investigate surface reactions. Curves like on the book's cover are a typical representative.
2. Electroactive substances in the electrolyte: the other major case is redox-active substances dissolved in the electrolyte solution. The famous "duck-shaped" curve showing up in some electrochemistry software or advertisement is the classical curve form.
3. Electrodeposition: in principle, already part of the previous cases, the resulting curves, and underlying understanding is slightly different. Examples are in Sections 9.2.3 and 9.2.7.

In the following, we discuss the first two use cases. Depending on your background (e. g., material science vs. electroanalysis), one of the two will be the "normal" cyclic voltammetry, while the other may seem more exotic.

5.3.3 Cyclic voltammetry of metal electrodes

Cyclic voltammograms of metal electrodes focus on aqueous electrolytes. Nobel metal electrodes are of high interest for both sensors and energy applications because of their excellent catalytic properties. Among them, platinum is the most widely used, e. g., as the working electrode in chemical and biosensors, a recording and stimulation electrode in neurotechnology, or to catalyze the reduction of oxygen in fuel cells. Other metals, e. g., iron or its alloys, play a role in the context of corrosion research. Here, often LSV can obtain all desired information; a detailed discussion of the method's specifics is in Section 8.2.2.

Platinum shows one of the clearest and feature-rich CV curves among the different metals and is, therefore, somehow the standard when discussing novel material composition in the form of alloys or composites. Figure 5.21(A) shows the CV of a platinum electrode in oxygen-free sulfuric acid.

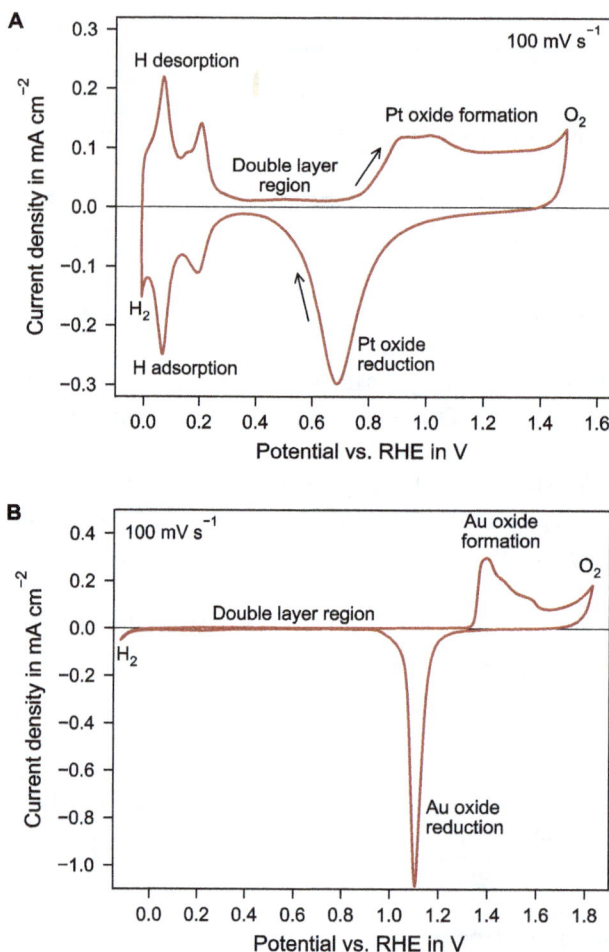

Figure 5.21: Cyclic voltammograms of noble metal electrodes in oxygen-free 1 N H_2SO_4: platinum (A) and gold (B). The gas evolution reactions (H_2, O_2) limit the potential window.

A broad region with a positive current starting at around 0.85 V indicates the platinum oxide formation in the upward scan. Calculating the charge density in this region suggests that close to two monolayers of PtO build up before molecular oxygen (O_2) evolution starts. In the beginning, the oxygen gas dissolves in the electrolyte. For higher potentials, beyond the potential range in this plot, gas bubbles form. Please note that the electrolyte was oxygen-free (degassed and during the measurement flushed with nitrogen as discussed in Section 4.4). That means the oxide formation is independent of molecular oxygen and purely relies on water as reactants.

Upon reversal of the potential, the current approaches zero until a strong reduction peak appears when the platinum oxide formed before gets removed. Once the platinum electrode is oxide-free, the current approaches zero again. Until here, the charge from the oxide reduction balanced the oxide formation. Depending on the surface status of the electrode and the purity of the electrolyte solution, the platinum oxide reduction might broaden. In this case, the current does not return and the following features set in without a gap in between.

The left part of the curve is the hydrogen-related region. In the downward scan, two distinct peaks are characteristic before the molecular hydrogen (H_2) evolution. Here, a monolayer of proton adsorbs onto the electrode. As strictly one monolayer forms (the process is fast but limited to one proton per platinum atom), the charge in this region allows calculation of the actual catalytic surface area and relation to the geometrical area to estimate the electrode's roughness (Figure 8.16).

In the upward scan, three characteristic peaks represent the proton desorption. The third peak, the very tiny one between the two more pronounced peaks, is often not visible (compare the arrow and discussion to Figure 5.23). Especially in neutral pH electrolytes, only two peaks are visible. Ramping to a lower potential would cause a more pronounced hydrogen gas formation. In this case, the oxidation of the just-formed gas causes another peak before the proton desorption.

The region between proton desorption and the onset of platinum oxide formation is named the *double layer region*. Here, the only current contribution is from charge-up of the double capacity:

$$i = c_{dl} \cdot \frac{dE}{dt} = c_{dl} \cdot v \tag{5.16}$$

The capacitive current always contributes (positive in the upward, negative in the reverse scan), but only in this region is it visible. Section 8.4 discusses the platinum surface reactions in more detail.

Figure 5.21(B) shows the CV of a gold electrode in oxygen-free sulfuric acid. The cyclic voltammogram has some features similar to platinum but has overall a very distinct appearance. Most characteristically is the very narrow gold oxide reduction peak[4] at 1.1 V. In contrast, different gold electrodes show a wide variety of signatures in the gold oxide formation. Notable is also the enormous double layer region.

---

4 Its strict dependency on the electrolyte pH (compare, e. g., Figure 2(D) in [45]) would even enable pH sensing. However, the effort to run CVs and the uncertainties with nonideal electrolytes makes it not a suitable sensor principle compared to, e. g., the pH glass electrode (Section 5.1.4) or metal oxides (Section 10.1.1).

While the overall curve shapes are more consistent with platinum than with gold, the exact form of the peaks in the hydrogen desorption from platinum reflects the different manufacturing processes and polishing. Also, the absolute values may differ between individual electrodes because of varying surface roughness. Different potentiostats do not play a significant role when the data acquisition method and current range are appropriate (Section 3.2.2). Switching a cell with a platinum electrode between potentiostats from different manufacturer produce the quantitatively identical curves.

Cyclic voltammograms, like the ones in Figure 5.21, represent the stationary case of the surface processes. Obtaining such clear curves in a reproducible way requires, apart from careful cleaning of the setup and high-grade chemicals, extensive cycling (at least 10, for publication grade curves better up to 100 cycles) to reach a stationary state. Upon cycling, the polycrystalline electrode surface reorients and exposes most of the active catalytic sites. At the same time, the continuous formation of an oxide layer and its removal clean the surface from possibly adsorbed contaminants. Intermediate exchange of the electrolyte reduces the risk of redeposition of foreign species just removed from the surface. Cyclic potential stepping between two extreme potentials could improve the run-in process. However, it is crucial that the pretreatment or initial cycling is not overdone to keep the electrode roughness in its original status.

! An important aspect is the careful selection of the other electrodes. The counter electrode is preferably the same material as the working electrode to avoid contamination by traces of metal released by dissolution or the potential cycling. The inner filling of the reference electrode should match the electrolyte, or a salt-bridge is needed. For example, an Ag/AgCl electrode and KCl solution as inner filling could ruin the experiment by Cl^- ions leaking into an H_2SO_4 electrolyte solution. Here, a bridge solution with a K_2SO_4 solution would improve.

Influence of scanrate and turning points

The scanrate has a minor influence on the qualitative curve shape within a reasonable range (e. g., between $10\,mV\,s^{-1}$ to $1\,V\,s^{-1}$). However, the impact on quantitative behavior is immense (Figure 5.22).

The charge in all three curves is roughly the same (the charge in the oxide formation region shows less than 1 % difference for $500\,mV\,s^{-1}$ and less than 10 % for $1000\,mV\,s^{-1}$ compared to $100\,mV\,s^{-1}$). Assuming a constant charge, it is evident that with a higher scanrate, the actual current (which is charge by time) is higher. A more detailed look reveals a slight shift of the platinum oxide reduction peak to the left with increasing scanrate (dashed line). Here, the potential already reached lower values for a faster scanrate before the oxide on the surface depletes because of the limit in the electron transfer. However, that might occur at an earlier time point with a faster scanrate because the platinum oxide reduction sees sooner a stronger driving force (lower potential).

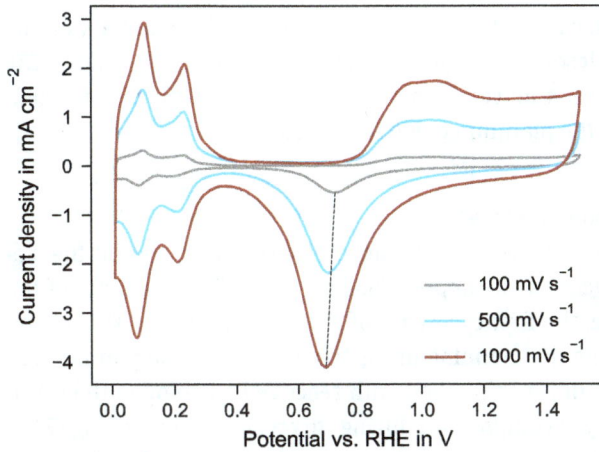

Figure 5.22: Cyclic voltammograms of a platinum electrode with different scanrates in oxygen-free 1 N H_2SO_4. The qualitative behavior is the same; the enclosed charge is also in the same range for all three curves.

The turning points of the applied waveform influence the cyclic voltammogram strongly. When one process is affected (e. g., less platinum oxide formation), the counter process (e. g., the platinum oxide reduction) diminishes accordingly. Therefore, the variation of the turning points is a helpful approach to figure out interrelation between the different features in the curve. Figure 5.23 shows the results with varying anodic boundaries in a platinum CV.

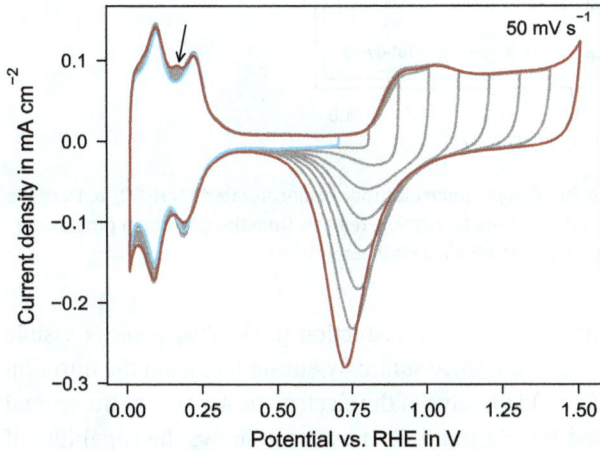

Figure 5.23: Cyclic voltammograms of a platinum electrode with different anodic boundaries in oxygen-free 1 N H_2SO_4. Those variations help to identify the interrelation between the processes.

A minor but strictly reproducible detail is the influence of anodic boundary on the tiny third peak in the hydrogen desorption region (marked with an arrow in Figure 5.23). With a lower anodic limit, the peak more and more disappears but gets visible again once the entire cycle, including the platinum oxide region, runs.

First cycles to trace the electrode's history

While in general, cyclic voltammetry with noble metal electrodes focuses on the stationary situation after many cycles, some applications require information about the electrode history and, therefore, the first cycles are of the highest interest. One example is the tracing of remains from the manufacturing process of thin-film platinum. In microfabrication, one can use a dry-etch method with reactive ions to open the passivation layer. Depending on the gas composition during this reactive ion etching (RIE), remains can stick to the electrode surface. Figure 5.24 shows the initial cycles of such an electrode directly after the cleanroom process.

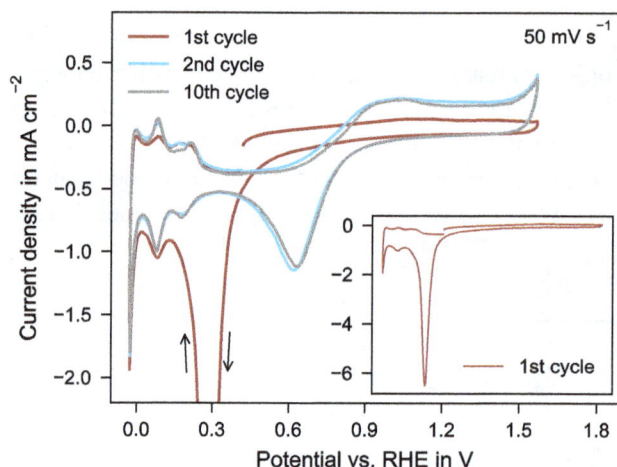

Figure 5.24: Cyclic voltammograms of a thin-film platinum electrode in air-saturated 1 N H_2SO_4 to trace the electrode's history. In the first cycle, directly after the fabrication, remains from the cleanroom process are visible. Fortunately, already one cycle is enough to obtain a clean electrode surface.

In the first cycle, there is a mighty and sharp reduction peak. This peak is visible only once during the CV after the fabrication. Assumingly, substances from the thin-film processing detach from the surface and dissolve in the electrolyte. Already in the second cycle, the curve looks as expected for platinum. This example shows the capability of cyclic voltammetry for both process evaluation, following up the fabrication, and using this method for electrode cleaning.

All noble metal electrodes should see a characterization by cyclic voltammetry after their fabrication. Commercial macroscopic electrodes usually undergo testing after production. In some cases, the manufacturer provides a certificate, including the cyclic voltammogram. For microfabricated electrodes, I strongly recommend a similar characterization and thereby cleaning and conditioning after cleanroom processing. For analytical measurements, one should do cyclic voltammetry additionally in the final setup before each experiment. The excellent sensitivity of the voltammetric methods makes it easily possible to see adsorption or desorption processes with much lower than monolayer resolution. But it also emphasizes the need to be aware of any trace amount of contaminations to ensure reproducible experiments.

Combination with redox-active substances in the electrolyte

There are also experiments possible in which both the surface reaction and redox-active substances from the solution play a role. An interesting example is the cyclic voltammetry of glucose at a platinum electrode. Here, the surface state of the electrode influences also the possible processes with glucose.

The addition of glucose causes five peaks not part of the platinum's cyclic voltammogram. The common understanding of the related processes are as follows (numbers are according to the labels in Figure 5.25):

1. In the hydrogen region, glucose adsorbs onto the bare platinum electrode, followed by deprotonation of the glucose molecule.
2. The adsorbed molecule from peak 1 gets oxidized. It is generally agreed that this process depends on OH_{ads} adsorbed to the electrode. This process diminishes when the OH_{ads} gets replaced at higher potentials.

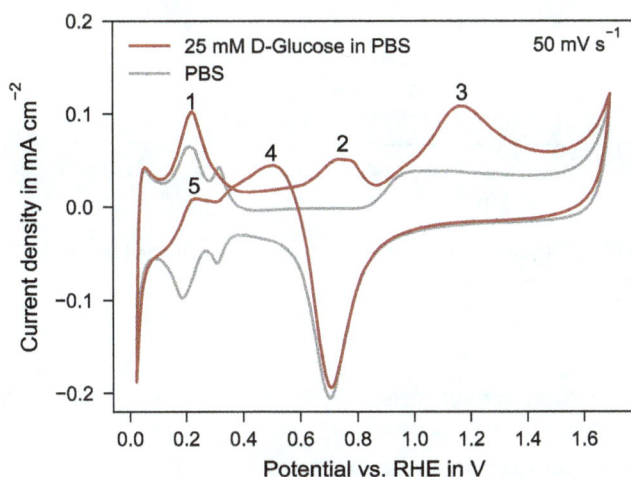

Figure 5.25: Cyclic voltammograms of glucose in phosphate-buffered saline (PBS) pH 7.4 at a platinum electrode. Five distinct peaks appear deviating from the regular curve shape caused by the surface reactions of the platinum electrode.

3. Reaching to the region of platinum oxide formation, the direct oxidation of glucose from the solution occurs. Assumingly, the platinum oxide film catalysis the process. Here, the reaction is diffusion controlled, which is in line with the observation of scanrate dependency for peak 3 compared to the previous peaks.
4. In the reverse scan, after the platinum oxide reduction, glucose gets oxidized on bare platinum.
5. The possibility to distinguish this peak from 4 depends on the solution composition (e. g., phosphate) and pH.

Glucose oxidation plays a role in nonenzymatic glucose sensors. Here, the focus is on the bulk process at peak 3. Further application is the biofuel cell using glucose and oxygen to generate electrical energy.

Another example of CV investigations is the reduction of molecular oxygen at a platinum electrode (Figure 5.26). The difference between the curve with and without oxygen in the electrolyte represents oxygen reduction, which only occurs when the electrode surface is oxide-free.

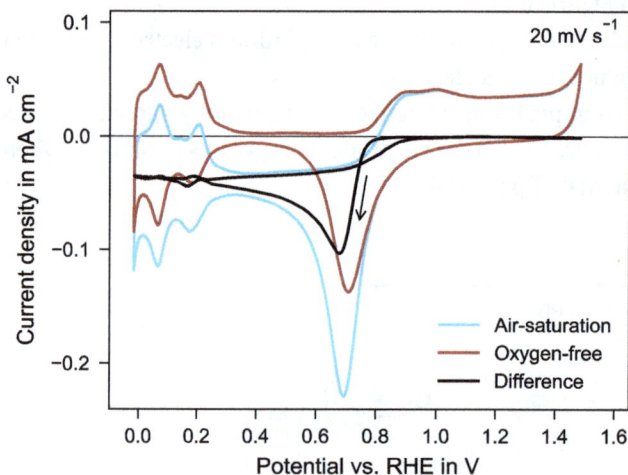

Figure 5.26: Cyclic voltammograms of a platinum electrode in air-saturated and oxygen-free $1\,N\,H_2SO_4$. The difference represents the current from the oxygen reduction.

5.3.4 Cyclic voltammetry of electroactive substances in the electrolyte

The second large use case (or from an electroanalytical perspective, the first) is the cyclic voltammetry with electroactive substances in the electrode. Ideally, reactions from the electrode material do not appear in the cyclic voltammogram, as is the case with carbon electrodes. But also noble metal or mercury electrodes are suitable considering the typically much higher current densities from the electroactive substance than from the electrode material itself.

The theory describing the obtained curves focuses on the prototypic reaction with O and R as used before (equation (2.7)). An extension of the complexity is possible, for example, by considering a downstream chemical reaction ("EC mechanism") or several coupled reactions ("ECEC mechanism").

Reversible systems (Nernstian systems)

We start our walk-through considering the oxidation reaction

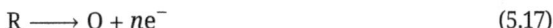

$$R \longrightarrow O + ne^- \tag{5.17}$$

with just R being in the solution before the experiment. Further, we consider an infinite source of R and the oxidation reaction as the only source of O. The electrode is planar in a stagnant solution, so we face a linear diffusion profile. The distance from the electrode is x. At the electrode ($x = 0$), the continuity equation describes the situation:

$$D_O\left(\frac{\partial[O]}{\partial x}\right)_{x=0} + D_R\left(\frac{\partial[R]}{\partial x}\right)_{x=0} = 0 \tag{5.18}$$

At first, we focus on a reversible system (compare Section 2.5.1), allowing the application of the Nernst equation as another boundary condition:

$$\frac{[O]}{[R]} = \exp\left(\frac{nF}{RT}(E - E^{0\prime})\right) \tag{5.19}$$

The equation requires the formal potential $E^{0\prime}$ instead of E^0 (Section 2.2.3), correcting the usage of concentrations instead of activities.

There is no analytical solution for the current response, but scientists developed many numerical approaches. Classically, numerical inversion of the Laplace transformation helped to solve the differential equation. With normalized "current functions," a lookup in tables provided the solution [47]. More recently, several simulators were realized [48–50].

Check the book's webpage to find links to free and commercial simulation programs. You can choose between Windows programs, Excel files with macros, or MATLAB/Python code upon your preference. Even some online tools are available, allowing the accurate simulation of simple systems. **!**

Ramping the potential from a low value at which the oxidation does not occur across the reaction's $E^{0\prime}$ to positive overpotentials results in a linear scan voltammogram, as shown in Figure 5.27.

As expected, the current rise when we approach $E^{0\prime}$ (zero overpotential). However, the maximum, E_p, which appears around 30 mV later, is less intuitive. A closer look a

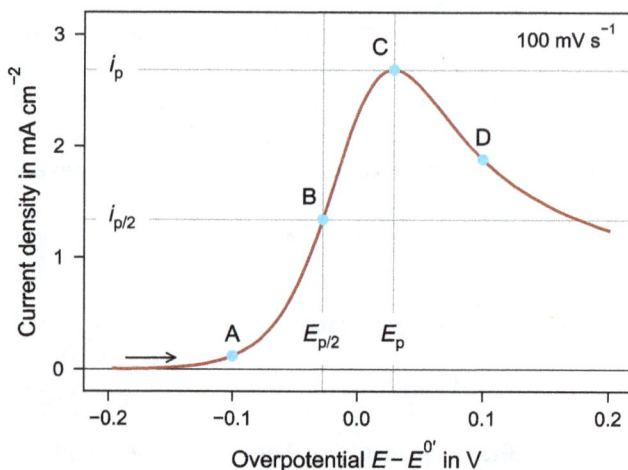

Figure 5.27: Linear scan voltammetry of a reversible system. See Figure 5.28 for concentration profiles at the points labeled with letters. Parameters: $n = 1$, $k_0 = 1\,cm\,s^{-1}$, $D = 1 \times 10^{-5}\,cm^2\,s^{-1}$, $c = 10\,mM$, and $v = 100\,mV\,s^{-1}$.

the diffusion situation before, at, and after the maximum explains the behavior (Figure 5.28). Initially (point A), a large amount of O is in the vicinity of the electrode. When the potential increases, a more extended diffusion gradient forms (B) with the concentrations of O and R balanced at zero overpotential, slightly before the peak (C). Once a linear gradient is established, the current drops (D).

At a given temperature, the curve shape depends solely on the number of electrons, measured by the distance between the peak potential and the potential at which the current reached half of its maximum:

$$|E_p - E_{p/2}| = 2.20 \cdot \frac{RT}{nF} \tag{5.20}$$

This parameter is 56.5 mV per electron at room temperature, which is a helpful tool to check quickly if the curve shape resembles a reversible system.

The *Randles–Ševčík equation* provides a general accepted numerical description of the peak current density:

$$i_p = \pm 0.446 \cdot nFc \sqrt{\frac{nFD}{RT} \cdot v} \tag{5.21}$$

More common is the expression at room temperature:

$$i_p = \pm 2.69 \times 10^5 \, As\, mol^{-1}\, V^{-0.5} \cdot n^{1.5} \cdot D^{0.5} \cdot c \cdot v^{0.5} \tag{5.22}$$

The sign depends on whether it is the peak for oxidation (i_{pa}) or reduction (i_{pc}). The diffusion constant D and the bulk concentration c relates to the reactant, so R for the ox-

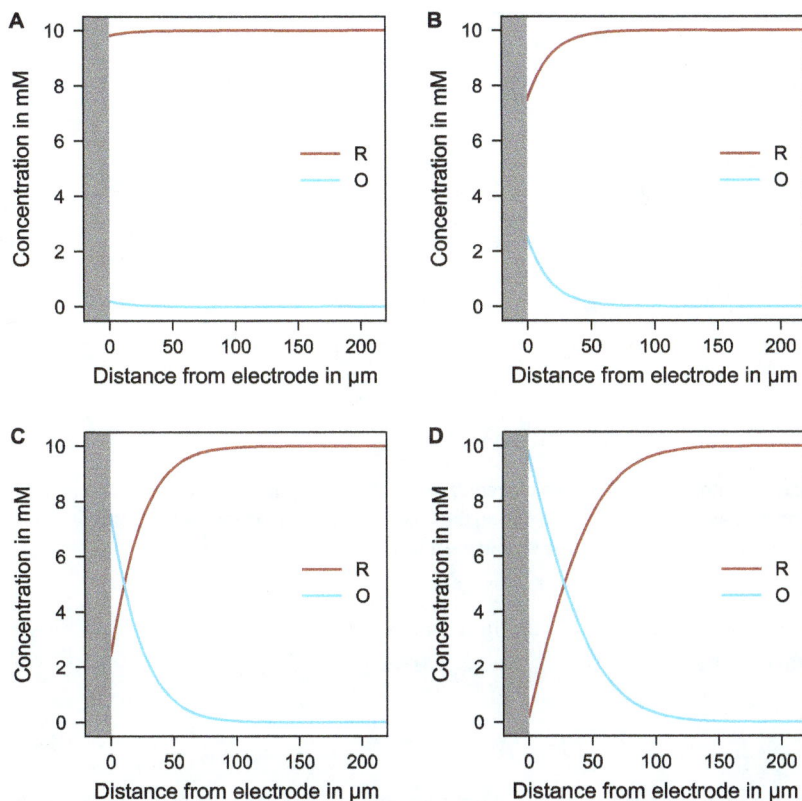

Figure 5.28: Concentration profiles for the linear scan voltammogram in Figure 5.27. The different subplots are before the wave (A), when the current reaches half its maximum (B), at the current's maximum (C), and after the maximum (D). Parameters: $n = 1$, $k_0 = 1\,cm\,s^{-1}$, $D = 1 \times 10^{-5}\,cm^2\,s^{-1}$, $c = 10\,mM$, and $v = 100\,mV\,s^{-1}$.

idation and O for the reduction reaction. Typically, the values of D_O and D_R are close, so $D = D_O = D_R$ is in most cases a reasonable assumption when considering both oxidation and reduction.

We continue our walk-through by expanding from linear scan to cyclic voltamme-try. Figure 5.29 shows the resulting curve when reversing the potential at its highest point. In the reverse scan, the just-formed O gets reduced back to R:

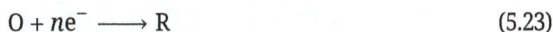

$$O + ne^- \longrightarrow R \tag{5.23}$$

The resulting curve looks roughly mirrored at the x-axis. With a closer look, the re-verse scan starts with a current above zero continuing from the endpoint of the upward scan – overall often described as "duck-shaped."

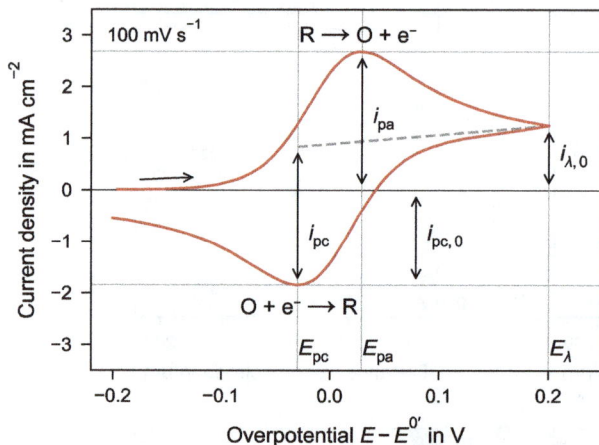

Figure 5.29: Cyclic voltammogram of a reversible redox system. In the beginning, only R is present in the electrolyte; during the reverse scan, O formed during the upward ramp gets reduced. Parameters: $n = 1$, $k_0 = 1\,\mathrm{cm\,s^{-1}}$, $D = 1 \times 10^{-5}\,\mathrm{cm^2\,s^{-1}}$, $[R] = 10\,\mathrm{mM}$, and $v = 100\,\mathrm{mV\,s^{-1}}$.

The turning point is E_λ (what is E_{max} when starting with the upward scan). A closer analysis reveals the peak potential E_{pa} and E_{pc} are separated by

$$E_{pa} - E_{pc} = 2.22 \cdot \frac{RT}{nF} \tag{5.24}$$

assuming a large enough scan window. This parameter is 57 mV per electron at room temperature. The following section discusses the exact influence of the turning point E_λ on the peak separation.

The peak current densities for a reversible system are symmetrical:

$$\left| \frac{i_{pa}}{i_{pc}} \right| = 1 \tag{5.25}$$

Please note that the peak current density in the backward scan (i_{pc}) refers to the curve extending the current in the reverse scan before the onset of the reduction (dashed line in Figure 5.29) rather than to the zero axes ($i_{pc,0}$). The *Nicholson equation* provides a linear approximation for this background current density:

$$\frac{i_{pc}}{i_{pa}} = \frac{i_{pc,0}}{i_{pa}} + \frac{0.485 \cdot i_{\lambda,0}}{i_{pa}} + 0.086 \tag{5.26}$$

The prerequisite for this approximation is a sufficiently high turning points, i. e., $E_\lambda - E_{pa} \geq 60/n\,\mathrm{mV}$.

Similar to the linear sweep voltammetry discussed before, the peak potentials are independent of the scanrate, and peak current densities show a square root dependency of the scanrate according to equation (5.21).

Seeing the curve shape in Figure 5.29 for the first time, the CV might seem drawn arbitrarily without a clear rationale. However, experiments match exactly the prediction. Considering the surface concentration of the species O and R at the electrode facilitates a better understanding of the situation within one cycle (Figure 5.30).

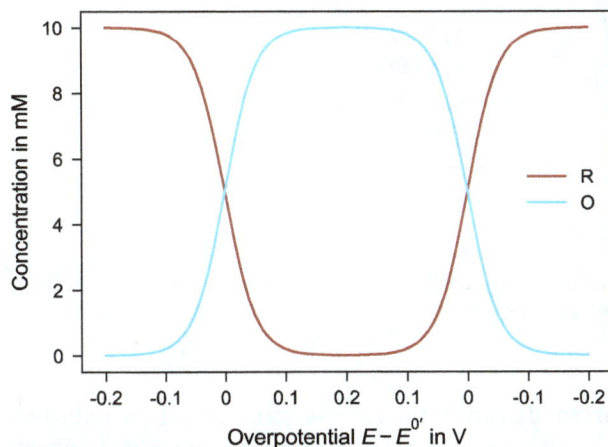

Figure 5.30: Surface concentrations during the cyclic voltammogram in Figure 5.29. Parameters: $n = 1$, $k_0 = 1\,\mathrm{cm\,s^{-1}}$, $D = 1 \times 10^{-5}\,\mathrm{cm^2\,s^{-1}}$, $[R] = 10\,\mathrm{mM}$, and $v = 100\,\mathrm{mV\,s^{-1}}$.

At the beginning of the experiment, the surface concentrations at the electrode resemble the bulk concentration ($[R] = 10\,\mathrm{mM}$, no O). R depletes at the surface during the upward scan and drops. The gradient propagates further into the solution, and the current increases until it reaches a maximum. After the peak potential (around 0.03 V), the decline of the surface concentration slows down. Eventually, it approaches zero until, in the backward scan, the driving force for the oxidation gets lower, and the surface concentration rises again because of R diffusing from the bulk. The behavior of O in the surface concentration plot is inverse.

Influence of the turning point
The turning points in cyclic voltammetry (E_λ) strongly influence the overall appearance of the reverse scan. Ideally, the applied waveform's upper and lower turning points should be as high/low as possible. However, because of the limitation due to the electrolyte's potential window, and because in reality, other processes may interfere, practical experiments may stay within the range of about ±100 mV from the peak potentials. Figure 5.31 illustrates the influence of different turning potentials.

The value of the turning point affects both the cathodic peak current density measured to the x-axis ($i_{pc,0}$) and the position of the cathodic peak potential accordingly. The Nicholson equation (equation (5.26)) accounts for the influence on $i_{pc,0}$, providing a

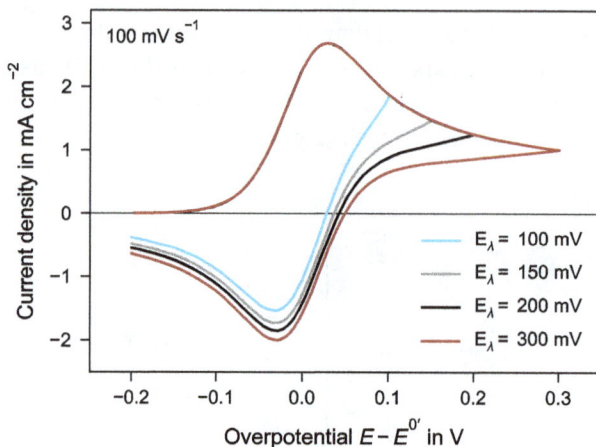

Figure 5.31: Influence of the upper turning point (E_λ) on the reverse scan. Parameters: $n = 1$, $k_0 = 1\,\mathrm{cm\,s^{-1}}$, $D = 1 \times 10^{-5}\,\mathrm{cm^2\,s^{-1}}$, $[R] = 10\,\mathrm{mM}$, and $v = 100\,\mathrm{mV\,s^{-1}}$.

linear approximation of the current density in the reverse scan without the reduction reaction. The peak separation slightly increases from the ideal 57 mV (equation (5.24)) for a turning point closer to the formal potential ($E^{0'}$). Table 5.2 summarizes some typical values.

Table 5.2: Peak separation: deviation from 57 mV per electron for finite turning potentials E_λ, adapted from [47].

$|E_\lambda - E^{0'}|/n$ in mV	$(E_{pa} - E_{pc})/n$ in mV
100	60.5
150	59.2
200	58.3
300	57.8

Irreversible systems

For an irreversible system, the charge transfer becomes dominant (Section 2.5.1), and the Nernst equation is not applicable anymore. Instead, the boundary condition should consider the reaction rates k_a or k_c depending on the scan direction:

$$D_R \left(\frac{\partial [R]}{\partial x} \right)_{x=0} = k_a \cdot [R] \tag{5.27}$$

$$D_O \left(\frac{\partial [O]}{\partial x} \right)_{x=0} = k_c \cdot [O] \tag{5.28}$$

Figure 5.32 shows a typical linear scan voltammogram of an irreversible system. While the overall shape resembles the reversible case for one scan direction, the peak potential remarkably shifts to higher overpotential along with a lower peak height.

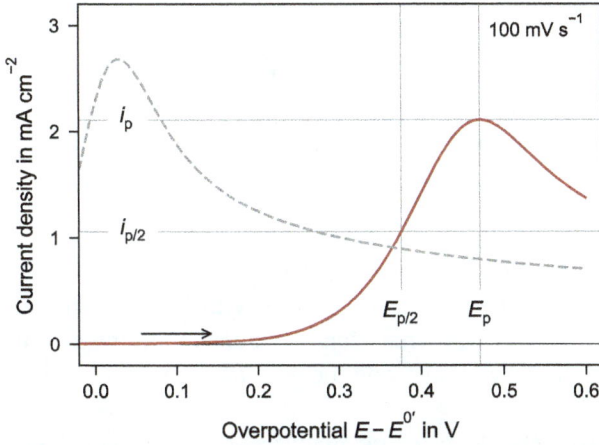

Figure 5.32: Linear scan voltammetry of an irreversible system. The dashed line is the reversible system from Figure 5.27 for comparison. Parameters: $n = 1$, $k_0 = 1 \times 10^{-6}$ cm s^{-1}, $D = 1 \times 10^{-5}$ cm^2 s^{-1}, [R] = 10 mM, $a = 0.5$, and $v = 100$ mV s^{-1}.

In contrast to the reversible system, the peak potential depends on many factors, including the scanrate:

$$E_p = E^{0\prime} \pm \frac{RT}{a_{a|c}nF}\left(0.780 + \ln\frac{\sqrt{D}}{k_0} + \frac{1}{2}\ln\left(\frac{a_{a|c}nF}{RT}\cdot v\right)\right)$$ (5.29)

For clarity, the equation shows the charge transfer coefficient as $a_{a|c}$. Depending on oxidation or reduction, it should be $a_a = 1 - a$ or $a_c = a$.

More simple, the peak potential relates to the decadic logarithm of the scanrate depending on the charge transfer coefficient and number of electrons only:

$$\left|\frac{dE_p}{d\lg v}\right| = \frac{29.5\,\text{mV}}{a_{a|c}n}$$ (5.30)

Many systems allow the approximation of a by 0.5, reducing the dependency to 15 mV per electron per decade change in scanrate.

The curve shape considers the charge transfer too and is generally wider than in the reversible case:

$$|E_p - E_{p/2}| = 1.857\cdot\frac{RT}{a_{a|c}nF}$$ (5.31)

This parameter is 47.7 mV/a_{alc} per electron at room temperature.

Analog to the Randles–Ševčík equation (equation (5.21)), the peak current density depends on the square root of the scanrate:

$$i_p = \pm 0.496 \cdot a_{alc} nFc \sqrt{\frac{a_{alc} nFD}{RT}} \cdot v \tag{5.32}$$

More common is the expression at room temperature:

$$i_p = \pm 2.99 \times 10^5 \, \text{As mol}^{-1} \, \text{V}^{-0.5} \cdot a_{alc} \cdot n^{1.5} \cdot D^{0.5} \cdot c \cdot v^{0.5} \tag{5.33}$$

Reversibility

So far, we met in our walk-through reversible systems ("Nernstian systems") in which mass transfer dominates, and the Nernst equation is applicable. While there is no exact analytical solution for the resulting curve shape, handsome equations allow the description of the peak current density, the curve shape, and the position of the peak relative to the formal potential.

On the other hand, we saw irreversible systems ("totally irreverible systems") with slow charge transfer dominating the situation. The numerical solutions depend on the charge transfer coefficient, and the peak position also changes with scanrate on the potential axis. Table 5.3 summarizes some essential aspects of reversible and irreversible systems.

Table 5.3: Comparison of reversible (Nernstian) and irreversible systems.

	Reversible system	Irreversible system
Limiting effect	Mass transfer	Charge transfer
Peak current density	$i_p \propto \sqrt{v}$	$i_p \propto \sqrt{v}$ and $i_p \propto a_{alc}$
Peak potential	E_p independent of v	$\left\|\frac{dE_p}{d\lg v}\right\| = \frac{29.5 \, \text{mV}}{a_{alc} n}$
Peak shape	$\|E_p - E_{p/2}\| = \frac{56.5 \, \text{mV}}{n}$	$\|E_p - E_{p/2}\| = \frac{47.7 \, \text{mV}}{a_{alc} n}$
Peak separation	$E_{pa} - E_{pc} = \frac{57.0 \, \text{mV}}{n}$	–
Peak ratio	$\left\|\frac{i_{pc}}{i_{pa}}\right\|$	–

There are also the quasi-reversible systems between reversible and irreversible system cases, sometimes called "kinetic systems." Many experiments end up in this regime. However, changing the conditions, most simply the scanrate, allows shifting a quasi-reversible system to the one or other case if desired.

The parameter Λ allows for classification of the different cases. The dimensionless value depends on the rate constant, the number of electrons, scanrate, and the diffusion constant assuming the same diffusion constant for O and R:

$$\Lambda = \frac{k_0}{\sqrt{\frac{nFD}{RT} \cdot v}} \tag{5.34}$$

The following values distinguish the different systems, while the exact numbers of the boundaries are disputable:

- Reversible systems: $\Lambda \geq 15$
- Quasi-reversible systems: $10^{-3} < \Lambda < 15$
- Irreversible systems: $\Lambda \leq 10^{-3}$

Figure 5.33 shows cyclic voltammograms for the three domains. The quasi-reversible system resembles the reversible case but wider peak separation and lower peak current density. For the irreversible system, the peak shift is much more prominent, and the backward scan's peak is largely delayed and suppressed. In actual experiments, the peak in the backwards scan may not be reached at all.

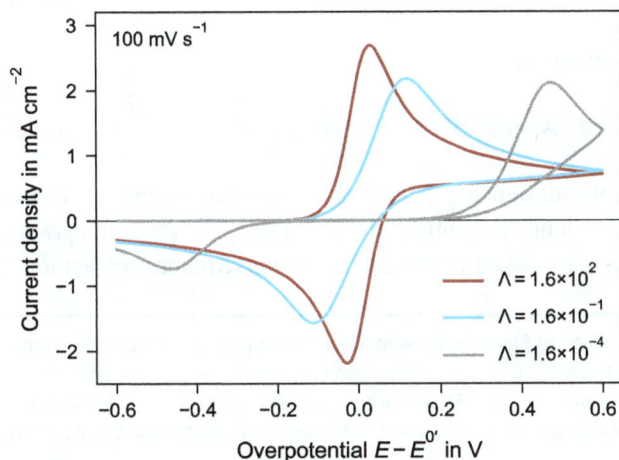

Figure 5.33: Comparison of reversible ($\Lambda = 1.6 \times 10^2$), quasi-reversible ($\Lambda = 1.6 \times 10^{-1}$), and irreversible system ($\Lambda = 1.6 \times 10^{-4}$). Parameters: $n = 1$, $k_0 = 1,1 \times 10^{-3}$, or $1 \times 10^{-6} \, cm \, s^{-1}$, $D = 1 \times 10^{-5} \, cm^2 \, s^{-1}$, $[R] = 10 \, mM$, $a = 0.5$, and $v = 100 \, mV \, s^{-1}$.

The peak separation in a quasi-reversible system depends on Λ, reaching values of more than 200 mV. Table 5.4 shows some typical values. However, using measurements of the peak separation at different scanrate to deduce the reversibility is difficult as parasitic effects like the uncompensated resistance (Section 3.2.4) may critically influence the measurement values. Using software that allows matching experimental data to simulated curve shapes is a more reliable option.

Table 5.4: Peak separation in quasi-reversible systems as a function of Λ. Adapted from [51].

Λ	$(E_{pa} - E_{pc})/n$	Λ	$(E_{pa} - E_{pc})/n$
> 15	57 mV	1.1	72 mV
11	61 mV	0.56	84 mV
3.9	63 mV	0.42	92 mV
3.4	64 mV	0.28	105 mV
2.8	65 mV	0.20	121 mV
2.3	66 mV	0.14	141 mV
1.7	68 mV	0.056	212 mV

The peak current density is different for reversible (equation (5.21)) and irreversible systems (equation (5.32)) but the equations has a similar form:

$$i_p = \pm 0.436 \cdot nFc \sqrt{\frac{nFD}{RT} \cdot v} \tag{5.35}$$

At room temperature, the expression is

$$i_p = \pm 2.63 \times 10^5 \, \text{As mol}^{-1} \, \text{V}^{-0.5} \cdot n^{1.5} \cdot D^{0.5} \cdot c \cdot v^{0.5} \tag{5.36}$$

Figure 5.34 illustrates various influencing factors on cyclic voltammograms of a reversible and a quasi-reversible system. With software to simulate cyclic voltammograms and change the parameter, one can easily understand the various influencing factors.

---

i ■■■ Task 5.1: Check the book's webpage to find links to various simulation programs (e. g., online tools free to use but sufficient to complete this task). Consider a simple redox system O + e⁻ ⇌ R and select a value for $E^{0\prime}$. Define a scan from −0.3 V below $E^{0\prime}$ to 0.3 V above and back. Assume only R ($c = 1$ mM) in the electrolyte. The diffusion constants of O and R are $D_O = D_R = 1 \times 10^{-5} \, \text{cm}^2 \, \text{s}^{-1}$, if needed use $\alpha = 0.5$. Start with $k_0 = 1 \, \text{cm s}^{-1}$.

1. Use different values for the scanrate (e. g., in the range between 10 mV s⁻¹ and 1 V s⁻¹). What is the influence on the peak position, the peak separation, and the peak current density?
2. Use different values for α (e. g., 0.3, 0.5, 0.7) with a scan rate of 100 mV s⁻¹. What is the influence on the peaks and the curve shape?
3. Repeat the above steps for a quasi-reversible reaction. Lower the reaction rate k_0 to a value so that Λ is between 0.01 and 1.

Multistep reactions

By now, we considered a simple one-step reaction. Especially with organic molecules, multistep reactions are common. A typical system consists of two coupled redox pairs:

$$A \underset{k_{AB}^0}{\rightleftharpoons} B + e^- \underset{k_{BC}^0}{\rightleftharpoons} C + e^- \tag{5.37}$$

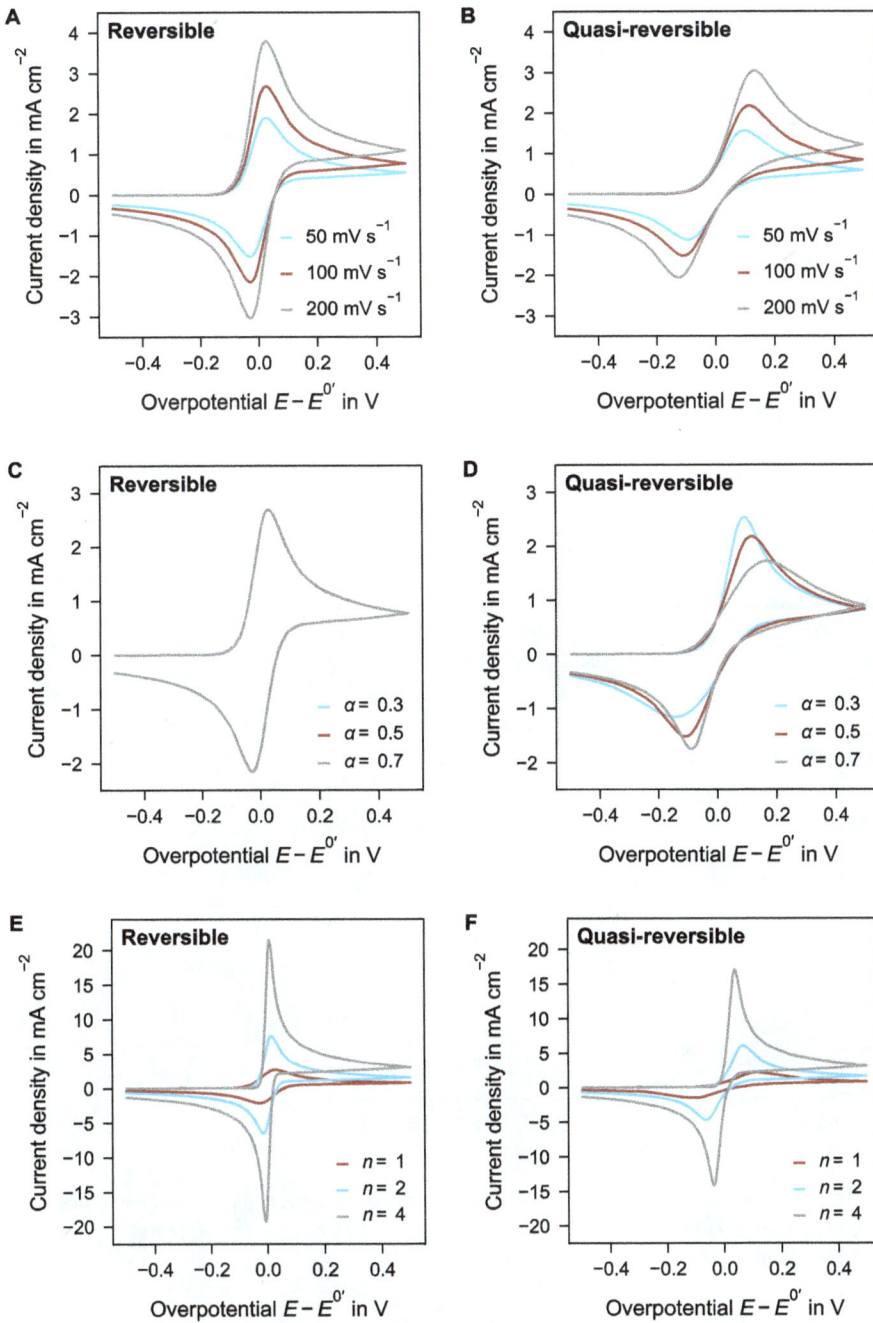

Figure 5.34: Influencing factors on cyclic voltammograms: scanrate v (A, B), symmetry factor α (C, D), and the number of electrons n (E, F) for a reversible system ($k_0 = 1\,\text{cm s}^{-1}$) and a quasi-reversible system ($k_0 = 1 \times 10^{-3}\,\text{cm s}^{-1}$). Parameters if not mentioned in the plots: $n = 1$, $D = 1 \times 10^{-5}\,\text{cm}^2\,\text{s}^{-1}$, $[R] = 10\,\text{mM}$, $\alpha = 0.5$, and $v = 100\,\text{mV s}^{-1}$.

Besides reversibility for each step, the distance between the two reactions' formal potential and which comes first in the initial scan direction matters. Figure 5.35(A) shows the example for $E_{AB}^{0\prime} < E_{BC}^{0\prime}$. Here, the first oxidation reaction fuels the second, and on the backward scan, it happens in reverse order. The other possible situation is Figure 5.35(B) showing only one wave with a higher current density.

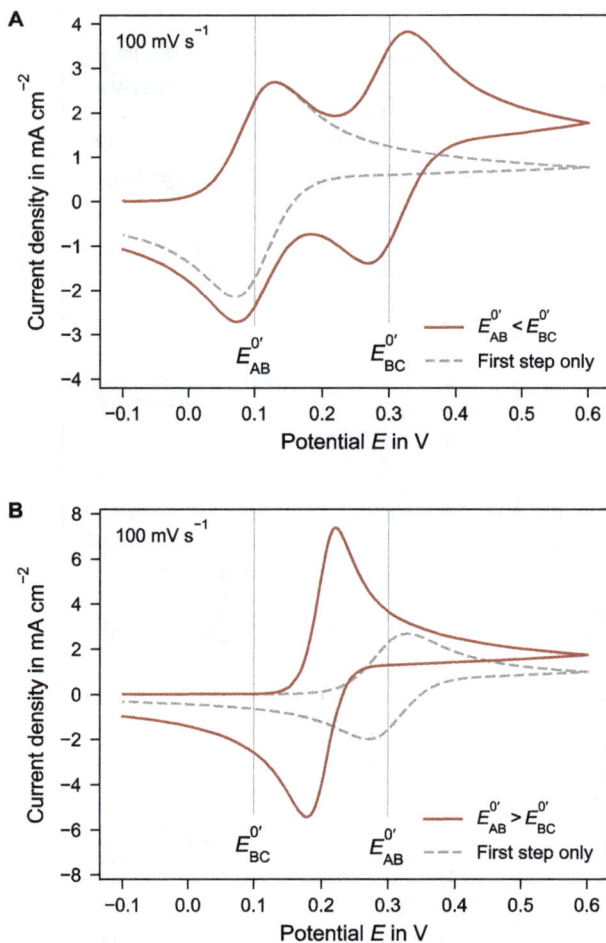

Figure 5.35: Cyclic voltammograms of a two-step reaction. The difference is in the order of the formal potentials of each step. Parameters: $E_{AB}^{0\prime} = 0.1\,\text{V}$ or $0.3\,\text{V}$, $E_{BC}^{0\prime} = 0.3\,\text{V}$ or $0.1\,\text{V}$, $k_{AB}^{0} = 1\,\text{cm s}^{-1}$, $k_{BC}^{0} = 0.1\,\text{cm s}^{-1}$, $D = 1 \times 10^{-5}\,\text{cm}^2\,\text{s}^{-1}$, $[A] = 10\,\text{mM}$, and $v = 100\,\text{mV s}^{-1}$.

Looking at the species' surface concentration provides better insights on how the curve shape forms. Figure 5.36 shows the surface concentrations from the two-step reaction with $E_{AB}^{0\prime} < E_{BC}^{0\prime}$ (Figure 5.35(A)). The system starts with the bulk concentrations (A=10 mM, no B, no C). First, the formation of B rises, reaching an equal concentration

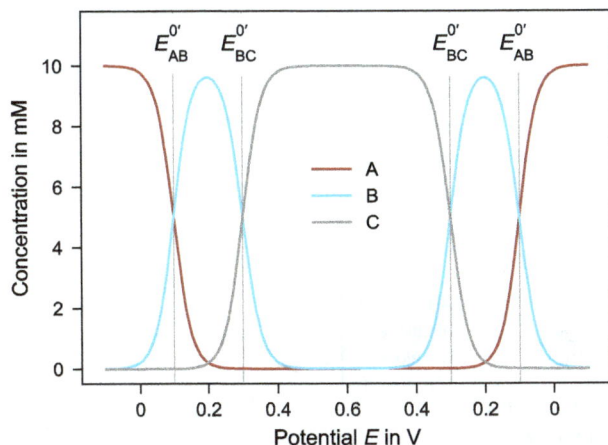

Figure 5.36: Surface concentrations during the cyclic voltammogram in Figure 5.35(A). Parameters: $E_{AB}^{0\prime}$ = 0.1 V, $E_{BC}^{0\prime}$ = 0.3 V, k_{AB}^{0} = 1 cm s^{-1}, k_{BC}^{0} = 0.1 cm s^{-1}, D = 1 × 10^{-5} cm^2 s^{-1}, [A] = 10 mM, and v = 100 mV s^{-1}.

of A and B at the corresponding formal potential (what is its definition). Second, the oxidation from B to C depletes the surface concentration of B. Similarly, the values for B and C are the same at their formal potential. The same situations occur in the reverse order on the backward scan (right-half of the surface concentration plot).

Multicomponent systems

Another case is a multicomponent system with independent reactions, e. g., two redox couples without common species:

$$A \overset{k_{AB}^{0}}{\rightleftharpoons} B + e^{-} \tag{5.38}$$

$$C \overset{k_{CD}^{0}}{\rightleftharpoons} D + e^{-} \tag{5.39}$$

Suppose no other mechanisms like adsorption or chemical reactions play a role. In that case, the systems' responses superimpose, as shown in Figure 5.37. Deconvolution allows interpretation of measurement curves from multicomponent situations but gets more difficult when the formal potentials are close, or the current ranges strongly differ.

The general appearance of a cyclic voltammogram makes it impossible to distinguish between a multistep reaction and a multicomponent system. The most straightforward approach in attempting to decipher the situation is the variation of the turning points of the applied waveform. The exact strategy depends on the different mechanisms involved and requires some experience. Matching the measured data to various simulated models may help.

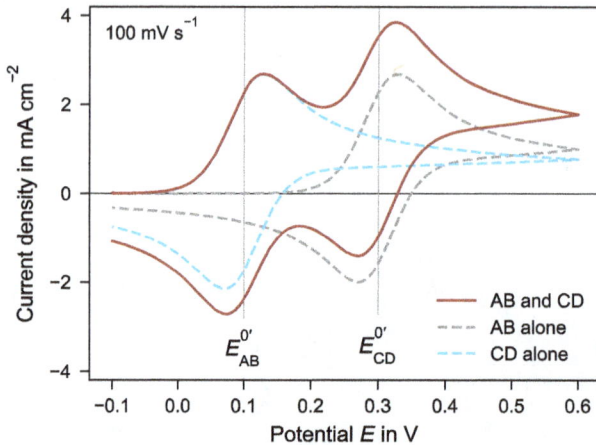

Figure 5.37: Cyclic voltammogram of a multicomponent system. Parameters: $E_{AB}^{0\prime} = 0.1\,V$, $E_{CD}^{0\prime} = 0.3\,V$, $k_{AB}^0 = 1\,cm\,s^{-1}$, $k_{CD}^0 = 0.1\,cm\,s^{-1}$, $D = 1 \times 10^{-5}\,cm^2\,s^{-1}$, $[A] = [C] = 10\,mM$, and $v = 100\,mV\,s^{-1}$.

Reaction mechanism including chemical reactions

Also, chemical reactions can contribute to the overall response. The most straightforward mechanism is the redox reaction followed by a chemical reaction ("EC mechanism") with products that are not involved in the redox reaction:

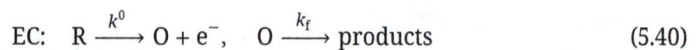

$$\text{EC:} \quad R \xrightarrow{k^0} O + e^-, \quad O \xrightarrow{k_f} \text{products} \tag{5.40}$$

Figure 5.38 shows an example of the cyclic voltammogram of an EC mechanism and the redox reaction alone for comparison. The chemical reaction consumes O continuously, depending on k_f the reduction peak in the backward scan is less pronounced. For higher rates, the diffusion situation during the forward scan gets affected, and the peak current density slightly increases along with a shift of the peak potential to the left.

The chemical reaction can also produce R in a catalysis reaction ("EC' mechanism"), or a second redox reaction can follow the just discussed combination of an electrochemical and chemical reaction ("ECE mechanism"):

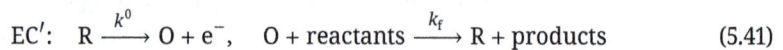

$$\text{EC':} \quad R \xrightarrow{k^0} O + e^-, \quad O + \text{reactants} \xrightarrow{k_f} R + \text{products} \tag{5.41}$$

$$\text{ECE:} \quad R \xrightarrow{k_1^0} O + e^-, \quad O \xrightarrow{k_f} S, \quad S \xrightarrow{k_2^0} T + e^- \tag{5.42}$$

Many other combinations with several electrochemical steps and chemical reactions before or after are possible. Further refinements include, e. g., the reversibility of the chemical reaction. An excellent overview of many different mechanisms and their influence on cyclic voltammograms is provided in Heinze [52].

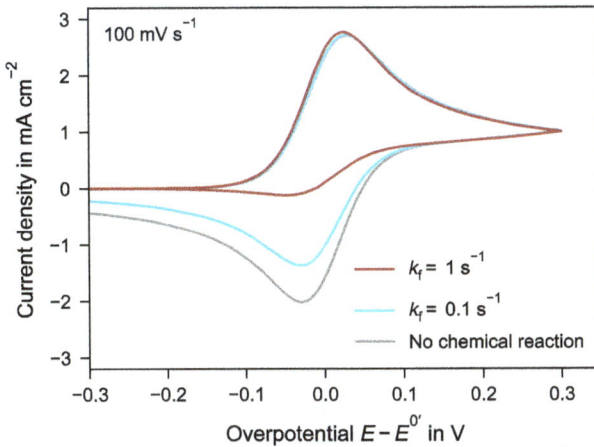

Figure 5.38: Cyclic voltammogram of a redox reaction followed by a chemical reaction ("EC mechanism"). k_f is the chemical reaction's rate constant. Parameters: $n = 1$, $k^0 = 1\,\mathrm{cm\,s^{-1}}$, $k_f = 1\,\mathrm{s^{-1}}$ or $0.1\,\mathrm{s^{-1}}$, $D = 1 \times 10^{-5}\,\mathrm{cm^2\,s^{-1}}$, $[R] = 10\,\mathrm{mM}$, and $v = 100\,\mathrm{mV\,s^{-1}}$.

Adsorption

Electrode reactions with an adsorption step play a role in many applications. The current contribution depends on the characteristics of the adsorption process. A simple case is a reversible adsorption (diffusion is not rate-limiting) as illustrated in Figure 5.39. Here, the labels are for adsorption of cations; the adsorption is then the reduction process in the downward scan. The oxidation and reduction curve is similar but mirrored at the *x*-axis. Changing the scan direction in between would cause the current to flip immediately to the opposite side.

The peak current density is proportional to the scanrate:

$$i_p = \pm \frac{n^2 F^2 \Gamma^*}{4RT} \cdot v \tag{5.43}$$

Γ^* is the surface coverage of the adsorbed species (e. g., in $\mathrm{mol\,cm^{-2}}$).

The curve shape, characterized by the width of the peak at half-height, depends at constant temperature on the number of electrons only:

$$\Delta E_{p/2} = 3.530 \cdot \frac{RT}{nF} \tag{5.44}$$

This parameter is 90.5 mV per electron at room temperature.

Variation of the scanrate is a simple but powerful diagnostic tool. With the dependency of the peak current density on the scanrate, it is possible to discriminate between an adsorption process ($\propto v$) or a redox reaction with freely diffusing species ($\propto \sqrt{v}$). In more detail, the dependency of the peak potential and peak separation allows for the identification of reversible, quasi-reversible, and irreversible processes.

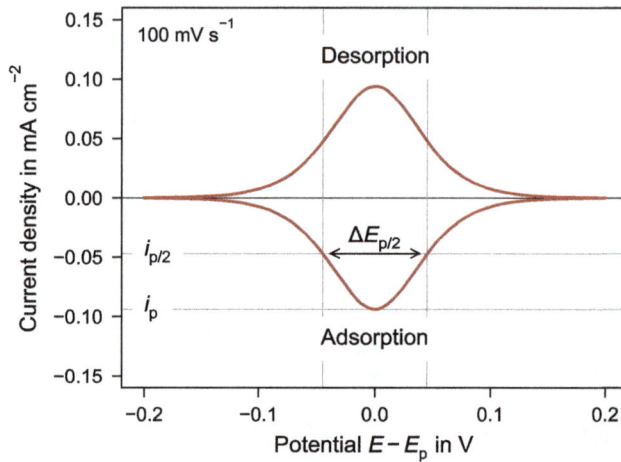

Figure 5.39: Cyclic voltammogram of an adsorption process. The labels refer to the adsorption of cations. Parameters: $\Gamma^* = 1 \times 10^{-9}$ mol cm^{-2}, $n = 1$, and $v = 100$ mV s^{-1}.

Nonidealities

Actual measurements of cyclic voltammograms suffer from additional contributions due to charging up the double layer capacity and the uncompensated resistance (Section 3.2.4). The capacitive current increases linearly with the scanrate. Therefore, its significance is highest at immense scanrates, e. g., during fast-scan voltammetry (Section 5.3.10). The uncompensated resistance matters more for larger current, i. e., larger electrode surface.

Figure 5.40 shows the effect of the double layer capacity and the uncompensated resistance. The blue curve represents a typical experimental situation that requires compensation for exact evaluations of the peak shift or peak position. Facing a CV like the grey curve with planar electrodes in a well-conducting electrolyte indicates an issue with the setup.

---

i ■□□ Task 5.2: Influence of the double layer capacity on voltammetry: You run a linear scan experiment with a scanrate of 100 mV s^{-1}. The double layer capacity of your electrode is 20 µF cm^{-2}. Which current density do you expect in the absence of any redox process?

Microelectrodes

In our walkthrough, we focused on linear diffusion only. However, with finite electrode dimensions, most prominently with microelectrodes, a hemispherical diffusion profile prevails (Figure 2.11). The before mentioned tables with classical solutions for cyclic voltammetry include data for spherical correction [47]. Many cyclic voltammetry simulation programs do contain the option to simulate hemispherical or disk-shaped electrodes. Especially with ultramicroelectrodes (Section 4.3.6), cyclic voltammograms look

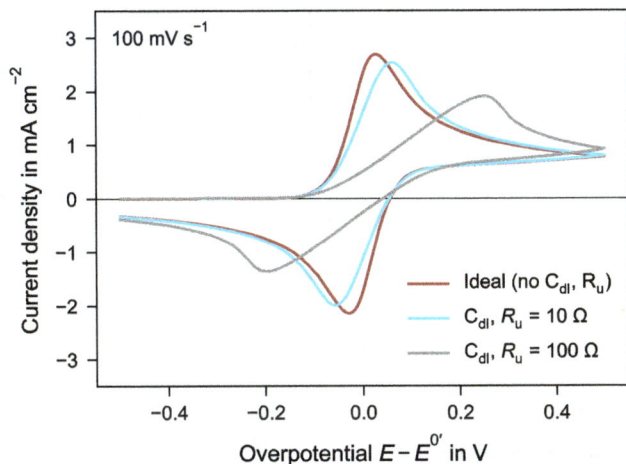

Figure 5.40: Influence of the double layer capacity and the uncompensated resistance on the CV shape. Parameters: $C_{dl} = 20\,\mu F\,cm^{-2}$, $n = 1$, $k_0 = 1\,cm\,s^{-1}$, $D = 1 \times 10^{-5}\,cm^2\,s^{-1}$, $[R] = 10\,mM$, and $v = 100\,mV\,s^{-1}$. The uncompensated resistance R_u relates to an electrode area of $1\,cm^2$.

much different for the previously discussed situation with a scanrate in the range of $100\,mV\,s^{-1}$.

Figure 5.41 shows cyclic voltammograms of a disk-shaped microelectrode with a $10\,\mu m$ diameter. The blue curve ($100\,mV\,s^{-1}$) is for the same parameters as the CV in Figure 5.29. However, the current density approaches a limiting value and returns nearly on the same path in the reverse scan because of the hemispherical diffusion profile. In contrast, the red curve ($10\,V\,s^{-1}$) resembles the known shape, which is the typical behavior for microelectrodes.

Please note the overall higher current density than Figure 5.29 in the range of one order of magnitude. This ratio also illustrates another advantage of microelectrodes. While parasitic effects like the charge-up of the double layer capacity contribute with the same current density for both electrodes, the much higher faradaic current density leads to a better signal-to-background ratio of the microelectrode than the electrode with a planar diffusion profile.

For the situation of low enough scanrate to reach a steady-state current I_{ss} exact solutions are available:

$$\text{Hemisphere:} \quad I_{ss} = 2\pi FrDc \tag{5.45}$$

$$\text{Disk:} \quad I_{ss} = 4FrDc \tag{5.46}$$

r is the electrode radius and c the bulk concentration. Heinze provides more details on cyclic voltammetry with ultramicroelectrodes in [38].

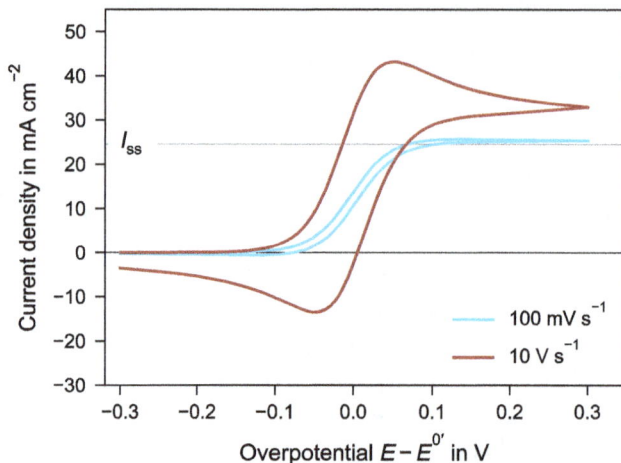

Figure 5.41: Disk-shaped microelectrode (10 μm diameter). Parameters: $n = 1$, $k_0 = 1\,\text{cm s}^{-1}$, $D = 1 \times 10^{-5}\,\text{cm}^2\,\text{s}^{-1}$, $[R] = 10\,\text{mM}$, and $v = 100\,\text{mV s}^{-1}$ versus $10\,\text{V s}^{-1}$.

5.3.5 Polarography

Polarography, the voltammetry with mercury electrodes (Section 4.3.2), is one of the oldest electroanalytic methods. Invented in the 1920s by Jaroslav Heyrovský, it is for sure the most classic classical method. Many conventions we saw for the other techniques in this chapter originate from what we treat today as the exceptional case, the situation of a mercury drop in an aqueous electrolyte. Still, especially for the electrochemical detection of metal ions, it is the golden standard any microsensor in this domain should compare to, e. g., its limit of detection, which is typically around μM and can reach values as low as 0.1 nM depending on the analyte and the used technique. The possibility of liquid metal at room temperature contributes to the features such as simple renewal of the electrode surface and adjusting the electrode area according to the experimental needs. However, successful measurements are only possible when contamination is avoided, even from oxygen in the air, which is more complicated than having a solid metal electrode.

Two significant disadvantages come along with mercury drops. Mercury, including its vapor, is toxic and dangerous for the environment when accidentally released and requires careful waste treatment. While applying polarography in analytical laboratories with appropriate precaution and waste disposal measures is considered safe, one should avoid using a mercury electrode in the field or with untrained staff. The second disadvantage is the low compatibility with microfabricated sensors. While there is the possibility to coat solid electrodes with a thin mercury film [27], environmental concerns remain while compromising on the features of the continuous renewal of the droplet.

The most common form of polarography is at the dropping mercury electrode (DME). The drop lifetime τ is in the seconds range depending on the setup and the mass

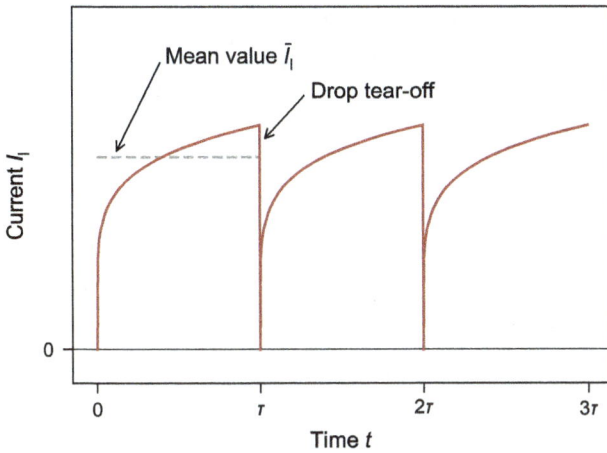

Figure 5.42: Dropping mercury electrode: during the drop lifetime τ, the surface area increases and the limiting current I_l accordingly. $\bar{I}_l$ is the mean current integrated over a drop's lifetime.

flow rate $\dot{m}$ of the mercury. In polarography, the analyte is historically called a "depolarizer" with its bulk concentration c. During the measurement, the electrolyte is at rest so that diffusion limits the faradaic current.

Figure 5.42 shows a typical raw signal at the DME. The diffusion limited current during one drop lifecycle depends on the increasing drop surface ($\propto t^{2/3}$) and the Cottrell behavior ($\propto t^{-1/2}$):

$$I_l(t) = \underbrace{4\pi \left(\frac{3\dot{m}}{4\pi\rho} \right)^{2/3}}_{\text{Electrode area}} nFc \sqrt{\frac{D}{\pi}} \cdot t^{1/6} \tag{5.47}$$

ρ is the density of mercury. Either synchronization of the current readout to a specific drop size or integration of the current over the entire drop lifetime provides the basis of the voltammogram:

$$\bar{I}_l = \frac{1}{\tau} \int_0^\tau I_l(t) dt \tag{5.48}$$

More common is the *Ilkovič equation* to describe the mean limiting current:

$$\bar{I}_l = 0.607 \cdot n \cdot D^{1/2} \cdot \dot{m}^{2/3} \cdot c \cdot \tau^{1/6} \tag{5.49}$$

The prefactor considers the numeric values of the constants and an additional correction factor. The current is in unit A with the diffusion constant D in $cm^2 s^{-1}$, the mass flow of the mercury $\dot{m}$ in $mg\, s^{-1}$, τ in s, and the bulk concentration c in $mol\, cm^{-3}$.

Figure 5.43 shows an idealized polarogram of three different metal ions: (Cu(II), Cd(II), and Zn(II)). The currents are the mean values $\bar{I}_l$ along a drop lifetime. Considering the step heights for each ion allows calculation of the concentration with the Ilkovič equation (equation (5.49)).

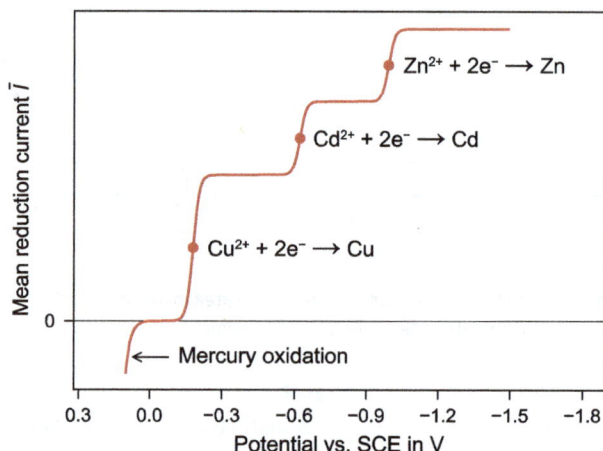

Figure 5.43: Dropping mercury electrode: Idealized polarogram of three different metal ions. The ratio of the step heights indicates that twice as much copper as the other two ions is in the solution. The half-wave potentials match a 1 M KCl electrolyte. For higher potentials (> 0 V), oxidation of the mercury occurs. This plot follows the polarographic sign convention (Figure 5.19), deviating from this book's standard.

Practical measurements suffer from unwanted, additional effects. Especially for lower analyte concentration, the capacitive and other background currents require attention and subtraction as a baseline. The diffusion situation may be disturbed by convective contributions and deviate from the Ilkovič equation. Therefore, calibration with standard solutions with known ion concentrations is often a better approach to obtaining accurate results.

The potential window of mercury limits the possible reactions mainly to the reduction side (see the arrow in Figure 5.43). At the lower end of the potential window, the reduction of the electrolyte (e. g., potassium ions) limits the useful range.

Possible analytes are elemental ions, mainly metals: Al(III), As(III), Ba(II), Be(II), Ca(II), Cs(I), Bi(III), Cd(II), Co(II), Cr(III), Cr(VI), Cu(II), Fe(II), Fe(III), Ge(IV), K(I), Li(I), Mn(II), Mo(VI), Na(I), Ni(II), Pb(II), Pd(II), Rb(I), Sb(III), Se(IV), Sn(II), Sn(IV), Sr(II), Ti(III), Ti(IV), Tl(I), U(VI), V(V), W(VI), Zn(II), Zr(IV). Also, the measurement of organic molecules is possible. Please note that some of the mentioned analytes require a specific electrolyte solution.

5.3.6 Pulse voltammetry

In the first-half of the last century, electroanalysis focused exclusively on mercury electrodes. In particular, with its outstanding property of a constantly new electrode surface, the DME posed a big challenge for synchronizing measurements with the droplet life cycle. Different voltammetry variants came up, both terminology and typical time scale matching the situation at the DME. Today, the various pulse techniques are mainly used at solid electrodes and offer similar and new benefits.

Normal pulse voltammetry

In *normal pulse voltammetry* (NPV), the electrode is at a base potential from which consecutively higher potentials get measured by short pulses. Figure 5.44 shows the time trace of the applied waveform. The measurement device samples the current at the end of each pulse, minimizing the contribution from capacitive effects. The base potential is preferable at a value without faradaic current. Using a longer time at the base potential allows restoration of the analytes diffusion situation at the electrode so that during the short pulses, the faradaic process runs strongly and forms a new concentration gradient. Apart from the higher current due to the pulsing, this method lowers the overall consumption of the analyte and, maybe most importantly, reduces the risk of inactivating the electrode by reaction products. However, the shorter timescale on which the faradaic process occurs may miss out, e. g., coupled reactions.

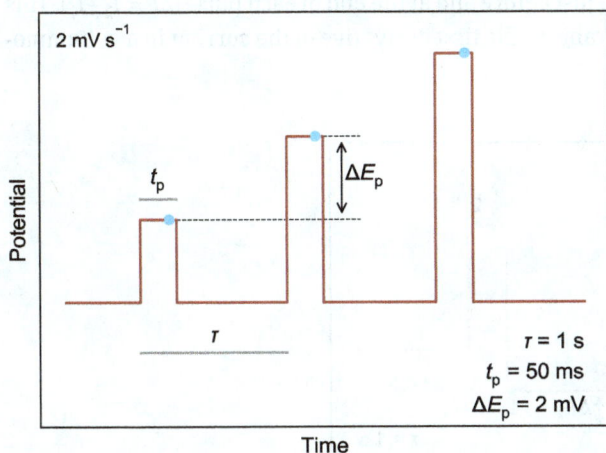

Figure 5.44: Normal pulse voltammetry (NPV): applied waveform and some typical values [53]. Please note that the durations are not to scale. The blue points indicate current sampling.

The normal pulse voltammogram is the current sampled at the end of each pulse versus potential curve (Figure 5.47(B)). The limiting current density depends linearly on the concentration with the parameter t_p setting the sensitivity:

$$i_{lim} = \pm nFc \sqrt{\frac{D}{\pi t_p}} \tag{5.50}$$

While it seems tempting to increase the sensitivity by decreasing t_p, for too low values, the capacitive currents dominate.

Developed for a mercury electrode, NPV also works well on carbon electrodes because of the absence of faradaic processes of the electrode. On metal electrodes such as platinum or gold, the surface reactions of the electrode material are also contributing to the signal raising the background. Especially for lower analyte concentrations, other methods suit better. An interesting example with a higher analyte concentration is the NPV study of methanol oxidation at a platinum electrode by Xu et al. [54].

A closely related method is the *reverse pulse voltammetry* (RPV), which resembles the NPV but with a base potential in the electroactive region, causing a diffusion-limited situation from which small pulses bring the system out of diffusion control.

Differential pulse voltammetry

In *differential pulse voltammetry* (DPV), the potential changes slowly in steps with a short pulse at the beginning of each step (Figure 5.45). The measurement signal is the difference between a current reading just before and at the end of each pulse ($\Delta I = I_2 - I_1$). This scheme leads to a curve comparable to the first derivative of the current in a voltammogram (Figure 5.47(C)).

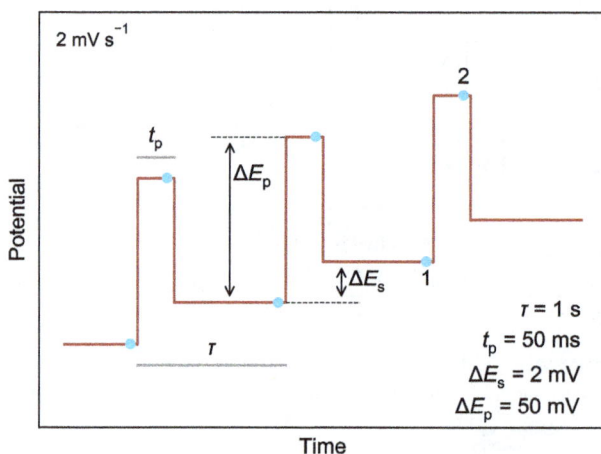

Figure 5.45: Differential pulse voltammetry (DPV): applied waveform and some typical values [53]. Please note that the durations and peak/step heights are not to scale.

The contribution from capacitive current is much lower compared to NPV because of the differential principle. The peak shape provides some diagnostic parameters, e. g., a symmetrical peak indicates a reversible reaction. Overall, this method allows a lower detection limit in many cases, typically in the range of two to three orders of magnitude below classical cyclic voltammetry.

The peak current density depends on both t_p and ΔE_p:

$$\Delta i_p = \pm nFc\sqrt{\frac{D}{\pi t_p}} \cdot \frac{1 - \exp(\frac{nF\Delta E_p}{2RT})}{1 + \exp(\frac{nF\Delta E_p}{2RT})} \tag{5.51}$$

A lower t_p allows for higher sensitivity. The influence of ΔE_p is ambivalent. With too low values, the peak disappears, while too high values make the compensation of the capacitive currents impossible and disturb the diffusion profile.

Staircase voltammetry

Also, staircase voltammetry, in this book, introduced first as an inferior technique in digital signal generation and data acquisition (Section 3.2.2), is a pulse voltammetry method with the feature of low contribution from capacitive effects. Current sampling is at the end of each step (Figure 5.46).

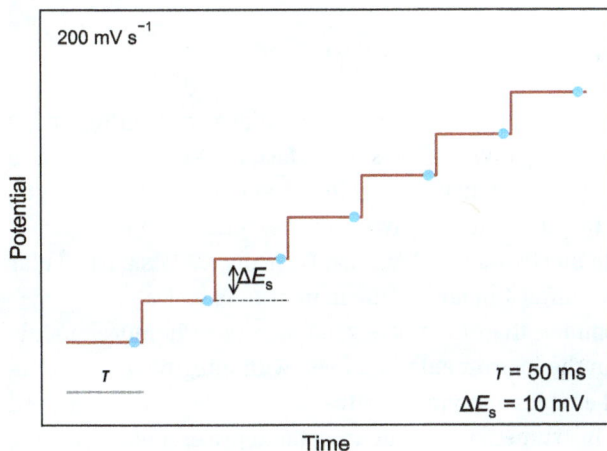

Figure 5.46: Staircase voltammetry: applied waveform and some typical values [53]. The blue dots indicate the current acquisition.

Data analysis

Figure 5.47 compares the resulting voltammograms schematically from the above-discussed methods. For NPV and staircase voltammetry, the output signal is current,

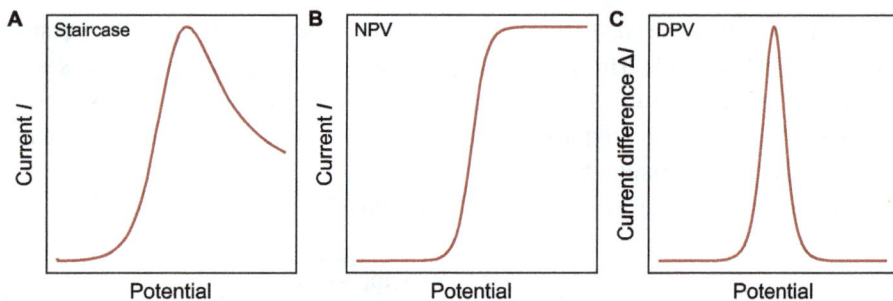

Figure 5.47: Voltammograms for the various pulse voltammetry methods: in staircase voltammetry (A) and normal pulse voltammetry (B), the y-axis is current; for differential pulse voltammetry (C), it is the current difference resembling the first derivative.

while for DPV, the current difference, which is comparable to the first derivative, is on the y-axis.

Comparing the sensitivity of the different methods, a clear ranking is possible:

$$NPV > DPV > staircase$$

Considering other features like the detection limits, the situation strongly depends on the electrochemical system.

5.3.7 Square-wave voltammetry

While the previously discussed pulse methods result in potential sweeps in the range of $10\,\text{mV}\,\text{s}^{-1}$, *square-wave voltammetry* (SWV) allows much faster measurements up to several $V\,\text{s}^{-1}$. Here, a square-wave signal superimposes the potential ramp (Figure 5.48). The higher sweep rate comes with quick analysis, lower analyte consumption, and fewer chances for reaction products to inactivate the electrode. Formally, SWV is a particular case of DPV, but the two methods differ fundamentally in their time scales.

The SWV became more common than the pulse voltammetry techniques, mainly because of the improvements in the instrumentation. Even with integrated circuit potentiostats, SWV is possible and enables applying an advanced electrochemical method in a sensor system with small instrumentation and acceptable price. A good starting point to learn more about the possibilities of this method is Osteryoung and Osteryoung [55].

In SWV, the difference signal consisting of the forward current (I_1) and backward current (I_2) is the primary output quantity. Further analysis may consider the two components, too. Figure 5.49 shows the square-wave voltammogram of a reversible system. While the individual components I_1 and I_2 depend on the scan direction, the difference is symmetrical concerning both the x-axis and the formal potential.

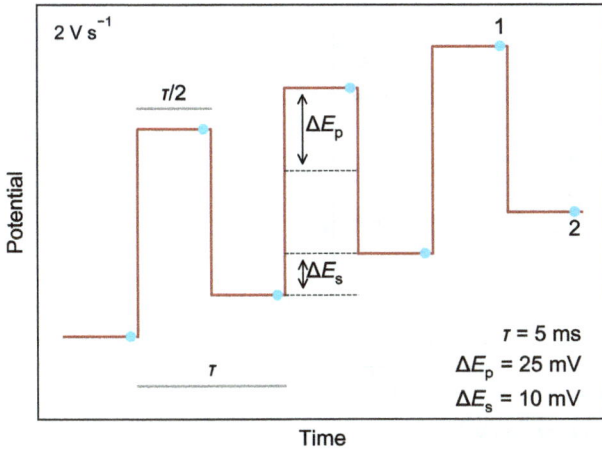

Figure 5.48: Square-wave voltammetry (SWV): applied waveform and some typical values [53].

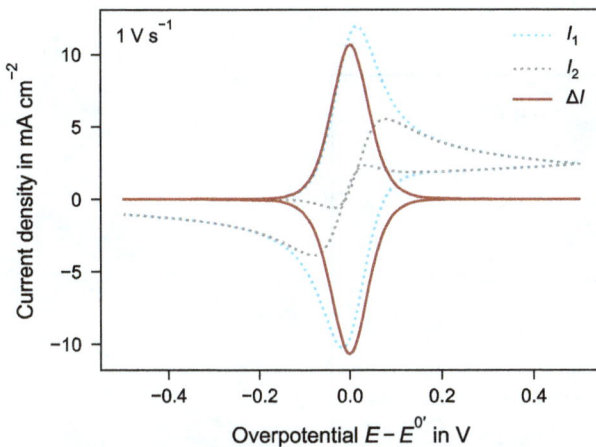

Figure 5.49: Square-wave voltammogram of a reversible system: the output signal (ΔI) is symmetrical. Parameters: $n = 1$, $k_0 = 1\,\mathrm{cm\,s^{-1}}$, $D = 1 \times 10^{-5}\,\mathrm{cm^2\,s^{-1}}$, $[R] = 10\,\mathrm{mM}$, $\Delta E_p = 25\,\mathrm{mV}$, $\Delta E_s = 5\,\mathrm{mV}$, and $\tau = 5\,\mathrm{ms}$.

In contrast, a quasi-reversible system deviates from the symmetry, and the difference signal is lower in comparison (Figure 5.50).

The power of SWV becomes visible with more complex systems, e. g., for multistep reactions or with multicomponent systems. The method's capability for a more precise peak separation compared to linear scan voltammetry makes it a valuable tool in stripping voltammetry (Section 5.3.9).

Figure 5.51 shows the square-wave voltammogram of the multicomponent system analyzed by CV in Figure 5.37. Here, as the system is reversible, the easily distinguishable

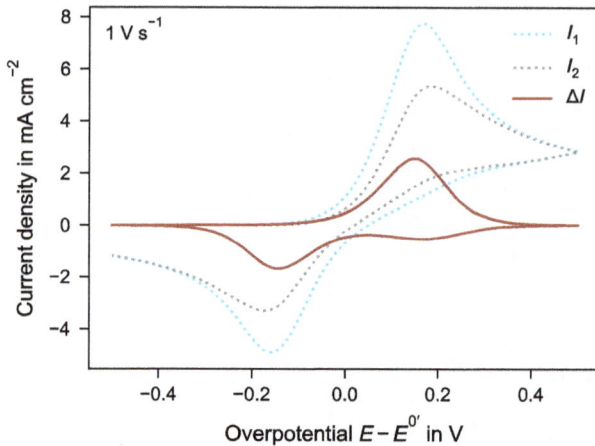

Figure 5.50: Square-wave voltammogram of a quasi-reversible system: in contrast to a reversible system, the output signal (ΔI) is lower and without symmetry. Parameters: $n = 1$, $k_0 = 1 \times 10^{-3}$ cm s^{-1}, $D = 1 \times 10^{-5}$ cm^2 s^{-1}, $[R] = 10$ mM, $\Delta E_p = 25$ mV, $\Delta E_s = 5$ mV, and $\tau = 5$ ms.

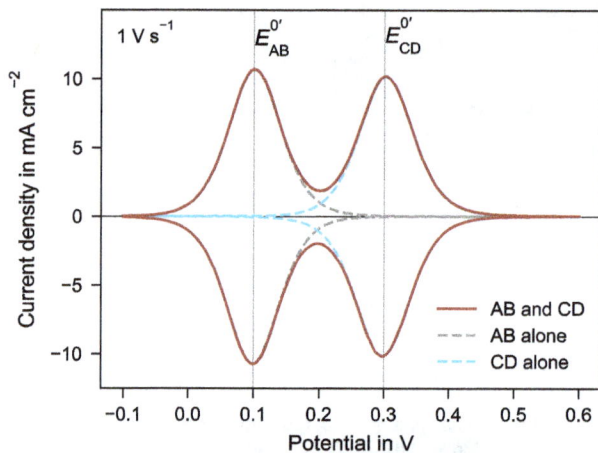

Figure 5.51: Square-wave voltammogram of the multicomponent system analyzed by CV in Figure 5.37. Parameters: $E_{AB}^{0\prime} = 0.1$ V, $E_{CD}^{0\prime} = 0.3$ V, $k_{AB}^0 = 1$ cm s^{-1}, $k_{CD}^0 = 0.1$ cm s^{-1}, $D = 1 \times 10^{-5}$ cm^2 s^{-1}, $[A] = [C] = 10$ mM, $\Delta E_p = 25$ mV, $\Delta E_s = 5$ mV, and $\tau = 5$ ms.

peaks are at the respective formal potentials. Apart from the clearer separation of the peaks, the results are obtained much faster with SWV than with CV.

Instead of the square-wave superimposed onto a potential sweep, also a sinusoidal waveform is possible. From the pattern, one might think of just the next step. However, the resulting method, called AC voltammetry, differs a lot in its principle and outcomes.

5.3.8 AC voltammetry

AC voltammetry is the superposition of a sinusoidal signal E_{AC} onto a DC voltage E_{DC}, typically swept linearly or cyclic:

$$E_{AC} = \hat{E}_{AC} \cdot \sin \omega t \quad \text{with } \omega = 2\pi f \tag{5.52}$$

The frequency of the AC signal is generally between f = 10 Hz and 100 kHz, with 100 Hz as typical value. Figure 5.52 shows the applied waveform schematically. The AC component I_{AC} of the current response has a sinusoidal waveform too, but possibly with a phase shift of ϕ:

$$I_{AC} = \hat{I}_{AC} \cdot \sin(\omega t + \phi) \tag{5.53}$$

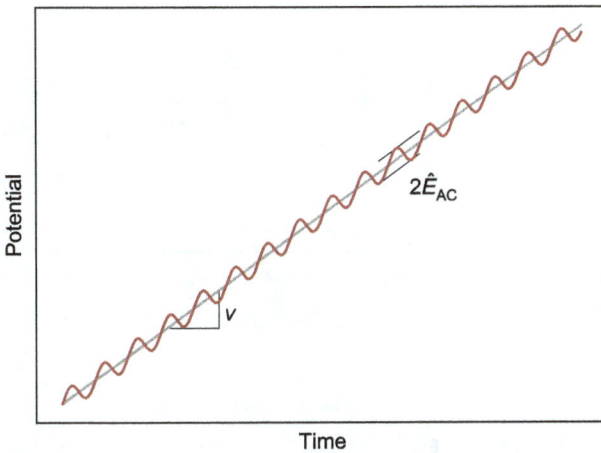

Figure 5.52: AC voltammetry: the applied waveform is the superposition of a DC sweep (scanrate v) with a small, sinusoidal perturbation (E_{AC}).

One important criterion for the parameter selection is the separation between the time domains of the AC signal and the DC sweep. Therefore, the scanrate should be much lower than the perturbation from the AC signal:

$$v \ll \hat{E}_{AC} \cdot \omega \tag{5.54}$$

This condition allows treating the diffusion situation of the DC and AC contributions independently. Depending on the amplitude of the AC voltage $\hat{E}_{AC}$, the system is linear or shows higher harmonics.

For a reversible system, the peak current density of the AC signal depends on the frequency and the applied amplitude:

$$i_p = \frac{n^2 F^2 c \hat{E}_{AC} \sqrt{\omega D}}{4RT} \tag{5.55}$$

With larger AC amplitude (e. g., 50 mV), higher harmonics become visible. Using a lock-in amplifier or digital signal processing with Fourier transformation allows separating the different harmonics. The fundamental (ω) contains both faradaic and capacitive contributions; higher harmonic ($2\omega, 3\omega, \ldots$) depend on the faradaic process only. Figure 5.53 compares AC voltammograms of a reversible and a quasi-reversible system. A reversible system shows a sharp peak at the formal potential in the fundamental and

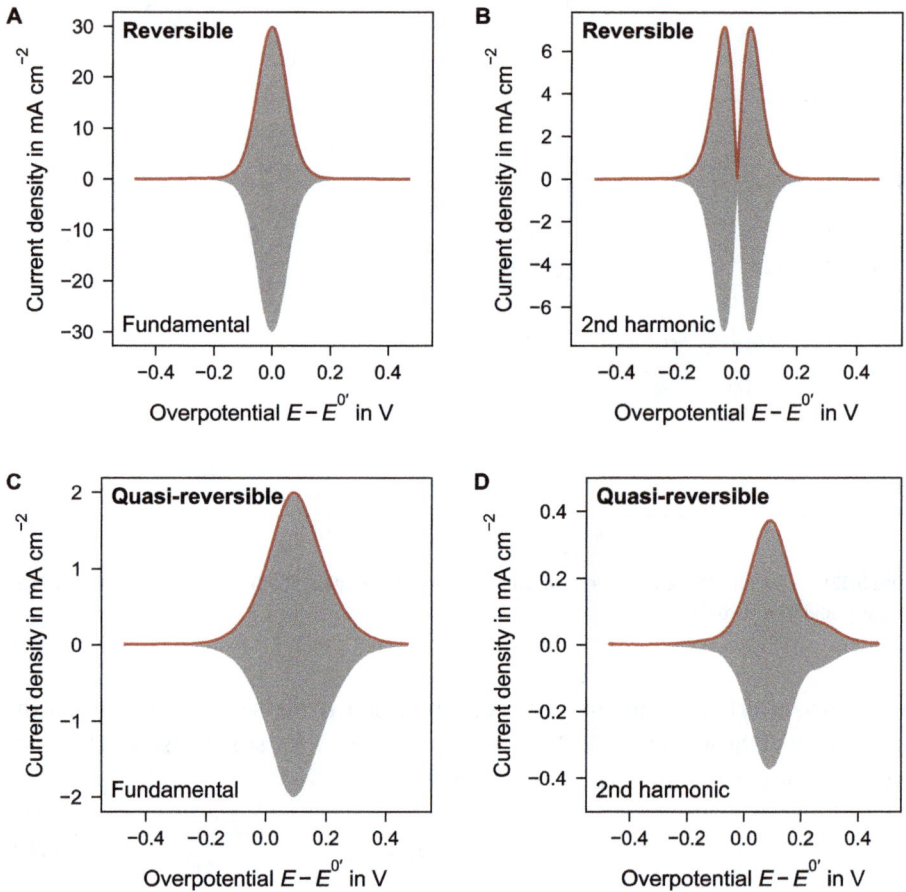

Figure 5.53: AC voltammetry: fundamental (A, C) and 2nd harmonic response (B, D) of a reversible (A, B) and a quasi-reversible system (C, D). Parameters: $k^0 = 1\,\mathrm{cm\,s^{-1}}$ versus $1 \times 10^{-3}\,\mathrm{cm\,s^{-1}}$, $\alpha = 0.5$, $n = 1$, $D = 1 \times 10^{-5}\,\mathrm{cm^2\,s^{-1}}$, $[R] = 10\,\mathrm{mM}$, $\hat{E}_{AC} = 50\,\mathrm{mV}$, $f = 100\,\mathrm{Hz}$, and $v = 0.1\,\mathrm{V\,s^{-1}}$.

a distinct double peak in the 2nd harmonic. A quasi-reversible system shows a broader and shifted peak in the fundamental with less characteristic behavior in the 2nd harmonic. The exact treatment of the higher harmonics is complex, and starting from the 3rd harmonic, the signals are often too weak for reliable analysis.

One convenient analysis is the deduction of the transfer coefficient α in quasi-reversible systems from a complete AC voltammetry cycle. The quasi-reversible system shows two separate peaks, so there is a crossing between the forward and backward scan curves (Figure 5.54). The potential at the intersection is the crossover potential E_{co}, which provides a simple method to determine the charge transfer coefficient α:

$$E_{co} = E_{1/2} + \frac{RT}{F} \ln \frac{\alpha}{1 - \alpha} \tag{5.56}$$

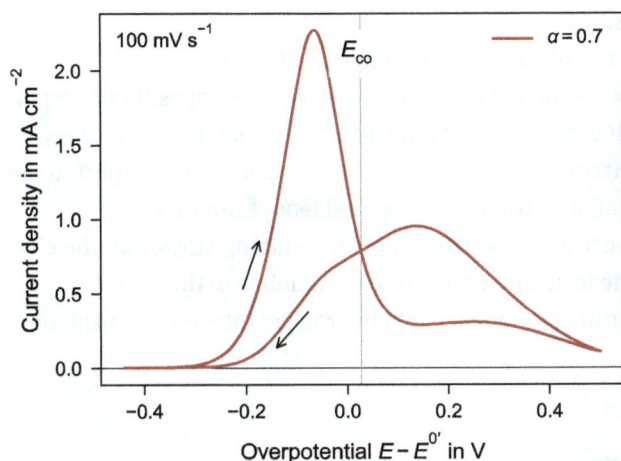

Figure 5.54: AC voltammetry of a quasi-reversible system: a complete cycle of the fundamental's envelope. The crossover potential allows the determination of the charge transfer coefficient α. Parameters: $\alpha = 0.7$, $n = 1$, $k^0 = 1 \times 10^{-3}$ cm s^{-1}, $D = 1 \times 10^{-5}$ cm^2 s^{-1}, $[R] = 10$ mM, $\hat{E}_{AC} = 50$ mV, $f = 100$ Hz, and $v = 0.1$ V s^{-1}.

Further kinetic studies are possible by analyzing the phase shift or higher harmonics of the current response. However, the detailed treatment is far beyond the scope of this book.

5.3.9 Stripping voltammetry

Stripping voltammetry combines voltammetric detection with a preconcentration phase. At the beginning of the measurement, the analyte electrodeposits onto the electrode. Afterward, the just-deposited material redissolves (gets "stripped") during a linear sweep, and the measured current allows quantification of the analyte. The duration of the electrodeposition defines the method's sensitivity; the potential at which the stripping occurs provides selectivity.

At the mercury electrode, the analyte is not on but dissolved into the electrode, why mercury film electrodes (MFE) are beneficial. While the potential window of the mercury electrode allows for attractively low deposition potential, solid electrodes, e. g., carbon and noble metals, became more common today.

Anodic stripping voltammetry

Anodic stripping voltammetry (ASV) is the most common variant. Here, a low potential drives the cathodic electrodeposition; the stripping occurs during an upward sweep by oxidation of the deposited material. ASV is most common for the detection of metal ions (Me^{n+}) in aqueous electrolytes.

A typical experimental protocol (Figure 5.55) comprises three steps:

1. Electrodeposition at a low potential: $Me^{n+} + ne^- \longrightarrow Me$. The deposition potential should be less than a few 100 mV below the lowest half-wave potential of the desired analytes. In this phase, stirring the solution enforces the material transport to the electrode. The typical duration is from one to several tens of minutes.
2. Rest period ("equilibration"): after the deposition, the stirring stops, and the electrolyte reaches steady state to ensure undisturbed conditions in the next step.
3. Anodic stripping: linear scan voltammetry into the range of the metal's oxidation: $Me \longrightarrow Me^{n+} + ne^-$.

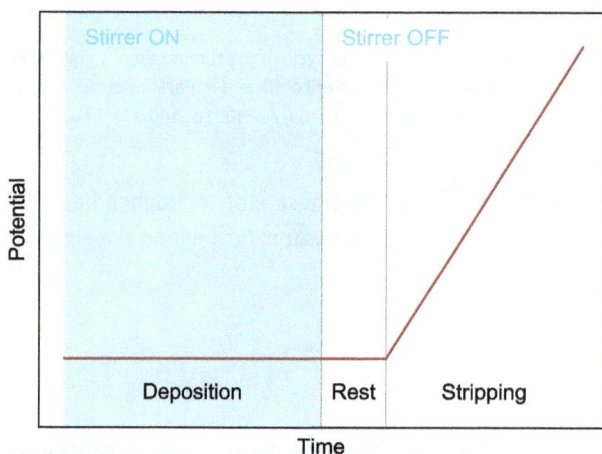

Figure 5.55: Anodic stripping voltammetry (ASV): the analytes electrodeposit onto the electrode at a low potential, the measurement ("stripping") is anodic.

Figure 5.56 shows an ideal stripping voltammogram with two different metals. The resulting curve provides qualitative information about the presence of different metals and quantification if peaks are separated clear enough to analyze or deconvolute.

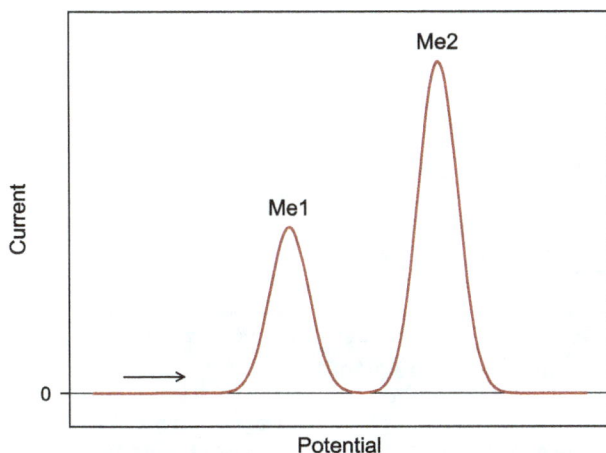

Figure 5.56: Ideal anodic stripping voltammogram: during the linear sweep, the deposited metals get oxidized. Metal Me1 has a lower half-wave potential than Me2. Assuming the same number of electrons in the stripping reactions, more Me2 was in the analyte.

Depending on the system and required detection limit, subtracting the blank signal from a run without the analyte is necessary. Other methods for stripping, e. g., square-wave voltammetry (Section 5.3.7), helps to lower the detection limit and increase the selectivity. With optimized parameters detection limits down to 1×10^{-10} M, for some analyte/electrode combinations even 1×10^{-12} M are possible.

Figure 5.57 shows a measured stripping voltammogram of four different metals with a HMDE. SWV and DPV did the stripping, the background was corrected by blanks in the same condition [56].

Cathodic stripping voltammetry
Similar to the ASV, the inverse method is possible either. In cathodic stripping voltammetry (CSV), the analyte gets deposited at a high potential by oxidation and stripped cathodically in a downward scan. Typical applications address the detection of halides anions or organic compounds.

Adsorptive stripping voltammetry
A related method is the adsorptive stripping voltammetry (AdSV). Instead of electrodeposition, adsorption of the species does the preconcentration. Some analytes require the addition of complexing agents to promote the adsorption.

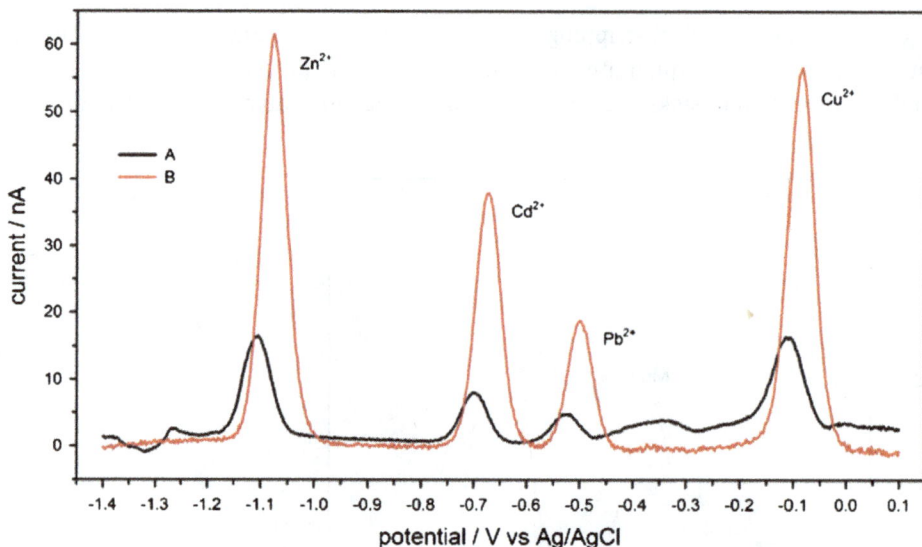

Figure 5.57: Anodic stripping voltammetry using DPV (A) and SWV (B). The analyte contained 2 µg/l of Zn(II), Cd(II), Pb(II), and Cu(II), which corresponds approximately to a concentration of 30 nM, 20 nM, 10 nM, and 30 nM, respectively. [56] Reprinted from Analytica Chimica Acta, 701, Rodrigues et al., Increased sensitivity of anodic stripping voltammetry at the hanging mercury drop electrode by ultracathodic deposition, 152–156, Copyright 2011, with permission from Elsevier.

5.3.10 Fast-scan cyclic voltammetry

Fast-scan cyclic voltammetry (FSCV) referred to cyclic voltammetry with scan rates of $100\,V\,s^{-1}$ or faster. Today FSCV nearly exclusively describes measurements with carbon-fiber microelectrodes for electroanalysis in the brain. The main target molecules are neurotransmitters, most prominently dopamine but also other catecholamines. In contrast to the waveform in the regular CV, the typical waveform in FSCV comprises a longer phase at a low potential, allowing the analyte to adsorb onto the electrode. During the potential sweep, the analyte gets oxidized back and becomes soluble again. In principle, FSCV is a cyclic form of stripping voltammetry with extremely fast stripping. Figure 5.58 shows the typical waveform commonly used for dopamine detection.

The obtained voltammogram from the marked region in Figure 5.58 has a strong background, e. g., because of capacitive effects. Therefore, FSCV measurement requires background subtraction in the voltammogram. Several factors lead to selective detection of different analytes. The long adsorption time with quick stripping favor for adsorbed species. Additionally, measurements with modified waveforms distinguish the various neurotransmitters. Section 10.3.4 deepens the discussion on neurotransmitter monitoring with FSCV.

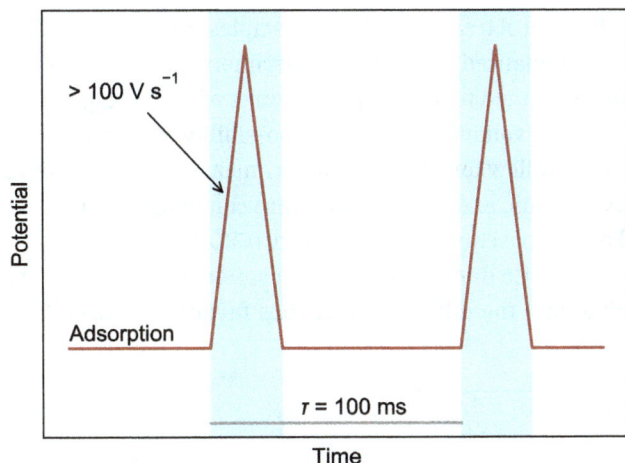

Figure 5.58: FSCV waveform for neurotransmitter detection, the marked regions generate the data for the voltammograms. Please note that the durations are not to scale.

5.4 Current-controlled techniques

In contrast to potential-controlled methods such as amperometry or voltammetry, galvanostatic control is less common. In electroanalysis, chronopotentiometry plays a role, while current ramps are used to characterize electrochemical energy conversion or storage systems.

5.4.1 Chronopotentiometry

Chronopotentiometry, directly opposing chronoamperometry (Section 5.2.3), is a method that uses a sequence of applied constant current phases to read the electrode potential. This method should be distinct from potentiometry (Section 5.1), the current-less potential measurement. Depending on the applied current, the required charge cannot sufficiently be provided by a specific process, and consequently, the electrode potential rises (positive current, oxidation) or falls (negative current, reduction). Depending on the electrochemical system, the potential may even leave the potential window of the electrolyte, causing its decomposition (electrolysis).

5.4.1.1 Reciprocal derivative chronopotentiometry

Reciprocal derivative chronopotentiometry (RDC) is a particular method applying a constant current (often a biphasic pulse with a negative and positive current phase). The resulting data is presented as dt/dE over E, leading to a curve shape resembling a cyclic voltammogram. The idea of RDC dates back to the 1950s, with the term in use since the late 1980s [57]. Later, theoretical treatment of the principles, e. g., for reversible [58] or irreversible processes [59], enhanced the original experimental approach of the method. However, it is still a niche method today. Distinct features of RDC, compared to voltammetric methods, such as cyclic voltammetry, are the possibility of measuring in low-conducting electrolytes [60] and allow for generally shorter measurement duration. When two or more subsequent cathodic and anodic phases are combined, the method is also called cyclic reciprocal derivative chronopotentiometry (CRDC) [61].

Depending on the application, pulse durations vary from sub-millisecond to a few ten-second range. Understanding that the different timescales favor processes differ-

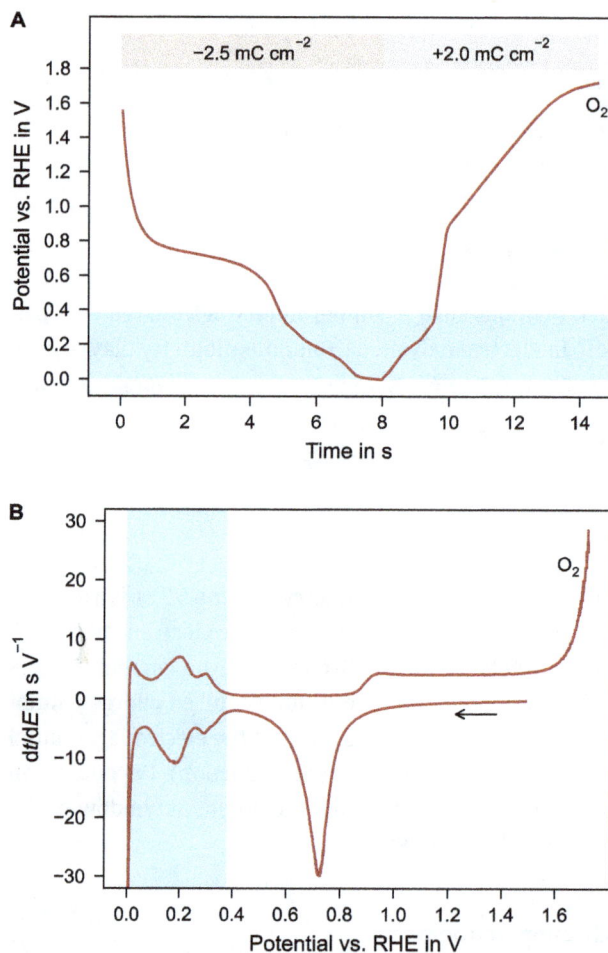

Figure 5.59: Reciprocal derivative chronopotentiometry of a platinum electrode in air-saturated phosphate-buffered saline (PBS) pH 7.4. A cyclic biphasic current pulse (cathodic first) results in a potential trace over time (A), which allows calculating the RDC curve (B) using the inverse of the potential slope. For illustration, the blue background indicates the hydrogen area and "O_2" the oxygen evolution reaction.

ently depending on kinetics or mass transport limitations is essential. While the charge in the cathodic and anodic phases of the current pulses needs to be different depending on the electrochemical system, the absolute value of the applied current should be the same in both phases to avoid confusing the scaling of the y-axis.

Most previous works focused on redox active species in the electrolyte. However, the method is also well suited to investigate reactions of the electrode material itself. Musa et al. applied RDC to platinum electrodes to relate the applied charge during stimulation (Section 10.3.3) with the underlying electrochemical reactions, especially the electrolysis of water. [62] Longer pulse durations in the second range help to separate the surface reactions without the dominant impact of the IR drop. Figure 5.59 shows the RDC results of a platinum electrode. The obtained curve resembles closely the typical platinum cyclic voltammogram (compare Figure 5.21A).

6 Combined methods

Combining the classical electrochemical methods with other actuation principles leads to better control of the mass transport (hydrodynamic methods) or provides spatial information (scanning methods). Exploiting other effects can yield more comprehensive analytical aspects, as discussed in the last section of this chapter.

6.1 Hydrodynamic methods

The previous chapter focused on the classical electrochemical methods, assuming that diffusion is the dominant mass transfer mechanism. While diffusion is always involved in the species transport to the electrode surface, forced convection may get the dominant role as exploited in the hydrodynamic methods. The most straightforward approach to forced convection is stirring. Although the flow regime is turbulent, reproducible steady-state conditions are achievable with careful control of the electrode's position in the beaker and the stirring parameters. Such setups can be a helpful tool to, e. g., calibrate amperometric sensors.

As introduced in detail in the next section, the rotating disk electrode combines both flow regimes: close to the electrode surface laminar conditions prevail while the bulk of the electrolyte is well mixed by forced convection driven by the electrode's rotation. Microfluidic flow cells are always laminar; increasing the fluid velocity to trigger turbulencies would cause a too large pressure drop.

Hydrodynamic electrodes are electrodes at which hydrodynamic methods define the mass transport conditions. Often the term means a rotating-disk electrode. Figure 6.1 compares the characteristic voltammograms between a diffusion-controlled regime (as discussed in Section 5.3) and a hydrodynamic electrode. However, high scanrates can lead to a peak formation even at a hydrodynamic electrode when the diffusion close to the electrode becomes dominant.

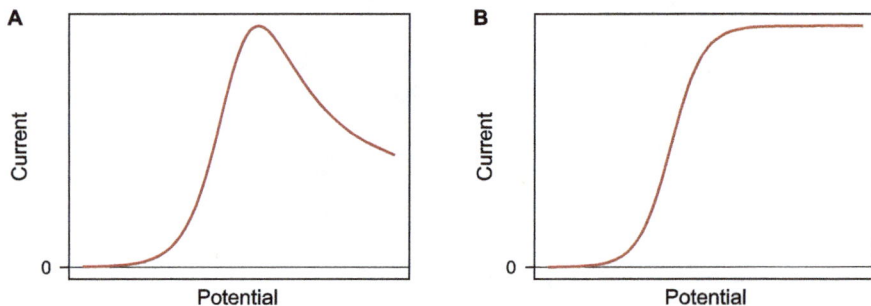

Figure 6.1: Typical voltammogram of an electrode reaction under diffusion control (A) or at a hydrodynamic electrode (B).

https://doi.org/10.1515/9783111488844-006

The zeta potential measurement is a hydrodynamic method, too, though the flow is not to control the mass transport conditions but to exploit electrokinetic phenomena.

The tendency of a flow for being turbulent depends on the relation of inertia to viscous forces. The Reynolds number Re is the ratio of the forces:

$$\text{Re} = \frac{\rho v L}{\eta} = \frac{vL}{v_k} \tag{6.1}$$

v is the flow velocity, η is the *dynamic viscosity* in Pa s (or Poise, $1\,\text{P} = 0.1\,\text{Pa s}$). In the context of hydrodynamic methods, the *kinematic viscosity* v_k in $\text{m}^2\,\text{s}^{-1}$ (or Stokes, $1\,\text{St} = 1\,\text{cm}^2\,\text{s}^{-1}$) is more common, here written with index k to avoid confusions with the scanrate v. The density of the medium relates the two viscosity definitions:

$$\eta = \rho \cdot v_k \tag{6.2}$$

The characteristic length L is, e. g., the diameter in a tube or the smaller distance between the walls in a rectangular channel. The critical Reynolds number is 2300. Lower Re values characterize conditions leading to a laminar flow; values above indicate turbulencies. However, depending on the geometry, formation of vortexes occurs at Re = 100 while smooth surfaces allow laminar conditions up to Re = 1×10^5.

6.1.1 Rotating disk electrode

The rotating disk electrode (RDE) is a powerful approach combining laminar flow at the electrode surface with a well-stirred bulk solution. The setup consists of a rotator turning the working electrode with speeds from below 100 up to several 1000 rpm. Brush or liquid mercury contacts electrically connects the electrode with the stationary setup.

A disk electrode in a cylindrical shaft rotates in the liquid electrolyte so that a radial, laminar flow develops created by the drag from the rotating electrode (Figure 6.2(A)). At the center, fresh electrolyte reaches the electrode vicinity. The *hydrodynamic boundary layer* characterizes the region of the laminar flow regime in front of the electrode:

$$\delta_H = 3.6 \sqrt{\frac{v_k}{\omega}} \tag{6.3}$$

The thickness δ_h depends on the kinematic viscosity v_k of the electrolyte and the rotation rate, here, expressed as ω, the angular frequency of rotation ($\omega = 2\pi \times$ rotation rate). A typical thickness is approximately 350 µm ($v_k = 0.01\,\text{cm}^2\,\text{s}^{-1}$, 1000 rpm).

The mass transport of the electroactive species toward the electrode occurs by diffusion from within the laminar flow region. The diffusion layer thickness

$$\delta = 1.61 \cdot \frac{D^{\frac{1}{3}} v^{\frac{1}{6}}}{\sqrt{\omega}} \tag{6.4}$$

A

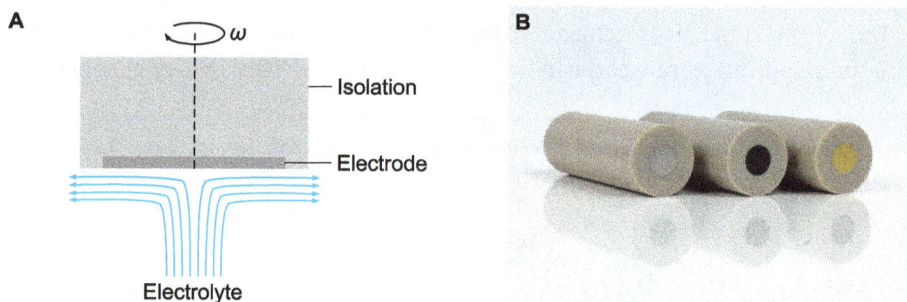

B

Figure 6.2: Rotating disk electrode (RDE): the rotation generates a radial flow in front of the electrode surface (A). Commercial RDE tips (B), reproduced with permission from Pine Research Instrumentation, USA.

is much lower than the hydrodynamic boundary layer δ_h. A typical ratio δ_h/δ is approximately 20 ($D = 1 \times 10^{-5}$ cm^2 s^{-1}, $v_k = 0.01$ cm^2 s^{-1}, 1000 rpm).

! A common source of confusion is the unit of ω. In this book, all equations with ω consider the angular frequency of rotation in rad s^{-1} given by $\omega = 2\pi \times$ rotation rate. The rotation rate in rpm divided by 9.55 converts the numbers to ω in rad s^{-1}. Some other formulations of the RDE equations exist considering ω in rpm and different prefactors accordingly.

Levich study – mass transport limitation

Most RDE measurements consider voltammetry, e. g., a line scan at different rotation rates. The potential range should cover the region from no reaction to the domination by mass transport limitation (Figure 6.3(A)). It is essential to choose a sufficiently low scanrate to ensure mass transport limitation. Then the current becomes constant, which is attributed as the limiting current density i_{lim}. The *Levich equation* relates the limiting current density to the rotation:

$$i_{lim} = \pm 0.620 \cdot nFcD^{\frac{2}{3}} v_k^{\frac{1}{6}} \cdot \sqrt{\omega} \tag{6.5}$$

c is the species' bulk concentration and D its diffusion constant. Depending on oxidation or reduction, i_{lim} is positive or negative.

The values for the limiting current density depend linearly on the square root of the rotation speed. Accordingly, in an i_{lim} versus $\sqrt{\omega}$ plot, the so-called *Levich plot* (Figure 6.3(B)), all data points are on a line with a slope allowing, e. g., determining the species' diffusion constant.

Diffusion constant measurement may suffer from the approximative nature of the Levich equation. The *Riddiford correction* leads to more precise results but also complicates the data evaluation. Instead of the constant factor 0.620 in equation (6.5), a term

A

B

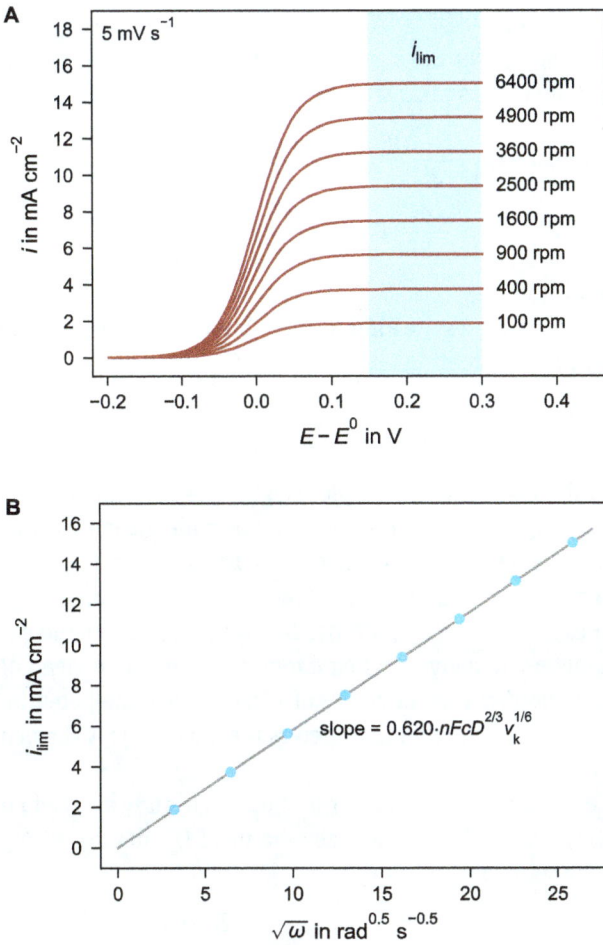

Figure 6.3: Levich study: line scan voltammetry at different rotation rates (A). Evaluation of the limiting current density in the Levich plot (B). Parameters: 5 mm electrode diameter, $n = 1$, $k^0 = 1\,\mathrm{cm\,s^{-1}}$, $D = 1 \times 10^{-5}\,\mathrm{cm^2\,s^{-1}}$, $v_k = 0.01\,\mathrm{cm^2\,s^{-1}}$, $c = 10\,\mathrm{mM}$, and $v = 5\,\mathrm{mV\,s^{-1}}$.

depending on the diffusion constant and kinematic viscosity attributes for fast diffusing species:

$$i_{\lim} = \pm \frac{0.554}{0.893 + 0.316 \cdot (\frac{D}{v_k})^{0.36}} \cdot nFcD^{\frac{2}{3}} v_k^{\frac{1}{6}} \cdot \sqrt{\omega} \tag{6.6}$$

A solution for the diffusion constant from the Levich plot considering the Riddiford correction requires an iterative data evaluation.

Koutecký–Levich analysis

Also, evaluation of the current response in the kinetically limited potential region is possible. The contributions from the electron transfer are

$$i_{k,a} = nFk^0 c \exp\left(\frac{(1-\alpha)F}{RT}(E - E^0)\right) \tag{6.7}$$

$$i_{k,c} = -nFk^0 c \exp\left(-\frac{\alpha F}{RT}(E - E^0)\right) \tag{6.8}$$

for oxidation and reduction, respectively.

The *Koutecký–Levich* equation combines the kinetic contribution with the limiting current density according to the Levich equation (equation (6.6)):

$$\frac{1}{i} = \frac{1}{i_k} + \frac{1}{i_{lim}} = \frac{1}{i_k} + \frac{1}{\pm 0.620 \cdot nFcD^{\frac{2}{3}} v^{\frac{1}{6}} \cdot \sqrt{\omega}} \tag{6.9}$$

Data analysis is possible in the *Koutecký–Levich plot* (Figure 6.4(B)) summarizing the values at different potentials as $1/i$ versus $1/\sqrt{\omega}$. The slopes provide the mass transport properties similar to the Levich study without the need for complete mass transport limitation, which might never be reached, e. g., because of overlapping reactions.

Additionally, extrapolation to $1/\sqrt{\omega} = 0$ (i. e., infinite fast rotation) allows the determination of the kinetic parameters, namely k^0 when n and α are known. Instead of reading $1/i_k$ for a single potential, plotting several values of $-\ln i_k$ against the potential again yields a linear relationship if the setup and parameters for the Koutecký–Levich study are correct.

In general, the Koutecký–Levich analysis is superior to the Levich study even when evaluating in the region of mass transport limitation. A zero value of $1/i_k$ in the limiting region confirms the validity of the setup and assumptions.

Experimental considerations

It is essential to choose a sufficiently low scanrate to ensure mass transport limitation over the whole range of rotation speeds (e. g., 5 mV s^{-1}). However, the total measurement duration may increase tremendously; that is why an automated setup to control the RDE speed from within the potentiostat software is favored.

The flow conditions limit the possible range of rotation rates. The lower limit comes from the requirement that the hydrodynamic boundary layer should not substantially exceed the electrode radius r:

$$\omega > 3.6^2 \cdot \frac{v_k}{r^2} \approx \frac{10 \cdot v_k}{r^2} \tag{6.10}$$

The upper limit is the requirement for a laminar flow regime, estimated by the Reynolds number (Re):

$$Re = \frac{\omega r^2}{v_k} < 2300 \tag{6.11}$$

A

B

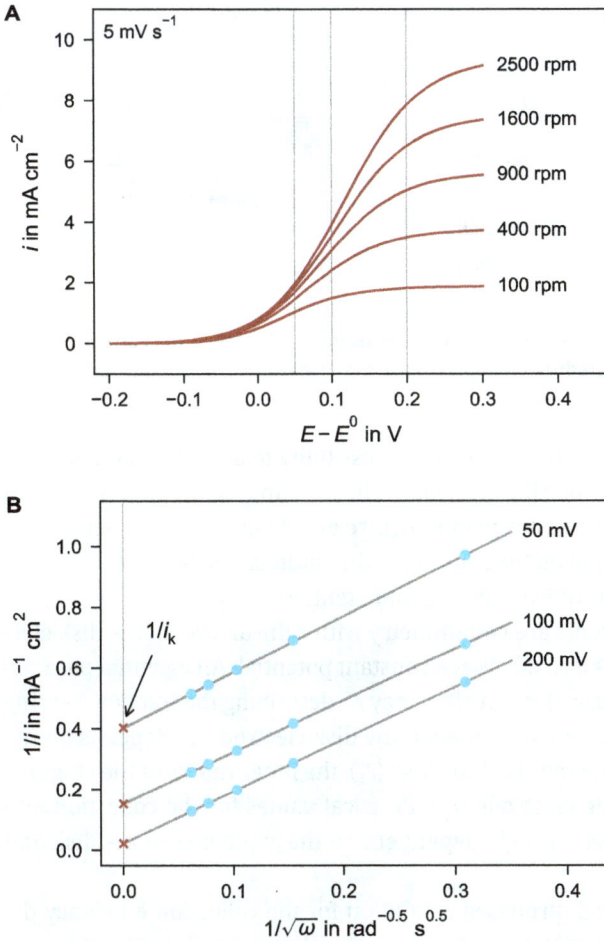

Figure 6.4: Koutecký–Levich study: line scan voltammetry at different rotation rates (A). Evaluation of the limiting current density in the Koutecký–Levich plot (B). Parameters: 5 mm electrode diameter, $n = 1$, $k^0 = 1 \times 10^{-3}$ cm s^{-1}, $D = 1 \times 10^{-5}$ cm^2 s^{-1}, $v_k = 0.01$ cm^2 s^{-1}, $c = 10$ mM, and $v = 5$ mV s^{-1}.

Deviations from ideal behavior in a Levich or Koutecký–Levich plot at extreme values of ω indicate a violation of one of these criteria. Additionally, practical limitations may play a role, e. g., RDE tips allow a certain maximal rotation speed only. Cheaper rotators are inaccurate at slow speeds or do not allow low rotation rates at all.

6.1.2 Rotating ring disk electrode

The rotating ring disk electrode (RRDE) is an enhancement of the RDE. Additionally, a ring-shaped electrode surrounds the disk electrode, separated by an isolating gap (Fig-

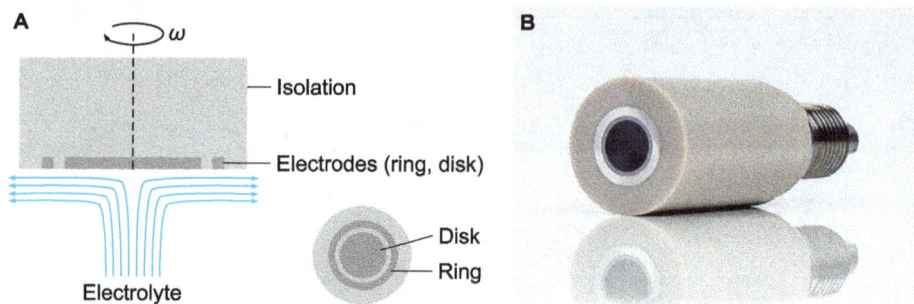

Figure 6.5: Rotating ring disk electrode (RRDE): a ring electrode surrounds the disk in addition to the RDE setup (A). Commercial RRDE tip (B), reproduced with permission from Pine Research Instrumentation, USA.

ure 6.5). The advantage of this arrangement is the possibility to analyze reaction products from the disk's reaction at the ring electrode. The disk and the ring electrode can be of different materials. RRDE measurements require an advanced rotator setup (two electrodes require a connection from the rotating to the stationary side), and the potentiostat needs to operate two working electrodes (bi-potentiostat).

Typically, RRDE measurements are voltammetry with a linear scan of the disk electrode's potential while the ring electrode is at a constant potential. An essential parameter for the data evaluation is the collection efficiency N, describing the fraction the ring can collect from the reaction products formed at the disk electrode. N depends on the electrodes' geometry, namely the radius of the disk (r_1), the inner radius of the ring (r_2), and the outer radius of the ring electrode (r_3). Practical values for the collection efficiency are between 20 and 40 %, strongly dependent on the gap between the disk and the ring electrode.

Albery and Bruckenstein [63] proposed a solution for the collection efficiency depending on the geometrical parameters r_1, r_2, and r_3, which is generally accepted:

$$N = 1 - F\left(\frac{a}{b}\right) + b^{\frac{2}{3}}(1 - F(a)) - (1 + a + b)^{\frac{2}{3}}\left(1 - F\left(\frac{a}{b}(1 + a + b)\right)\right) \tag{6.12}$$

$$a = \left(\frac{r_2}{r_1}\right)^3 - 1 \quad \text{and} \quad b = \left(\frac{r_3}{r_1}\right)^3 - \left(\frac{r_2}{r_1}\right)^3 \tag{6.13}$$

$$F(x) = \frac{\sqrt{3}}{4\pi} \cdot \ln \frac{(1 + x^{\frac{1}{3}})^3}{1 + x} + \frac{3}{2\pi} \cdot \arctan\left(\frac{2x^{\frac{1}{3}} - 1}{\sqrt{3}}\right) + \frac{1}{4} \tag{6.14}$$

---

i ■□□ Task 6.1: Consider an RRDE tip with a 5 mm diameter disk electrode and a ring electrode with 6 mm inner and 8 mm outer diameter. How large is the gap between the disk and the ring electrode? What is the collection efficiency?

Another approach is the experimental determination of the collection efficiency with a stable redox couple. The potential selection of the disk and ring must ensure mass transport limitation for the forward and backward reaction. The ratio of the disk (D) and ring (R) limiting currents lead to the collection efficiency (assuming same number of electrons involved in the two reactions):

$$N = -\frac{I_{\text{lim,R}}}{I_{\text{lim,D}}} \tag{6.15}$$

Measurement of reaction products

Commonly RRDE measurements comprise voltammetry at the disk and constant polarization to detect reaction products at the ring. Figure 6.6 shows a typical RRDE measure-

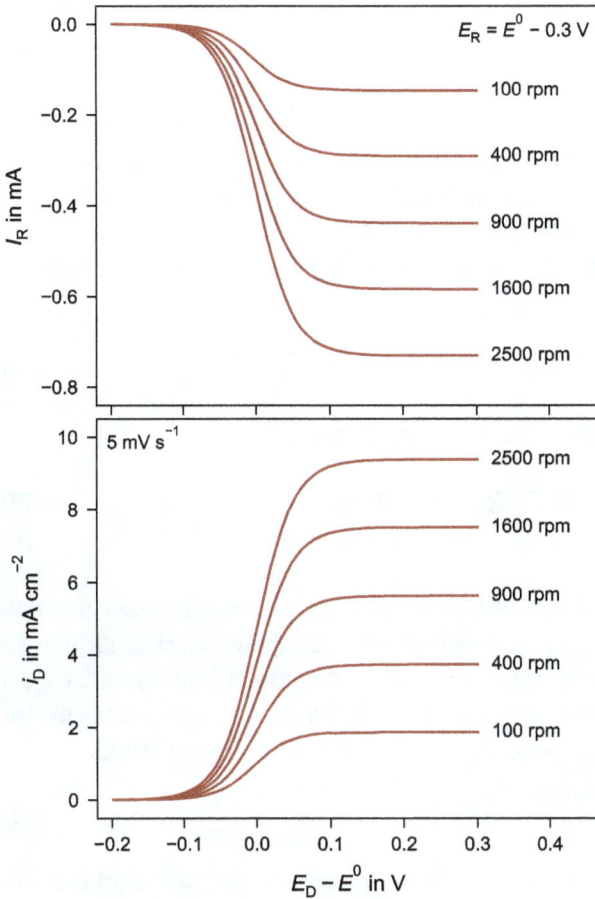

Figure 6.6: Rotating ring disk electrode (RRDE) measurement: The disk potential ramps up, and an oxidation reaction occurs (lower plot). The reaction product gets reduced at the ring with a fixed potential below the reaction's equilibrium (upper plot). Electrode geometries: $r_1 = 2.5\,\text{mm}$, $r_2 = 3\,\text{mm}$, and $r_3 = 4\,\text{mm}$. Parameters: $n = 1$, $k^0 = 1\,\text{cm s}^{-1}$, $D = 1 \times 10^{-5}\,\text{cm}^2\,\text{s}^{-1}$, $v_k = 0.01\,\text{cm}^2\,\text{s}^{-1}$, $c = 10\,\text{mM}$, and $v = 5\,\text{mV s}^{-1}$.

ment result. At the disk, an oxidation reaction occurs:

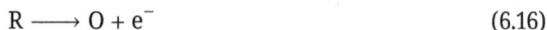

$$R \longrightarrow O + e^-$$ (6.16)

The oxidized form of the species O gets transport to the ring, which is constantly polarized at potential negative to the reaction's equilibrium so that the backward reaction runs:

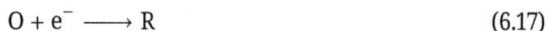

$$O + e^- \longrightarrow R$$ (6.17)

The plot in Figure 6.6 shares a common x-axis with the scanning potential of the disk. While current density is meaningful for the disk, the y-axis for the ring is in current. The exact values depend on the RRDE's geometry and, accordingly, the collection efficiency.

Example: oxygen reduction reaction
The most famous example of an RRDE measurement is investigating the oxygen reduction reaction (ORR). Knowledge of the ORR mechanism is highly relevant for fuel cells, especially during the development of novel platinum-based and other catalysts. RDE and especially RRDE are the standard methods to study the ORR.

In an aqueous electrolyte, the ORR can follow two pathways, the direct reduction in a four-electron process:

$$O_2 + 4H^+ + 4e^- \longrightarrow 2H_2O$$ (6.18)

or the indirect pathway with H_2O_2 as intermediate species:

$$O_2 + 2H^+ + 2e^- \longrightarrow H_2O_2$$ (6.19)
$$H_2O_2 + 2H^+ + 2e^- \longrightarrow 2H_2O$$ (6.20)

When both reactions (equation (6.19) and equation (6.20)) are completed, the indirect pathway is indistinguishable from the direct reduction. However, at an RRDE with the disk electrode polarized for the oxygen reduction, the intermediate species (H_2O_2) formed in equation (6.19) gets transported to the ring and shows an oxidation current assuming appropriate polarization (e. g., $E_R = 1.2\,V$ vs. RHE at a platinum ring):

$$H_2O_2 \longrightarrow O_2 + 2H^+ + 2e^-$$ (6.21)

A typical representative for the direct oxygen reduction is a platinum working electrode, while at a gold electrode, the indirect pathway occurs.

Under some circumstances, the indirect pathway can occur at platinum too. Figure 6.7 shows the results of a study on the influence of Cl^- ions on the ORR at a fuel cell catalyst in an acidic electrolyte [64]. In the presence of chloride in the electrolyte,

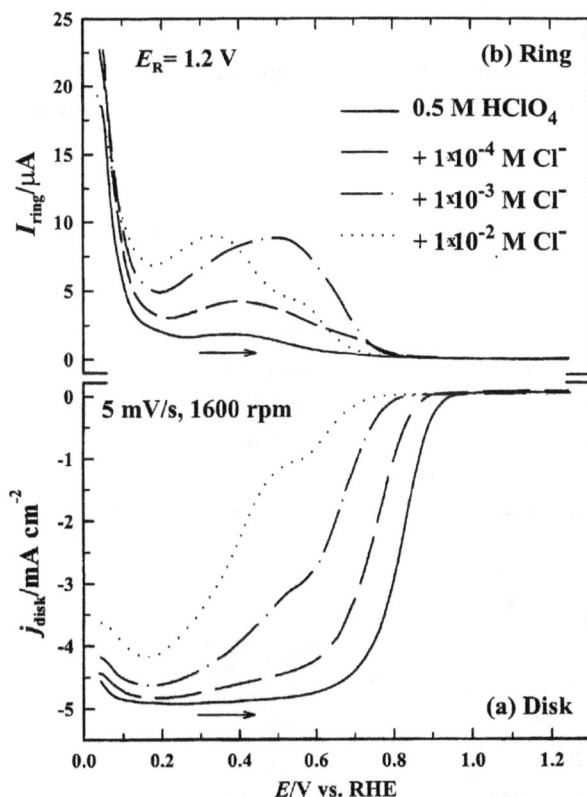

Figure 6.7: RRDE measurement of the ORR at a platinum-based catalyst in acidic electrolyte [64]. Depending on chloride ions, H_2O_2 peroxide was detected at the ring, suggesting the indirect ORR pathway. Reprinted from the Journal of Electroanalytical Chemistry, 508, Schmidt et al. The oxygen reduction reaction on a Pt/carbon fuel cell catalyst in the presence of chloride anions, 41–47, Copyright 2001, with permission from Elsevier.

a signal from the H_2O_2 oxidation becomes visible at disk potential around 0.4 V versus RHE, indicating that the ORR partially follows the indirect pathway.

Measurement of reaction products' stability

As introduced above, the collection efficiency depends on the RRDE geometries only. However, when the product of the disk reaction is unstable (e. g., a radical), the species may already degrade in the short passage time from the disk to the ring. In this case, the apparent collection efficiency depends on the rotation speed of the RRDE.

The collection efficiency's dependency on the rotation rate allows evaluating the stability of intermediate species by comparing the obtained values to the ideal collection efficiency based on geometry. Here, a larger gap between disk and ring can be beneficial depending on the species' lifetime.

An example for such a stability measurement is the study of Herranz et al. [65]. The authors investigate the stability of the superoxide radical in aprotic lithium-air battery electrolytes. At the disk, the ORR occurs in such electrolytes as a one-electron process producing superoxide:

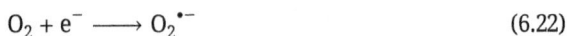

$$O_2 + e^- \longrightarrow O_2^{\bullet -} \tag{6.22}$$

At the ring, the superoxide gets oxidized back.

Figure 6.8 shows the collection efficiency from the RRDE measurements at different rotation rates. For the ionic liquid electrolyte (Pyr14TFSI), the observed collection efficiency resembles the value expected based on the geometry indicating superoxide stability. In contrast, in propylene carbonate (PC), the measured collection efficiency is lower and depends on the rotation rate, showing instability of the superoxide.

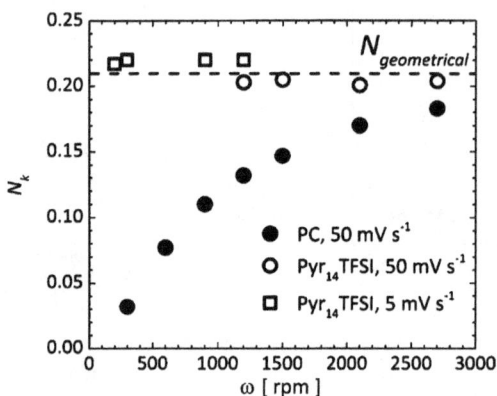

Figure 6.8: RRDE measurement of the ORR in aprotic electrolyte [65]: the observed collection efficiency indicates the stability of the reaction product formed at the disk (superoxide). Reprinted with permission from Herranz et al., Using Rotating Ring Disc Electrode Voltammetry to Quantify the Superoxide Radical Stability of Aprotic Li–Air Battery Electrolytes, J Phys Chem C. 116 (2012) 19084–19094. Copyright 2012 American Chemical Society.

6.1.3 Flow cells

Another approach to ensure reproducible mass transport is using flow cells with a microfluidic, ensuring laminar flow across the electrode surface. Figure 6.9(A) shows a typical microfluidic flow cell comprising several electrodes, including counter and reference. Practically the latter two are downstream to ensure that material from the reference electrode or reaction products formed at the counter electrode does not influence the working electrode reaction. Such flow cells play an essential role in microsystems for electroanalysis of products from a separation column or an enzyme reaction (Section 10.1.6).

Additionally, flow cells allow measurement comparable to the RRDE by using two consequent working electrodes (Figure 6.9(B)). WE1 would act as the disk, with WE2 as the ring electrode. In contrast to the RRDE, there is no comprehensive theoretical treatment, but estimates from experience or flow simulations considering the actual geometry.

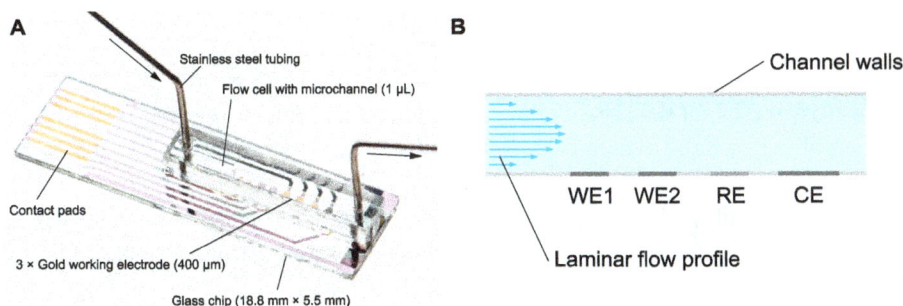

Figure 6.9: Flow cells: microfluidic flow cell mounted a glass-chip with thin-film electrodes (A). Illustration of a flow cell with electrodes at the bottom and a laminar flow profile (B). A: reprinted with permission from Uka et al., Electrochemical Microsensor for Microfluidic Glyphosate Monitoring in Water Using MIP-Based Concentrators. ACS Sensors 2021, 6 (7), 2738–2746. Copyright 2021 American Chemical Society.

6.1.4 Zeta potential measurement

Zeta potential measurement is a hydrodynamic method to characterize solid surfaces. The measured quantity is the streaming potential caused by a pressure-driven flow in a capillary or slit. Figure 6.10(A) illustrates a typical setup in which the samples form a slit with a height in the range of 100 µm. A practical realization of such a setup capable of also measuring porous films was described by Yaroshchuk and Luxbacher [66]. Within the slit, the pressure-driven flow causes charge separation along the shear plane (compare Figure 6.10(B) and Figure 2.24).

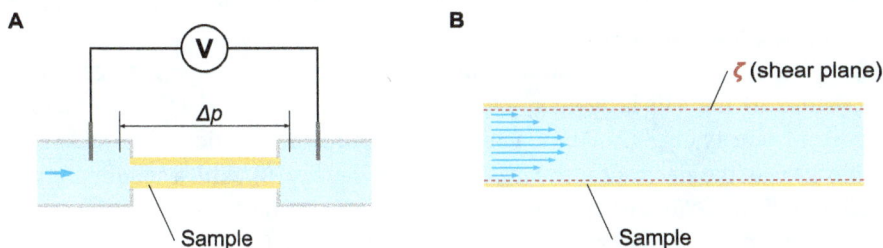

Figure 6.10: Zeta potential measurement: the pressure-driven flow in a small slit between the sample surface causes a streaming potential.

Measurement of the pressure drop Δp and streaming potential E_{str} along the slit provides the zeta potential ζ. Alternatively, the streaming current I_{str} caused by the flow can be measured. The *Helmholtz–Smoluchowski equations* relate the measured quantities to the zeta potential:

$$\zeta = \frac{\partial E_{str}}{\partial \Delta p} \cdot \frac{\eta}{\epsilon_r \epsilon_0} \cdot \frac{L}{A} \cdot \frac{1}{R} \tag{6.23}$$

$$\zeta = \frac{\partial I_{str}}{\partial \Delta p} \cdot \frac{\eta}{\epsilon_r \epsilon_0} \cdot \frac{L}{A} \tag{6.24}$$

A is the cross-section of the rectangular slit with length L. R is the electrical resistance of the sample slit. η is the dynamic viscosity, ϵ_0 the permittivity of free space, and ϵ_r the relative permittivity.

Other setups allow the measurement of zeta potential in particle suspensions or emulsions. Here, the zeta potential is a measure for the stability of the dispersion.

6.2 Scanning methods

Scanning methods are standard in surface analysis, such as mechanical or laser-scanning profilometry, atomic force microscopy (AFM), or the scanning tunneling microscope (STM). The electrochemical method equivalent to profilometry is scanning electrochemical microscopy (SECM) using redox reactions to probe the surface.

The electrochemical atomic force microscopy (EC-AFM) and the electrochemical scanning tunneling microscope (ESTM) are AFM and the STM applied to study electrochemical reactions in situ.

6.2.1 Scanning electrochemical microscopy

In scanning electrochemical microscopy (SECM), an ultramicroelectrode tip (Section 4.3.6) approaches the substrate with a piezo actuator at different positions. Usually, a bipotentiostat controls the potential of the electrode tip and the substrate. The electrolyte contains a redox couple with the tip electrode polarized for oxidation or reduction and the substrate to a potential driving the backward reaction.

Depending on the distance from the substrate and its conductivity, different cases occur (Figure 6.11): If the tip is far away from the substrate, hemispherical diffusion limits the current density to i_∞ (A). Approaching a nonconducting fraction of the substrate causes blockage of the electrode lowering the current density (B). With a conducting position on the substrate (and its appropriate polarization), the backward reaction at the substrate causes positive feedback with the reaction at the tip (C).

Figure 6.12 shows typical SECM approach curves. When reaching the close vicinity of the substrate, the tip current density deviates from the case of undisturbed hemi-

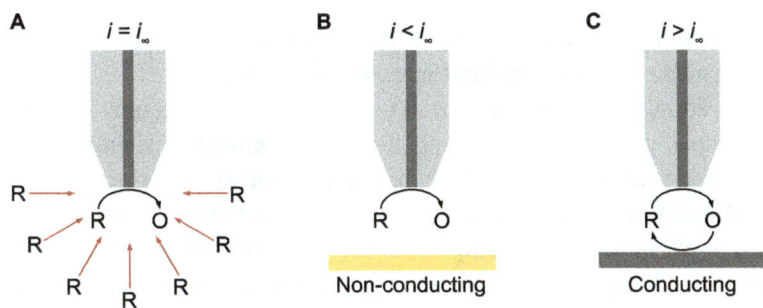

Figure 6.11: Scanning electrochemical microscopy (SECM): different cases occur depending on the conductivity of the substrate. Far away from the substrate, hemispherical diffusion dominates (A). Depending on the conductivity, the tip current drops (B) or rises by positive feedback (C).

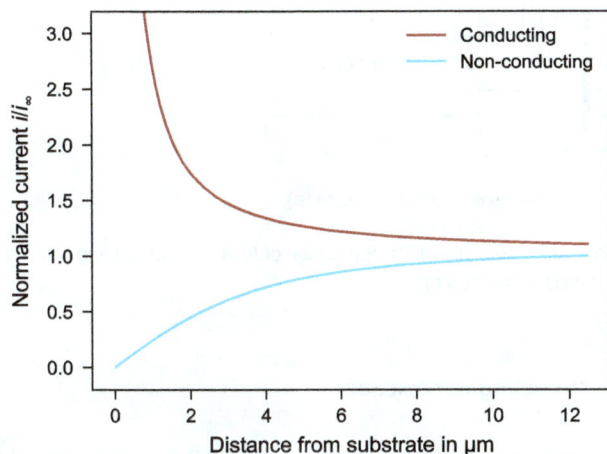

Figure 6.12: SECM approach curves: characteristic tip currents when approaching the substrate surface. The values for the distance refer to an electrode with 5 µm diameter.

spherical diffusion (i_∞). For a nonconducting place, the current density drops, while at a conducting part, positive feedback occurs. Further discussion on the different operation modes is, e. g., provided by Bard et al. [67].

Approaching many positions on the substrate in a scanning manner allows mapping the substrate's conductivity (redox activity). Apart from applications for different micro and nanostructured surfaces, SECM measurements are possible even with biophysical systems [68], including cells [69–71].

6.2.2 Electrochemical atomic force microscopy

In atomic force microscopy (AFM), a cantilever with a tip scans over the surface to investigate. A laser points to the cantilever so that the deflection of the cantilever is detected.

AFM allows measurement of surface topology in contact mode or with an oscillating cantilever in tapping mode. Additionally, force measurement is possible.

Electrochemical atomic force microscopy (EC-AFM) is the application of AFM on a conducting sample, which is the working electrode of an electrochemical cell (Figure 6.13). As the AFM principle does not include electrical current, the electrochemical actuation and the AFM measurement are independent (apart from disturbance of the mass transport situation by the AFM tip). This combination allows investigation of the electrode surface during the electrochemical reaction, e. g., electrodeposition or corrosion phenomena.

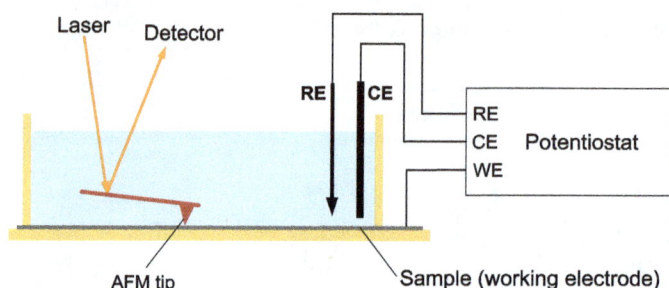

Figure 6.13: Electrochemical atomic force microscopy (EC-AFM): application of AFM on a conducting sample, which is the working electrode of an electrochemical cell.

6.2.3 Electrochemical scanning tunneling microscope

Similarly, the electrochemical scanning tunneling microscope (ESTM) was introduced [72]. As the scanning tunneling microscope (STM) principle depends on electrical current (the tunneling current), the instrumentation needs to consider both the electrochemical reaction and the tunneling current.

6.3 Other measurement methods

6.3.1 Electrochemical quartz crystal microbalance

The electrochemical quartz crystal microbalance (EQCM) allows measuring an electrode's mass change in the range from less than a monolayer to film-thickness of several tens of micrometers.

A quartz crystal microbalance (QCM) independent of electrochemistry detects a change of the crystals' mass by observing its resonance frequency in an oscillator circuit. AT-cut quartz crystals show a longitudinal piezoelectric oscillation mode, which is the preferred orientation for QCM application. Metal electrodes on both sides of a thin

A

B

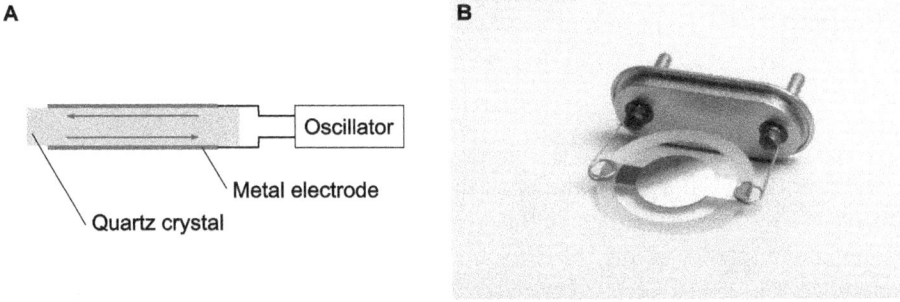

Figure 6.14: Quartz crystal microbalance (QCM): an AT-cut quartz crystal with metal electrodes on both sides shows a longitudinal piezoelectric oscillation mode (A). Commercial QCM crystal, also used in the electrochemical quartz crystal microbalance (B).

crystal disk connect to the oscillator circuit (Figure 6.14). The resonance frequency (e. g., 6 MHz for a crystal with approximately 280 μm thickness) shifts with the mass change[1] Δm. For moderate Δm, the frequency changes according to the *Sauerbrey equation*:

$$\Delta f = -\frac{2f_0^2 N_h}{\sqrt{\rho_q \mu_q}} \cdot \Delta m \tag{6.25}$$

f_0 is the fundamental resonance frequency, N_h the harmonic number, $\rho_q = 2.648 \, \text{g cm}^{-3}$ the density of quartz, and $\mu_q = 2.947 \times 10^{11} \, \text{g cm}^{-1} \, \text{s}^{-2}$ the shear modulus of quartz. The constants may be grouped as the sensitivity factor C_f, which is $81.5 \, \text{Hz} \, \mu\text{g}^{-1} \, \text{cm}^2$ for the fundamental mode of a 6 MHz crystal.

The Sauerbrey equation implies that the mass change occurs uniformly on the crystal surface, the deposited material is rigid (e. g., a metal), and mass change does not significantly affect the crystal's thickness. The rule of thumb $\Delta f < 0.02 f_0$ addresses the latter.

For larger deposits ($\Delta f > 0.02 f_0$), the material properties are not negligible. Lu and Lewis [73] proposed what we today call the *Z-match method* to calculate the change in mass accurately:

$$\Delta m = \frac{v_q \rho_q}{2\pi Z (f_0 + \Delta f)} \tan^{-1}\left(Z \tan\left(-\frac{\pi \Delta f}{f_0} \right) \right) \tag{6.26}$$

v_q is the wave speed for quartz:

$$v_q = \sqrt{\frac{\mu_q}{\rho_q}} = 3.336 \times 10^{-5} \, \text{cm s}^{-1} \tag{6.27}$$

---

1 The mass change Δm in EQCM theory usually has the units of mass per area (g cm^{-2}). This book follows this approach accepting some inconsistencies for the variable m.

The factor Z relates the properties of the deposited film to the quartz:

$$Z = \sqrt{\frac{\rho_q \mu_q}{\rho_f \mu_f}} \qquad (6.28)$$

ρ_f is the film's material density, and μ_f its shear modulus. Especially for the latter, it is difficult to find values depending on the deposited film. With $Z = 1$ (deposited material has the same properties as quartz), equation (6.26) simplifies to the Sauerbrey equation (equation (6.25)).

Conventionally, a QCM is installed, e. g., in a vacuum vapor deposition machine for metals to detect the mass of deposited thin films. Other uses include gas sensor applications to monitor adsorbed molecules. The Butterworth-Van Dyke model (Figure 6.15) describes the QCM. Here, the upper path with L, C_1, R depends on the piezoelectric movement (motional branch), while C_0 is a parasitic capacity. EQCM instrumentation sometimes allows adjustment of compensation for C_0 with a variable capacitor.

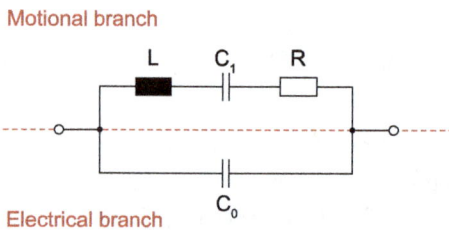

Figure 6.15: Butterworth-Van Dyke equivalent circuit to describe the QCM: L is proportional to the mass, C_1 relates to the crystal's stiffness, and R accounts for the dissipative losses. C_0 is a parasitic capacity.

In the EQCM, the electrolyte is on one side of the crystal, and one of the metal electrodes used in the oscillator circuit additionally connects to a potentiostat as the electrochemical cell's working electrode (Figure 6.16). The liquid phase of the electrolyte causes more substantial damping compared to air, but still measurements are possible. The presence of a liquid electrolyte accounts for an additional frequency shift, according to Kanazawa and Gordon [74]:

$$\Delta f = -\sqrt{\frac{\eta_1 \rho_1 f_0^3}{\pi \rho_q \mu_q}} \qquad (6.29)$$

η_1 and ρ_1 are the viscosity and density of the liquid electrolyte.

While it is debatable whether the Sauerbrey equation is applicable in liquids from a theoretical point of view, accurate measurements are possible by calibration and ensuring that the electrolyte surface is sufficiently far away from the crystal to avoid reflections. Instead of relying on the fundamental resonance frequency, overtones allow for

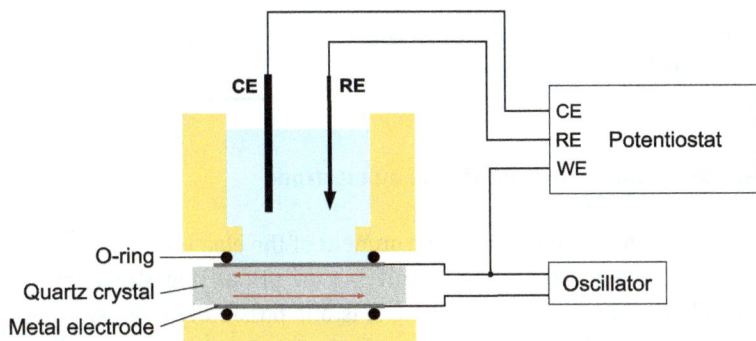

Figure 6.16: Electrochemical quartz crystal microbalance (EQCM): one side of the quartz crystal is in contact with the electrolyte using one of the metal electrodes on the quartz as the working electrode.

even higher sensitivity (only odd harmonics are possible). However, with the increasing number of harmonics (N_h), the signal gets weaker. Another approach to increase sensitivity is crystals with higher resonance frequency, e. g., 10 MHz. However, those crystals are thinner and, accordingly, more fragile.

Another practical consideration is the temperature sensitivity of EQCM measurements. The resonance frequency of a quartz crystal can change with a few $Hz\,K^{-1}$ in air and even more in contact with the liquid electrolyte. Therefore, especially for longer measurements, precise temperature control is essential.

Figures 8.15 and 9.7 show some typical curves obtained from EQCM measurements. Apart from observation of surface reaction and monitoring of electrodeposition, EQCM experiments can support corrosion studies. However, nonuniform effects like pitting corrosion violate the assumptions for exact quantitative treatment.

Viscoelastic properties of the deposited material

A further possibility is the additional consideration of dissipation caused by material deposited on the crystal. This effect is of interest, e. g., with organic molecules, which may only partially follow the crystal's lateral movement. The quality factor of the QCM is typically around 10^6, indicating a sharp peak at the resonance frequency. However, dissipation effects broaden the peak (a lower quality factor). The dissipation factor, which measures those effects, is the inverse of the quality factor and provides information about the viscoelastic properties of the deposited material and the electrolyte.

■■□ Task 6.2: Consider a 6 MHz AT-cut EQCM crystal. Assume a frequency resolution of your measurement device of 0.07 Hz.

1. What is the minimum change in mass (in $g\,cm^{-2}$) that the device can resolve? How much mass per area has a monolayer (ML) of lead on gold? Assume that the formation of 1 ML Pb requires a charge of 300 $\mu C\,cm^{-2}$ in a two-electron process. The molar mass of lead is 207.2 $g\,mol^{-1}$. What fraction of a monolayer can be resolved?

2.   While monitoring the electrodeposition of silver, what is the maximal layer thickness for which the Sauer-brey equation is applicable? The density of silver is 10.49 g cm^{-3}.

6.3.2 Sensor access to the microenvironment of an electrode

Many redox reactions also change the microenvironment of the electrolyte at the electrode, e. g., consuming dissolved oxygen (O_2), modifying the pH, or releasing ions and other species such as H_2O_2. The SECM (Section 6.2.1) is one possibility to assess those changes; for some cases, an RRDE (Section 6.1.2) can do the job. Discrete sensors placed or moved in front of the electrode are generally more robust, e. g., to describe the microenvironment in a corrosion study. Both optical and electrochemical sensor principles are in use, with the latter requiring special care that there is no electrical interference or conflict due to different ground levels between the sensor and the electrode to be investigated. Figure 6.17 shows the pH distribution next to a corrosion experiment with steel and nickel. The used electrode was a potentiometric pH sensor using a proton-selective ionomer. This work also measured the current distribution, providing an overall picture of the corrosion situation [75]. Other ionomers have been used, too, e. g., to map Na^+, Cl^-, or Mg^{2+} ions in front of a corroding electrode [76, 77].

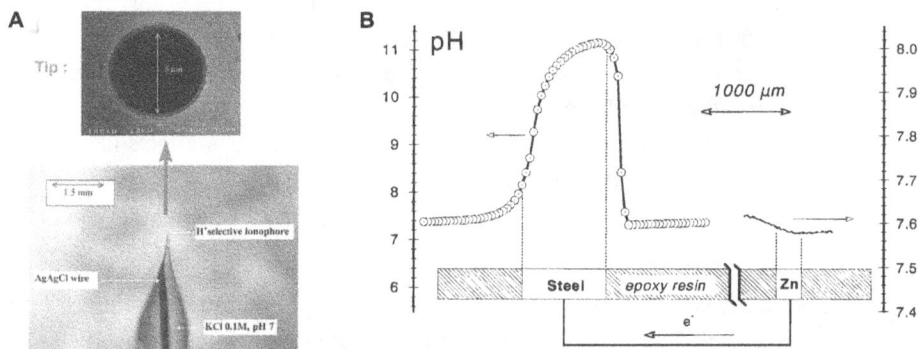

Figure 6.17: pH measurement of the microenvironment in a corrosion setup. The pH value was measured using a tiny pH electrode (A) scanning above the setup with connected steel and nickel in 0.5 M NaCl (B) [75]. © The Electrochemical Society. Reproduced by permission of IOP Publishing Ltd. All rights reserved.

Respirometry

An approach closely related is respirometry, which monitors the overall consumption of oxygen gas (O_2) or formation of hydrogen gas (H_2) by the reduction process in a corrosion experiment (compare Section 8.1.1). Strebl et al. proposed using an optical oxygen sensor combined with volumetric or manometric principles to deduce the amount of formed hydrogen [78–81]. Figure 6.18 shows a setup to combine electrochemical corrosion measurement with respirometry.

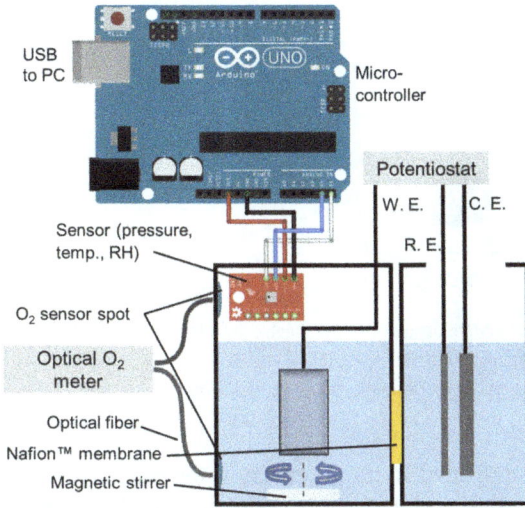

Figure 6.18: Respirometry measurement using an optical oxygen sensor and a pressure sensor. The H-cell allows the separation of the counter electrode into the right chamber connected by an ion conductive membrane to avoid disturbing the respirometry measurement [81]. Reprinted from Electrochimica Acta 412, M.G. Strebl et al., Coupling Respirometric HER and ORR Monitoring with Electrochemical Measurements, 140152, Copyright (2022), with permission from Elsevier.

6.3.3 Electrochemical noise analysis

Electrochemical noise analysis (ENA) is the measurement of stochastic fluctuation in potential or current. Generally, two causes of the noise exist:
1. Microfluctuations based on the thermal motion of particles lead to noise energy in the range of the thermal energy (equilibrium noise).
2. Macrofluctuations show energy much higher than thermal energy (nonequilibrium noise).

Especially the latter is of interest in ENA, mainly in the context of corrosion. Another noise source is the instrumentation itself, sometimes bringing ENA in a questionable context because of reproducibility issues. However, careful design of the setups and statistical analysis can prevent those pitfalls [82, 83].

Figure 6.19 shows two modes of noise monitoring that are common in corrosion analysis. The most simple yet powerful method is monitoring the open circuit potential (OCP, Section 3.4) by a high-impedance voltmeter (A). A zero-resistance amperemeter (ZRA, Figure B.4 in the Appendix) shorts two nominally identical working electrodes, and the observed current noise relates stochastical processes at the individual electrodes (B). Also, classical potentiostatic or galvanostatic control in a three-electrode setup is possible. An overview of the different setups and models provides Xia et al. [84]. The most common setup is as in Figure 6.19B for the current noise with the additional

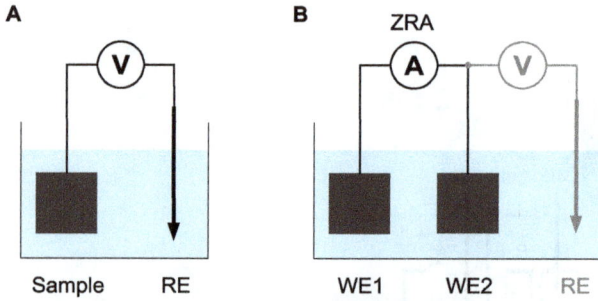

Figure 6.19: Electrochemical noise analysis (ENA): Measurement of the electrochemical potential noise (EPN) of a sample material (A) and the electrochemical current noise (ECN) with a zero-resistance am-peremeter (ZRA) between two identical samples as the working electrodes and optional the potential (B).

monitoring of the potential noise as indicated by the grey extension for the voltage recording. Typically, a bandpass filter limits the frequency range to, e. g., 1 mHz up to 1 kHz or even a more narrow band.

Different data analysis methods are established for ENA to assess corrosion. In the time domain, statistical analysis provides the standard deviations σ_E for the potential and σ_I for the current. The noise resistance R_n as the quotient of these standard deviations

$$R_n = \frac{\sigma_E}{\sigma_I} \tag{6.30}$$

provides a measure for the corrosion rate analog to the concept of the polarization resistance R_p (Sections 8.2.2 and 8.2.3) [84, 85].

Frequency domain analysis uses the power spectral densities Ψ_E of the potential and Ψ_I of the current noise providing the spectral noise impedance R_{sn}:

$$R_{sn} = \sqrt{\frac{\Psi_E}{\Psi_I}} \tag{6.31}$$

The zero-frequency impedance relates to the corrosion rate while the spectrum provides information about the corrosion type [82, 85].

Most complex information is obtained by the frequency-time domain analysis by short-time Fourier transform (STFT) or decomposition of the signal into different frequency bands ("crystals") using the Wavelet transformation. The latter provides inherent trend removal and computationally efficient, continuous information on the mechanisms, e. g., pitting versus uniform corrosion (Section 8.1.2) [86–89].

6.3.4 Spectroelectrochemistry

Spectroelectrochemistry is the combination of electrochemical methods with spectroscopy yielding very powerful in situ analysis possibilities. Different spectroscopic principles are possible, from ultraviolet-visible (UV/Vis) absorption, infrared, or Raman spectroscopy to nonoptical techniques like X-ray, nuclear magnetic resonance, or electron paramagnetic resonance spectroscopy. A great overview provides Lozeman et al. [90].

Figure 6.20 shows the setup of a spectroelectrochemical experiment for UV-Vis absorption spectroscopy. The working electrode needs to be optically transparent (e. g., indium-tin-oxide (ITO) on a glass substrate) or designed as a mesh or honeycomb structure, allowing a substantial fraction of the light to pass through. Such an arrangement enables the detection of reaction products while the electrochemical method, e. g., cyclic voltammetry, runs. With intransparent electrodes also measurements in reflective mode are possible.

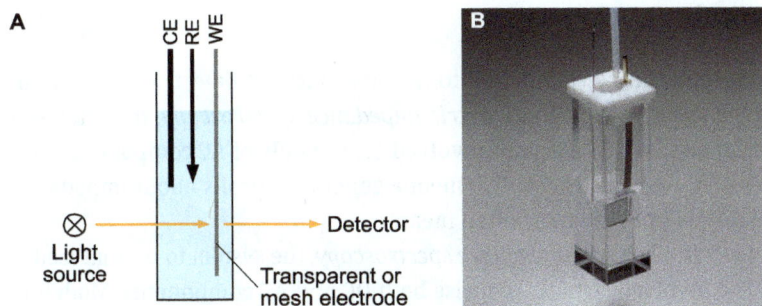

Figure 6.20: UV/Vis absorption spectroelectrochemistry: the working electrode is in the optical path of the spectrometer (A). Commercial spectroelectrochemical cell with a mesh electrode and inert gas connection (B). B: reproduced with permission from ALS, Japan.

7 Electrochemical impedance spectroscopy

Electrochemical impedance spectroscopy (EIS) comprises many powerful electrochemical methods for viewing a system from a more global point. Typically, the way of thinking in this field is more related to electrical engineering than to chemistry. Therefore, the formulation of the concepts differs significantly from other electrochemical methods. For this reason, EIS is in a separate chapter.

EIS provides insights into the electrochemical mechanism and allows material characterization. Accordingly, there are many different applications: fundamental research of reaction kinetics and catalysis, investigation of electrodeposition, corrosion analysis, battery research, and fuel cells. The seemingly simple approach of the EIS leads to a considerable amount of scientific work with wrong conclusions and overinterpretations. To a certain extent, those works cast a bad light on EIS. Therefore, this chapter strongly emphasizes the assumptions made for EIS and the limitations in modeling.

7.1 Fundamentals

Impedance spectroscopy can be classified into two approaches: dielectric and electrochemical impedance spectroscopy. In *dielectric impedance spectroscopy*, the analyzed system is coupled capacitively. The signals involved consist only of AC components, and the setup requires only two electrodes. If someone generally speaks about impedance spectroscopy, this is the approach most often meant.

In contrast, in *electrochemical impedance spectroscopy*, the system to be analyzed is coupled directly. The involved signals comprise both DC and AC components. Modeling considers the AC contributions only, while the DC component often prescribes the domain (e. g., electrode status) in which the system works. Therefore, an EIS setup usually comprises a three-electrode configuration and requires a potentiostat circuit.

7.1.1 Mathematical formulation and assumptions

The electrochemical system is considered as a "black box" with its terminals as the only possibility to get access. For the investigation of the system, a harmonic excitation is applied to probe the system's response, as indicated in Figure 7.1:

$$E(t) = E_0 \cdot \sin \omega t \tag{7.1}$$

A voltage $E(t)$ with a sinusoidal waveform with frequency f or angular frequency $\omega = 2\pi f$ and an amplitude E_0 is applied. Optionally, the applied signal contains a DC component. For a linear, time-invariant (LTI) system, the current response $I(t)$ is a signal with a sinusoidal waveform as well, related by the absolute value of the system's impedance

https://doi.org/10.1515/9783111488844-007

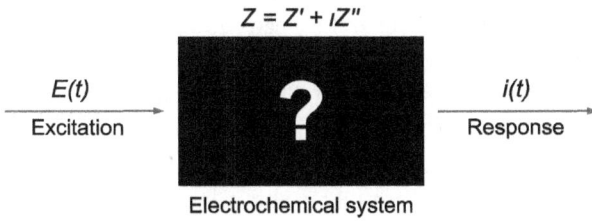

$$Z = Z' + \iota Z''$$

E(t)
Excitation

?

i(t)
Response

Electrochemical system

Figure 7.1: Schematic representation of electrochemical impedance spectroscopy (EIS): the electrochemical system is considered a "black box," which is investigated by probing with a harmonic excitation $E(t)$ and analyzing the current response $i(t)$.

$|Z|$ to the input signal. The electrochemical system might cause a phase shift ϕ between excitation and response.

$$I(t) = \frac{E_0}{|Z|} \cdot \sin(\omega t + \phi) \tag{7.2}$$

Assuming an LTI system and causality, an equivalent circuit model represented by its impedance allows the description of the system's behavior:

$$Z = Z' + \iota Z'' = |Z| \cdot \exp(\iota \phi) \tag{7.3}$$

The measurement hardware typically provides the absolute value $|Z|$ and phase ϕ. Alternatively, the impedance is represented by its real and imaginary parts. Simple relations allow switching between the two notations:

$$Z' = |R| \cdot \cos\phi \tag{7.4}$$

$$Z'' = |R| \cdot \sin\phi \tag{7.5}$$

$$|Z| = \sqrt{Z'^2 + Z''^2} \tag{7.6}$$

$$\phi = \arctan\left(\frac{Z''}{Z'}\right) \tag{7.7}$$

Plots with absolute value and phase over frequency or negative imaginary versus real part of the impedance are typical to visualize the data (Bode plot, Nyquist plot, Section 7.1.4).

With complex numbers, the prime symbol $'$ after a variable's name indicates the real part and $''$ the imaginary **i** part. The greek letter iota ι is used for the imaginary unit to avoid confusion with the current density i or the species flux density j.

For different DC values of the applied potential, the electrode surface could be in a different state (e. g., oxidized), and different redox reactions might occur. Voltage as excitation with current as the response allows explicit control of the electrode's status. Therefore, applying voltage is preferred over current excitation in most cases.

In EIS, the electrochemical systems are probed with different excitation frequencies. A typical range is between 100 kHz and 10 mHz. Numerical optimization allows matching the values of the circuit elements in an equivalent circuit with the measured data. It is essential to consider the assumptions carefully to obtain meaningful results.

Linearity

Electrochemical systems are typically highly nonlinear. Only for small excitation amplitudes linearity can be assumed. This assumption is most critical and often ignored, maybe because it is tempting that with higher excitation amplitude, smoother signals can be obtained. However, the assumed equivalent circuits do not match appropriately, and the approach leads to wrong results. Reasonable amplitudes of the AC signal are in the range of 10 mV.

Time-invariance or steady state

Another critical assumption is the time-invariance, which means the electrochemical system needs to be at a steady-state for the complete duration of the spectroscopy measurement. However, the excitation itself can change the electrode by blocking the surface or accumulation of reaction products. Shortening of the measurement duration by reducing the number of cycles per frequency, the frequency range, the number of measurement points, or even by measuring several frequencies at the same time (a so-called multisine approach, Section 7.1.3) allow achieving time-invariance also for less stable systems.

Causality

From the perspective of theory, it seems clear that the electrochemical system's response needs to be causal. In practice, many influencing factors need to be considered and kept constant to ensure that the system responds to the excitation signal only. Important factors are temperature, electrolyte composition, and flow conditions. For some applications, light exposure or electromagnetic influence could play a role. Besides careful planning, all experiments should be repeated sufficiently often to identify any noncausal contribution.

7.1.2 Measurement methods

Previously, matching AC bridge circuits was the commonly used method to measure impedance spectra. Therefore, the electrochemical cell was arranged with an assumed equivalent circuit in a full bridge, as shown in Figure 7.2. In the example, an RC parallel circuit with adjustable components R_x and C_x allows the matching of the cells' behavior.

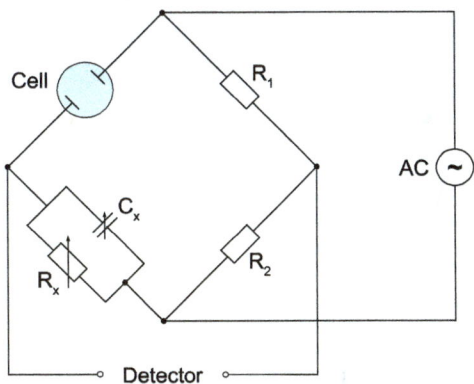

Figure 7.2: AC bridge with an RC parallel circuit with variable components R_x and C_x to match the electrochemical cells' characteristic. The elements need an adjustment for each measured frequency.

The operator needs to adjust for each requested frequency making the generation of broad spectra very cumbersome. Additionally, only a two-electrode setup for the electrochemical cell is possible. The AC bridge is, therefore, of minimal value for most applications and more of historical relevance.

Today's state-of-art method is the combination of a potentiostat with an AC source and a phase-sensitive detector. The bandwidth of the potentiostat needs to be sufficiently high to allow for the required frequencies. Optimally is an adjustable bandwidth to compromise between measurement at higher frequencies and stability. Practical realization can be a specific frequency response analyzer (FRA), a component in a modular potentiostat device, a discrete arrangement of several instruments, or an integrated circuit allowing for EIS (e. g., the example in Figure 3.16).

Today's potentiostats with built-in impedance spectrometers typically display only the AC component of the current or the impedance. The analog levels of the potentiostat circuit itself are not accessible via the software's user interface. If the signal consists mainly of a DC with only a small AC contribution, the noise may distort the intended output. Additionally, instabilities of the potentiostat circuit cannot be seen at all. Therefore, it is highly recommended to use the analog outputs if available and check the signal integrity with an oscilloscope (check with both DC and AC coupling).

Figure 7.3 illustrates a simplified setup around the previously introduced adder potentiostat (Section 3.2.1) with a lock-in amplifier as a phase-sensitive detector to extract the AC signals in terms of absolute value and phase of the impedance. A second voltage source connects to one input of the adder circuit providing the small AC signal superimposing to the standard DC source of the potentiostat. At the output of the current follower, a filter splits the signal into its AC and DC components. The lock-in amplifier uses the applied alternating voltage and the AC response to provide the absolute value and the phase of the impedance.

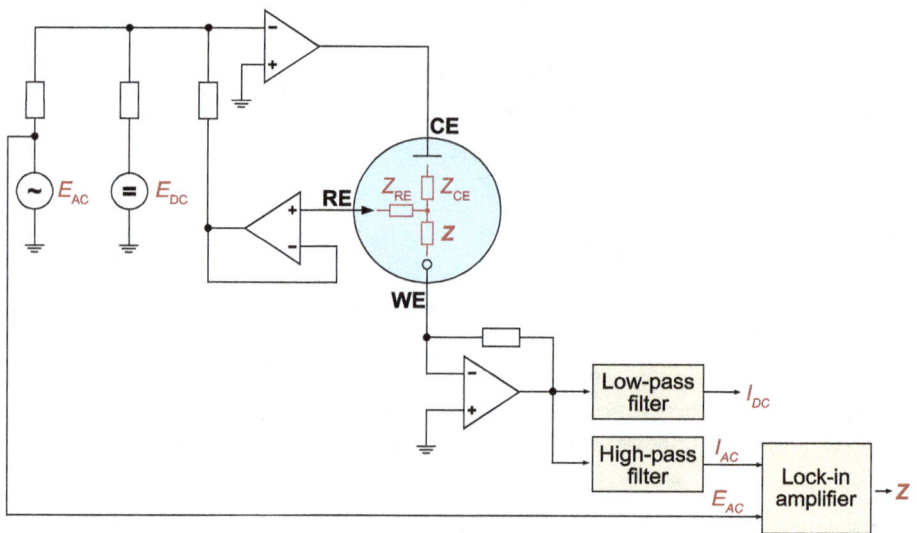

Figure 7.3: Illustration of an adder potentiostat enhanced by an alternating voltage source and phase-sensitive detector for electrochemical impedance spectroscopy (EIS).

---

i Some practical considerations for successful EIS measurements: Finding the right instrument settings is mainly about ensuring that the electrochemical cell is an LTI system. Linearity requires a low AC amplitude of the applied potential, e. g., 10 mV. The value of the previously read OCP is a good guess for the DC component of the potential. Alternatively, the selection of a different value addresses a specific regime or state of the electrode. Frequencies should be in the range of 100 kHz to 0.1 Hz with a logarithmic distribution of the individual values. Studies of corrosion phenomena or characterization of batteries might require even lower frequencies down to 1 mHz. Another parameter is the integration, which defines the number of periods measured per frequency. Higher integration time results in better SNR with the downside of longer total measurement time. A too long duration might violate the assumption of time-invariance.

The three-electrode setup with a potentiostat ensures that only the working electrode's impedance Z (including the uncompensated resistance) plays a role. In Figure 7.3, the different impedances in the electrochemical cell are indicated. The potentiostatic circuit hides the impedance Z_{CE}. Also the impedance of the reference electrode Z_{RE} shows up in parasitic effects, e. g., stability issues only.

7.1.3 Multisine approach

The premise of an LTI-system can also help to shorten the measurement duration. Especially with an electrochemical system that changes its properties, a largely reduced measurement time allows to approximate time-invariance. Instead of applying several frequencies sequentially in time, several values are superimposed, assured that the chosen values do not share harmonics. In an example with five simultaneous frequencies, the fundamental frequency f_0 is superimposed by $3f_0$, $5f_0$, $7f_0$, and $9f_0$.

The multisine approach enables time-resolved EIS measurements with up to more than 20 frequencies at the same time. The gain in measuring speed is affected by the lower accuracy with an increasing number of simultaneously measured frequencies. In all cases, cross-checking the linearity assumption is crucial (Section 7.3.1).

7.1.4 Data presentation

Many electrochemical systems show complex behavior in EIS experiments. Possibly visible effects comprise redox reactions and its kinetic properties, mass transport phenomena, and contributions from the interfacial regions. An essential role for the correct interpretation of the results plays the presentation of the data. Nyquist plots and Bode plots visualize impedance results for different frequencies. Lissajous figures allow online observation of the system's behavior at a single frequency.

Nyquist plot

For newbies to EIS, the Nyquist plot might seem a bit like a mystery. Once used to the typical curve shapes in this presentation form, it is easy to grasp the characteristic behavior of an electrochemical system with a glance at the plot. The negative imaginary part of the impedance is plotted over the real part. Although the curve contains data for all measured frequencies, the information on what is the frequency of a specific point gets lost. In general, data points for higher frequencies are to the left. Sometimes some of the data points are labeled with the corresponding frequency value. Figure 7.4(A) shows an example of a Nyquist plot for a simple RC parallel circuit. In diagrams with an equal scale on the vertical and horizontal axis, this circuit causes a hemispherical curve. The intersection with the horizontal axis is the resistance value. I strongly recommend getting used to the representation of EIS data in a Nyquist plot, as this facilitates the selection of the correct model.

Bode plot

The Bode plot shows the absolute value and phase of the impedance over the frequency. The horizontal axis is typically scaled logarithmically. The two curves can be in the same diagram with two vertical axes or two subplots, as shown in Figure 7.4(B). From the Bode plot, information about the specific values at a particular frequency can be read. Understanding of the overall system's behavior is less convenient. Optimally, EIS data is presented as a Nyquist plot and a Bode plot in parallel as it is done throughout this chapter.

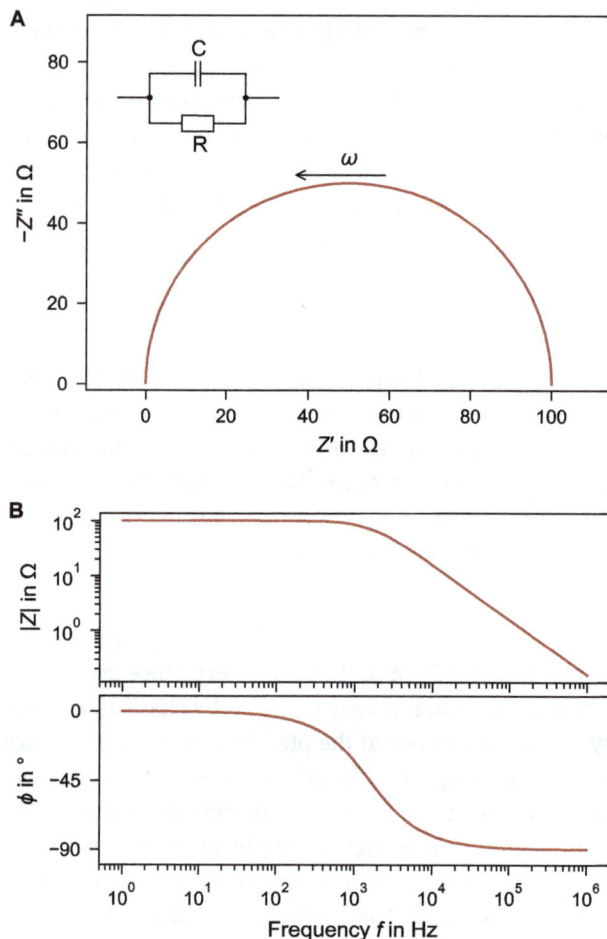

Figure 7.4: Nyquist plot (A) and Bode plot (B) of a RC parallel circuit. $R = 100\,\Omega$ and $C = 1\,\mu F$. The Nyquist plot provides a quick impression on the system's behavior; the Bode plot shows more details of the frequency dependency.

Lissajous figures

Lissajous figures might be known to you from physics or electronics experiments when connecting two waveforms to an oscilloscope in X-Y mode. In EIS, Lissajous figures are used to get an online presentation of the impedance characteristic at a particular frequency. The AC component of the applied potential connects to the horizontal axis, the obtained current response to the vertical axis. Figure 7.5 shows the Lissajous figures of impedance signals with different phase values. The diagram directly indicates the amplitude of the voltage (E_0) and the current (I_0). The intersection with the axes represents the phase. Besides the general estimation of the phase characteristics, this presentation allows getting a first impression on the validity of assuming an LTI system (Figure 7.6).

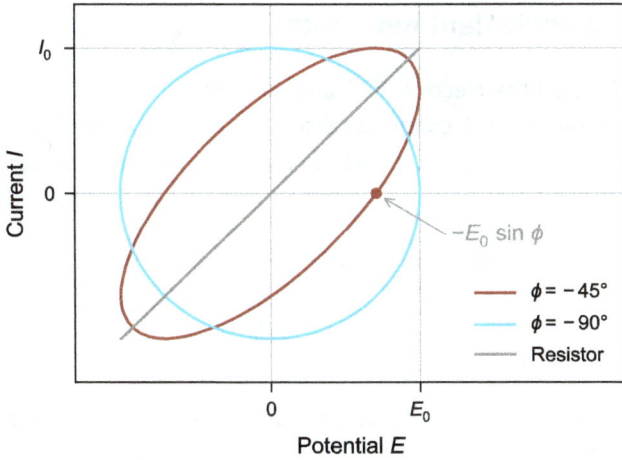

Figure 7.5: Lissajous figures for different phase values: A pure capacitive system ($\phi = -90°$) results in a circle in contrast to a line for a resistor ($\phi = 0°$).

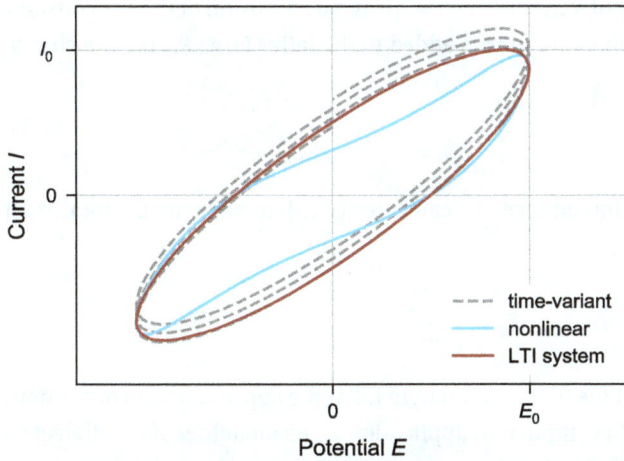

Figure 7.6: Representation of nonidealities in Lissajous figures: nonlinearity causes an unsymmetric curve, time-variant behavior a wandering of the signal. In comparison, an ideal LTI system ($\phi = -30°$).

Lissajous figures are a mighty tool for instant cross-check of the validity of the assumptions for EIS. Although modern hardware and software often hide those basic approaches, everyone serious about EIS should connect an oscilloscope in X-Y mode with AC coupling to the potentiostat if analog outputs are available or check if the EIS software provides a live view at Lissajous figures.

7.2 Circuit elements and equivalent networks

The goal in EIS analysis is to model the electrochemical system with an appropriate equivalent network. An equivalent network comprises discrete circuit elements, some are well known from electrical engineering (R, C, L), while others are particular to electrochemical systems.

7.2.1 Basic elements

Resistor: R
The most simple element is an ideal resistor R with its resistance R. The impedance of a resistor consists of a real part only:

$$Z_R = R \tag{7.8}$$

Please note the ambiguity of R in this context. To avoid confusion of the gas constant R with the resistance, preferable an index is added to the latter (e. g., R_{ct} for the charge transfer resistance).

Capacitor: C
In contrast to the resistor the impedance of a capacitor C with its capacity C contains an imaginary part only:

$$Z_C = \frac{1}{\iota \omega C} \tag{7.9}$$

The most prominent example of this element in EIS is the approximation of the double layer by a capacitor. Another important application is the modeling of the dielectric properties of a coating in the field of corrosion analysis.

Constant phase element: Q
As you might guess from the previous description of the electrical double layer (Section 2.6.1), a simple capacitor is often not an appropriate representation. The constant phase element (CPE) Q considers two parameters Y_0 and α:

$$Z_Q = \frac{1}{Y_0(\iota \omega)^\alpha} \quad \text{typically } \alpha = 0.8 .. 0.9 \tag{7.10}$$

The CPE is like an "imperfect capacitor," which represents in many electrochemical systems the contribution of the double layer better than a pure capacity. For $\alpha = 1$, the

CPE is an ideal capacitor, and the admittance parameter Y_0 represents the capacity. The parameter α directly relates to the phase shift:

$$\phi = -\alpha \cdot 90°$$ (7.11)

In Figure 7.7, the CPE with different parameter α is compared to an ideal capacitor in a parallel circuit with a resistor. In the Nyquist plot of experimental data, the deviation from semicircular behavior indicates the need for a CPE instead of a capacitor.

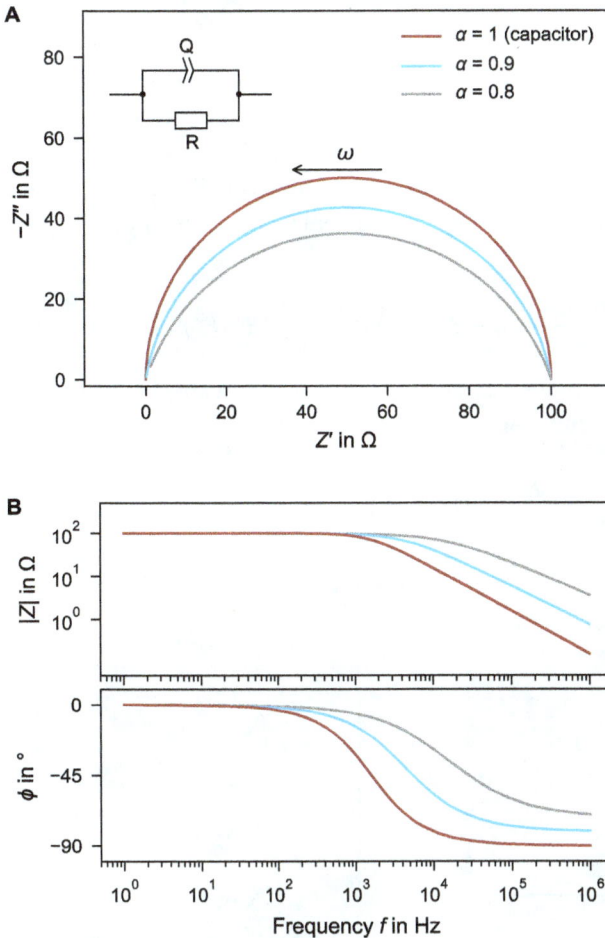

Figure 7.7: Example of a CPE in a parallel circuit with a resistor: Nyquist plot (A) and Bode plot (B). $R = 100\,\Omega$, $C = 1\,\mu F$ or Y_0 with the respective unit, and different values of α. For $\alpha = 1$, the CPE behaves like an ideal capacitor.

Inductor: L

The last fundamental element is the inductor L with its inductance L. Its impedance contains an imaginary part only:

$$Z_{\mathrm{L}} = \iota\omega L \tag{7.12}$$

For most applications, this element plays a minor role in EIS modeling. However, with corrosion phenomena such as pit formation or generally localized corrosion, inductive behavior can occur at low frequencies ($f \ll 1\,\mathrm{Hz}$) [91, 92]. Similarly, it was observed in proton exchange membrane fuel cells [93] or proton exchange membrane electrolyzers [94, 95]. Additionally, inductance comes into the arena when nonidealities of the setup play a role, e. g., the inductance of cables.

7.2.2 Charge transfer resistance

The charge transfer in electrochemical reactions is generally nonlinear. The Butler–Volmer equation (Section 2.3, Figure 7.8) relates the current (density) to the applied voltage:

$$i(E) = i_0\left(\exp\left(\frac{(1-\alpha)nF}{RT}(E - E^0)\right) - \exp\left(-\frac{\alpha nF}{RT}(E - E^0)\right)\right) \tag{7.13}$$

For small changes ΔE of the applied potential the relationship $\exp(x) \approx 1 + x$ allows linearization of the exponential functions:

$$\Delta i = \frac{i_0 nF}{RT} \cdot \Delta E \tag{7.14}$$

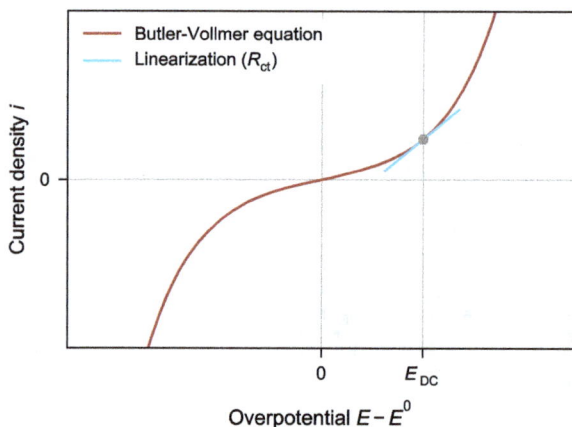

Figure 7.8: Reaction kinetics: to model the charge transfer in a redox reaction as an LTI system, a linear function approximates the generally nonlinear situation (Butler–Volmer equation) for a small change in applied voltage ΔE around a fixed DC potential E_{DC}. The inverse of its slope is the charge transfer resistance (R_{ct}).

In EIS with the valid assumption of linearity, the ΔE corresponds to the amplitude E_0 of the voltage stimuli's AC component. With the transition from current density Δi to current ΔI by multiplying with the electrode area A and rearranging for $\frac{\Delta E}{\Delta I}$, we get the charge transfer resistance:

$$R_{ct} = \frac{RT}{nFAi_0} \tag{7.15}$$

The charge transfer resistance depends on the individual redox reaction by the number of electrons n and the exchange current density i_0. In the context of corrosion analysis, R_{ct} is called polarization resistance R_p (Section 8.2.3).

■□□ Task 7.1: Linearize the Butler–Volmer equation for small changes ΔE around a DC potential E_{DC}. Start with $i(\Delta E)$ and apply the linearization to get Δi. You should end up with equation (7.14). From there you can easily get the definition of R_{ct} (equation (7.15)).

7.2.3 Randles circuit

The application of the charge transfer resistance to model the reaction of an electrochemical system under kinetic control needs a few more components. Most importantly, the double layer represented by its capacity C_{dl} or an appropriate CPE contributes in parallel to the charge transfer reaction. Additionally, the uncompensated resistance adds to the situation at the electrode. This model is the *Randles circuit* (Figure 7.9) suggested by Randles already in 1947 to described situations without mass transfer control.[1]

The Nyquist plot of the Randles circuit (Figure 7.9(A)) shows the characteristic semicircle of an RC parallel circuit with an offset in the horizontal direction by the uncompensated resistance. The charge transfer resistance is visible at the side of lower frequencies.

■■□ Task 7.2: Write down and simplify the complex impedance Z of the Randles circuit shown in Figure 7.9(A). Convert the impedance into absolute value $|Z|$ and phase ϕ. What is the time constant of this system?

7.2.4 Mass transport control – Warburg impedance

Once the electrochemical system is under mass transport control modeling using classical elements only is not possible anymore. Emil Warburg, a German physicist, derived

---

1 In his original work [96], Randles suggested another capacitor in series to R_{ct}, which is not a common element in nowadays "Randles circuit."

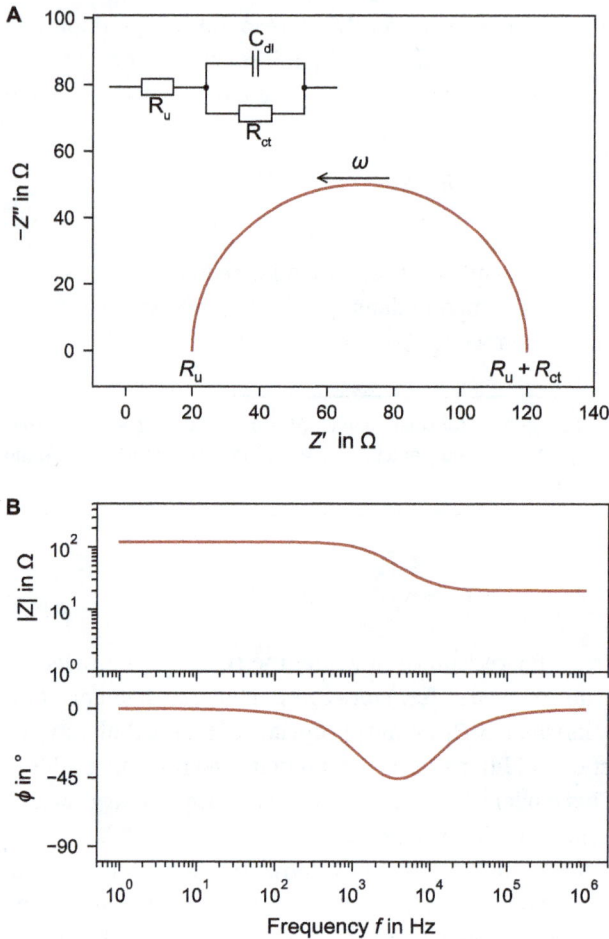

Figure 7.9: Randles circuit to model an electrochemical system under kinetic control: the charge transfer resistance (R_{ct}) is in parallel with the double layer capacity (C_{dl}) or. The uncompensated resistance (R_u) adds to the situation at the electrode.

that diffusion limitation causes an impedance with similar real and imaginary parts. The corresponding component is attributing to him called *Warburg element.* In the case of semiinfinite diffusion above the electrode, its impedance, the Warburg impedance Z_W, is inverse proportional to the square root of the frequency:

$$Z_W = \frac{A_W}{\sqrt{\omega}}(1 - \iota).\tag{7.16}$$

The Warburg constant A_W, sometimes also labeled σ, relates fit values to the diffusivity D and concentration c of the involved species, number of electrons n, and the electrode area A:

$$A_{\mathrm{w}} = \frac{RT}{n^2 F^2 Ac\sqrt{2D}}$$ (7.17)

Sometimes the Warburg impedance is defined with $1/\sqrt{\iota}$ instead of $(1 - \iota)$. In this case, A_{w} needs to be adjusted by multiplying with $\sqrt{2}$.

The Warburg element comes into the Randles circuit in series with the charge transfer resistance. For lower frequencies, diffusion limitation plays a role, and the Warburg element sets in. Fast changes in the applied potential do not translate into a change in mass transfer, and the Warburg element is conductive. In the Nyquist plot (Figure 7.10(A)), the Warburg element causes a diagonal line disturbing the semi-circular curve at lower frequencies. In the Bode plot (Figure 7.10(B)), the mass transfer limitation causes an increase in the absolute value of the impedance at low frequencies, along with a phase shift to more negative values.

In the situation of finite-length diffusion, depending on the nature of the boundary, two formulations are common: the Warburg impedance Z_{ws} for the "short-circuit" case with a transmissive boundary condition and Z_{wo} for the "open-circuit" case with a reflective boundary condition:

$$Z_{\mathrm{ws}} = \frac{A_{\mathrm{w}}}{\sqrt{\omega}}(1 - \iota) \cdot \tanh\left(\frac{\delta}{\sqrt{D}} \cdot \sqrt{\iota\omega}\right)$$ (7.18)

$$Z_{\mathrm{wo}} = \frac{A_{\mathrm{w}}}{\sqrt{\omega}}(1 - \iota) \cdot \coth\left(\frac{\delta}{\sqrt{D}} \cdot \sqrt{\iota\omega}\right)$$ (7.19)

The reflective boundary condition represents an impermeable boundary, e. g., a channel wall. A permeable boundary, as it is the case in rotating disc electrode experiments (Section 6.1.1), requires the transmissive boundary condition. In both cases, the Nernst diffusion layer thickness δ modulates the degree of deviation from the semi-infinite diffusion situation with its Warburg impedance Z_{w}.

The Nyquist plot (Figure 7.11(A)) shows the effect of finite-length diffusion with the two different types of boundary conditions. For an impermeable boundary, the imaginary part dominates in comparison to semiinfinite diffusion; for a permeable boundary, the real part is higher.

A more generalized treatment of the Warburg impedance in finite-length diffusion for different boundary conditions may require the modeling with a transmission line.

7.2.5 Coatings

An essential application of EIS is the characterization of dielectric coatings or paints used to passivate metals. The nondestructive nature of the method enables the moni-

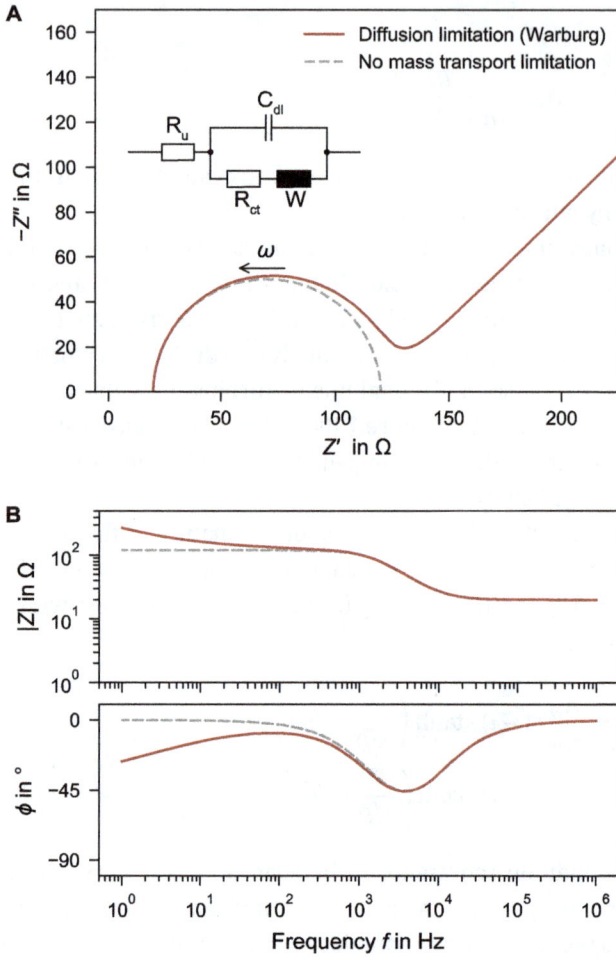

Figure 7.10: Mass transport limitation by diffusion: enhanced Randles circuit with the Warburg element W placed in series with the charge transfer resistance R_{ct}.

toring of failures over time. In comparison to other nondestructive methods, EIS works with the same physicochemical principle as it occurs in corrosion.

A perfect coating applied on metal (the working electrode) behaves like a series connection of the coating's capacity C_c and the uncompensated resistance R_u (Figure 7.12). Measurement of a perfect coating by EIS is not useful as such but can be the starting point in degradation monitoring.

Once the dielectric layer has defects (pinholes, pores, scratches), redox reactions can occur at the metal electrolyte interface. The typical model of this scenario (Figure 7.13) considers the coating's capacity in parallel with a Randles cell representing the situation within the defect(s).

A

B

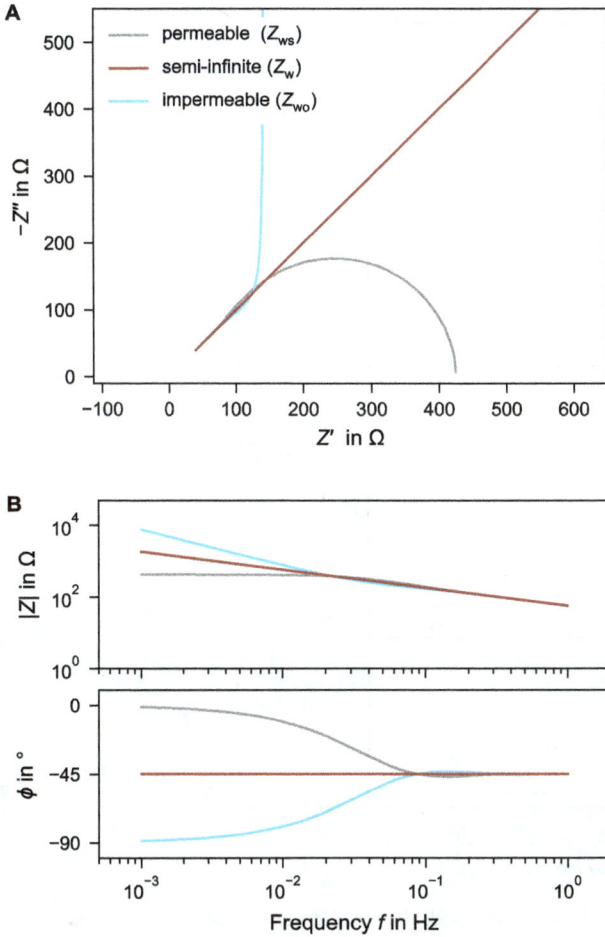

Figure 7.11: Comparison of the different Warburg impedances: semiinfinite diffusion (Z_w) and finite-length diffusion with reflective (Z_{ws}, impermeable) and transmissive (Z_{wo}, permeable) boundary condition.

R_d is the solution resistance within the hole(s). With EIS, it is not possible to discriminate between many small pinholes or a more substantial defect with the same area. The double layer capacity C_{dl} at the metal electrolyte interface within the defect may require a CPE to match the real situation. Depending on the electrochemical system, mass transfer limitation might occur, especially for lower frequencies. A Warburg element in series with the charge transfer resistance R_{ct} would attribute for it. The Nyquist plot for a coating with defects may show two characteristic semicircles. Depending on the parameters, both of them might not be as clearly visible as in Figure 7.14(A).

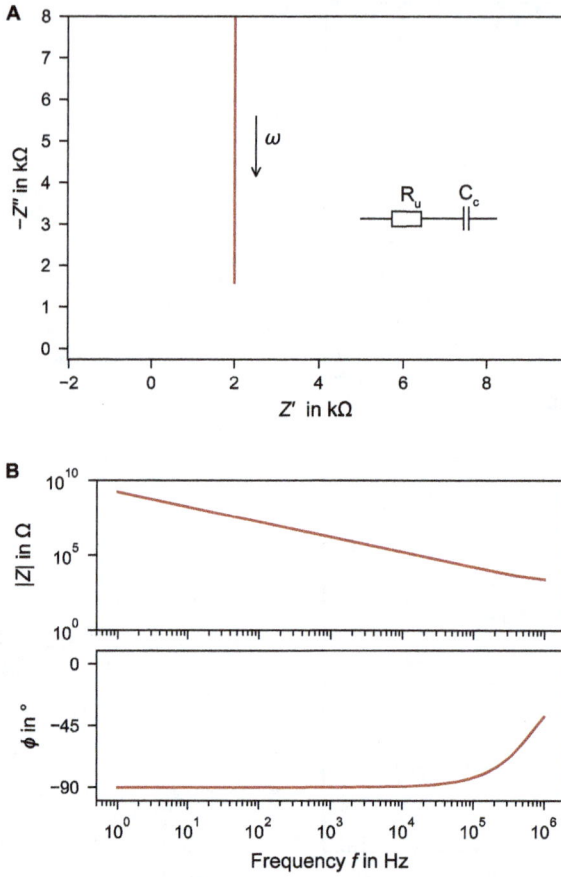

Figure 7.12: Perfect coating: an RC serial circuit is the model for a dielectric layer (paint) without any defects.

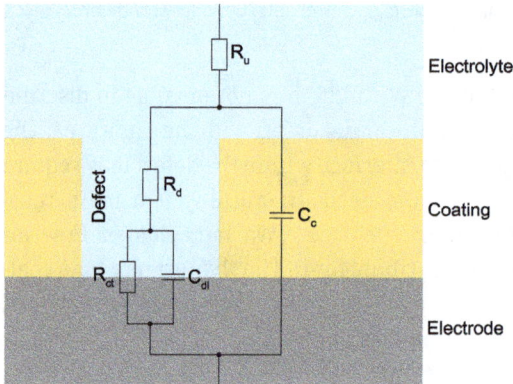

Figure 7.13: Schematic representation of a dielectric coating (paint) on a metal with a defect exposing the bare metal surface to the electrolyte.

A

B

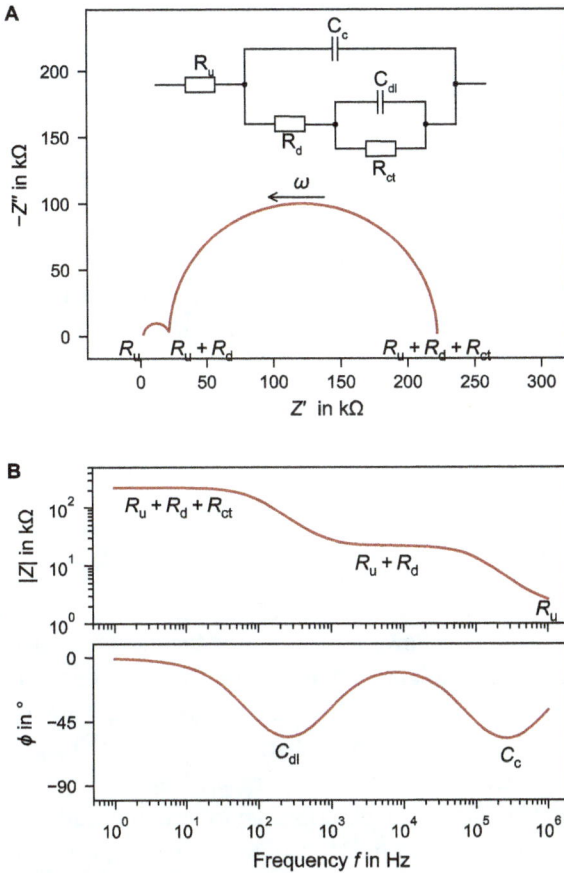

Figure 7.14: Coating with defect: The circuit of a perfect layer (R_u in series with the coating's capacity C_c) is enhanced by the electrolyte resistance in the defect (R_d) and a Randles circuit (R_{ct} in parallel to C_{dl}) to describe the metal-electrolyte interface.

Typical measurements of coatings start with an intact layer and observe the change either in the spectrum or at a single frequency over time to monitor degradation. In practice, this causes some experimental challenges:

- An electrode with a perfect coating does not show a defined OCP. The bare metal or a sample with a huge scratch allows measurement of the OCP in a prior experiment providing a good guess value for the applied DC potential in the EIS.
- The overall currents will be low, causing bad SNR. Besides sensitive instrumentation large sample size is recommended. Larger surface area also allows for better averaging of individual defects. Another solution would be to increase the amplitude of the AC excitation to values around 100 mV. In this case, verification of linearity is even more crucial once the coating degraded.

7.3 Toolbox

7.3.1 Kramer–Kronig test

As you might have recognized from the previous sections, an essential part of EIS analysis is understanding and ensuring the validity of the underlying assumptions. Most crucial is it in the case of linearity. The Kramers–Kronig test is a method to check the linearity formally. This test uses the Kramers–Kronig relations, which are the Hilbert transform applied to impedance. The functional principle of this test is most easily understood with the underlying equations. In a linear system, the real and imaginary parts of a function depend on each other:

$$Z'(\omega) = \frac{2}{\pi} \int_0^\infty \frac{x \cdot Z''(x)}{\omega^2 - x^2} \, dx \qquad (7.20)$$

$$Z''(\omega) = -\frac{2}{\pi} \int_0^\infty \frac{\omega \cdot Z'(x)}{\omega^2 - x^2} \, dx \qquad (7.21)$$

Knowing the real part for all frequencies allows reconstructing the imaginary part and vice versa. Practical limitations are unavailability of all values f up to infinity and the knowledge of the analytical function of Z' or Z''. The Kramers–Kronig test applies one of the relations to a fit model for the finite amount of discrete values of Z', respectively, Z''. A comparison of the reconstruction to the data for Z'' resp. Z' allows judgment whether the linearity assumption is valid. Boukamp described a practical realization of the test with a series of RC parallel circuits as a fit model [97], which was improved by Schönleber et al. [98]. Typically, it is sufficient to trust the Kramers–Kronig test implementation in the used EIS software.

7.3.2 Software

EIS software serves two purposes: simulation and fitting. Simulation means calculating impedance spectra (shown as a Nyquist plot or a Bode plot) with fixed parameters for the elements in an equivalent circuit model. The primary intension is the visualization of a specific circuit and investigation of the influence from different settings.

More important for the analysis of experimental data is the fitting part. Based on a guess for the equivalent circuit model, the software matches the parameter of the circuit elements to the measurement results. Visual inspection of the fit model plotted over the data points allows checking how appropriate the model was and adjusting the circuit to obtain a good fit. Either by visual evaluation of the fit quality or by considering error estimates of the fitting algorithm may help to find well-matching models for the experimental data set.

Once you publish impedance data, it is essential to show a plot of the fitted curve over the raw data as it **!** will help the reader to understand how appropriate the model is. In case the fit only partially matches the measurement results, it is crucial to understand in which region deviation occurs (what would not be visible with formal error estimates only).

Most potentiostat manufacturers provide quite useful software alongside their EIS hardware. A significant benefit is an optimal match to the specific properties of the particular equipment and the file format of the data set. Some software even reads in data in the file format of a different brand. Besides, there are several stand-alone software tools. Alternatively, there are Python libraries for EIS simulation and fitting. You can find an overview and links on the book's webpage. In the end, it depends on the scientist or engineer, which is the favored approach. Following the coding way with Python may result in a better understanding and provides more flexibility with less common fit models.

■■■ Task 7.3: Simulating different basic equivalent circuits and changing the parameter helps to get a **i** sound understanding of EIS models. Choose an EIS simulation tool of your choice and investigate different circuits:
1. Generate a Nyquist plot and a Bode plot for a simple RC parallel circuit representing the charge transfer resistance and double layer capacity.
2. Enhance the model with a CPE instead of the capacitor and use another element for the uncompensated resistance.
3. Finally, incorporate a Warburg element to consider mass transport limitation (you can assume semiinfinite diffusion).
4. Change all parameters of the circuit elements one by one to get a feeling of its influence on the EIS spectrum.

A possible solution to this task, indicating reasonable starting values for your exploration and a Jupyter Notebook showing how to implement in Python, are available on the book's webpage.

7.3.3 Challenges in modeling

Nonconverging fits are a big issue with most real-world EIS data. The exact strategy to obtain meaningful fit results depends on the used software and the electrochemical system.

Another challenge in modeling is the ambiguity of equivalent circuits, especially if more than one time-constant is involved. Nyquist plots with two hemispherical circles are an indication of a system with two time-constants. Figure 7.15 shows the Nyquist plot and Bode plot of a series connection of two RC parallel circuits and an uncompensated resistance. However, measured data would equally well fit the equivalent circuit of a coating with defect (Figure 7.13). Therefore, it is essential to consider that even if a model matches nicely experimental data, there might be no physical correspondence. In my

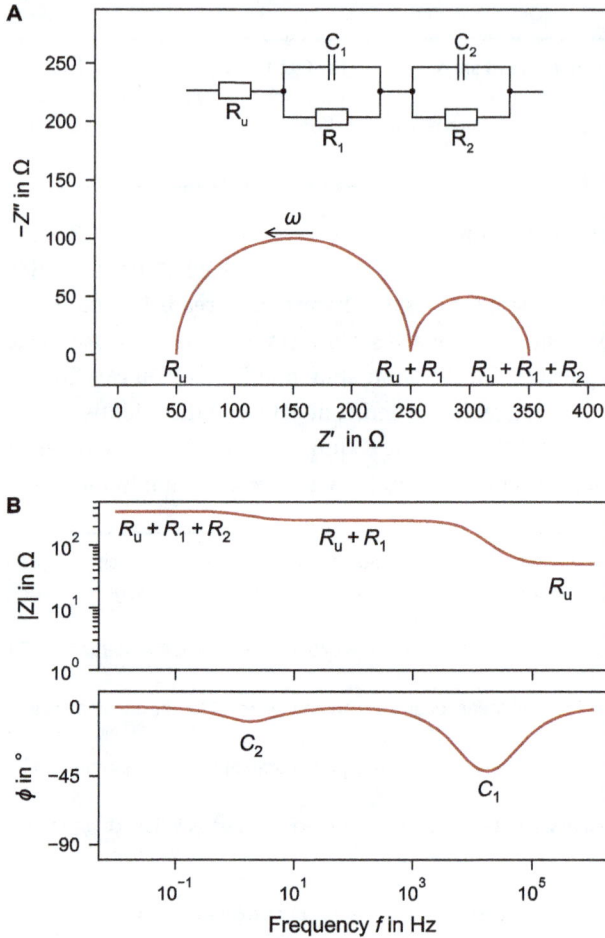

Figure 7.15: Ambiguity in modeling: Nyquist plot (A) and Bode plot (B) of a system with two time-constant. Please be aware that experimental data matching this model would equally well fit the equivalent circuit of a coating with defects shown in Figure 7.13. $R_u = 50\,\Omega$, $R_1 = 100\,\Omega$, $C_1 = 1000\,\mu F$, $R_2 = 200\,\Omega$, and $C_2 = 0.1\,\mu F$.

opinion, such tempting overinterpretation of successful matches is the reason for the majority of the wrong conclusions drawn in EIS research.

---

i Some practical hints to deal with nonconverging fits:
- Start with reasonable guess values for all parameters.
- Fix some of the parameters in the model to enforce convergence while optimizing the critical parameters.
- Reduce the complexity of the model initially (e. g., use a capacitor instead of a CPE). In a further run, use the values obtained with the simple model as start values for the full equivalent circuit.

Part III: **Applications**

"I believe that water will one day be employed as fuel, that hydrogen and oxygen which constitute it, used singly or together, will furnish an inexhaustible source of heat and light, of an intensity of which coal is not capable."

Jules Verne in The Mysterious Island (1874)

8 Selected aspects: material science

Material science touches in many areas electrochemical methods. Analysis of material properties benefits from electroanalysis, especially concerning metals, metal alloys, and carbon-based materials. Also, in the prediction and avoiding of failure, electrochemistry helps. Apart from the analysis, electrochemical methods contribute to the synthesis of new materials, e. g., nanomaterials.

Here, the focus is on two selected aspects of broader interest: corrosion and platinum electrochemistry. Corrosion analysis and prevention have a substantial economic impact. While consequences occur on a large scale (think of a pipeline leakage), the causes (onset of pitting corrosion) happens in the micro- and nanoworld. The scientific community around classical corrosion research seems independent from people working on electrochemical sensors or biomedical applications. However, the description of corrosion introduces concepts (e. g., mixed-potential or Pourbaix diagrams) that are helpful for other areas, too. The analysis methods are great examples of how the different electrochemical techniques can work together. Understanding the corrosion mechanism in steel helps interpret the degradation of sensor or implant electrodes made of noble metals in the biological environment (although the effects are different).

The second selected aspect is platinum electrochemistry for two reasons: platinum and its alloys are state-of-art materials for implant electrodes, in chemo- and biosensors, and as catalysts. Additionally, a sound understanding of platinum electrochemistry and its characterization facilitate applying the electrochemical methods to other platinum-group metals and material composites commonly used in fuel cells and electrolysis.

8.1 Corrosion

Corrosion is a term used in many scientific and technical disciplines, but also in everyday language, most people have a basic idea of what corrosion is. Outside the field, rust would be the first association or a mechanism to cause a catastrophic failure.

The international standard ISO 8044[1] defines corrosion as "physicochemical interaction between a *metallic material* and its environment that results in changes in the properties of the metal, and that may lead to significant impairment of the function of the metal, the environment or the technical system, of which these form a part" [99].

In a more general understanding, corrosion is a chemical or electrochemical process at the surface of a solid in contact with a fluid. Besides metals, solids can be semiconductors, ionic crystals, glass, or polymers. Fluids can be liquids such as water, maybe

---

1 The international standard ISO 8044, the most recent issue from 2020, replaces the German standard DIN 50900 as DIN EN ISO 8044. ASTM G193 is the equivalent US standard.

https://doi.org/10.1515/9783111488844-008

just in the form of humidity or, e. g., oil. The result of the corrosion process can be material loss or conversion into an insoluble compound. The solid forms the electrodes, and the liquid is the electrolyte of an electrochemical cell.

8.1.1 Fundamentals

Two half-reactions describe the corrosion of metal. Anodic half-reaction (oxidation) is the dissolution of the metal or the formation of insoluble products:

$$Me \longrightarrow Me^{n+} + ne^- \tag{8.1}$$

In the presence of molecular oxygen, the cathodic half-reactions (reduction) are

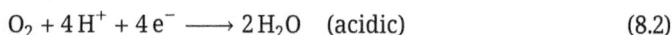

$$O_2 + 4\,H^+ + 4\,e^- \longrightarrow 2\,H_2O \quad \text{(acidic)} \tag{8.2}$$

or

$$O_2 + 2\,H_2O + 4\,e^- \longrightarrow 4\,OH^- \quad \text{(alkaline, neutral)} \tag{8.3}$$

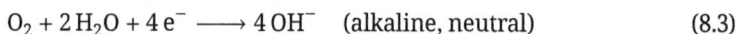

depending on the pH.

Thermodynamically less favored are the reduction reactions resulting in molecular hydrogen, which occurs in oxygen-deficient aqueous systems:

$$2\,H^+ + 2\,e^- \longrightarrow H_2 \quad \text{(acidic)} \tag{8.4}$$
$$2\,H_2O + 2\,e^- \longrightarrow H_2 + 2\,OH^- \quad \text{(alkaline, neutral)} \tag{8.5}$$

Here, it is not only the anodic metal dissolution but also the cathodic reaction that can cause the failure of the material due to hydrogen embrittlement, as observed with steel.

Local element

In contrast to the classical understanding of a galvanic cell as introduced in Section 1.3 no spatial separation of the two half-cells is required in case of corrosion. The two half-reactions occur at the same material (but different active sites) and form short-circuited galvanic cells, each a so-called *local element* (Figure 8.1).

The electrical potential of the local element in equilibrium, called corrosion potential E_{corr}, establishes based on the two involved half-reactions. Its equilibrium potential characterizes each half-cell. Under standard conditions, the cell potential is the standard potential E^0. For activities deviating from unity, the cell potential is the reversible cell potential as calculated by the Nernst equation (equation (2.10)).

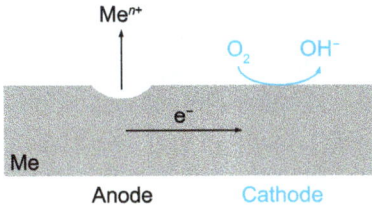

Figure 8.1: Local element: two half-reactions at neighboring sites contribute to the mixed potential. The anodic reaction causes the dissolution of the metal or the formation of an oxide. At the cathode, oxygen gets reduced.

The ability of each half-reaction pulling E_{corr} toward its own reversible cell potential is given by its exchange current density (Section 2.3.2). That means the corrosion potential is always in between the two reversible cell potentials of the two half-reactions and closer to the process with the higher exchange current density (assuming that none of the reactions depends on mass transport limitation).

When the anode and cathode sites of the local element randomly fluctuate across the metal surface, uniform corrosion occurs. However, localized corrosion starts as soon as a predominance for a position towards one of the half-cell reactions occurs (e. g., because of crack, a grain boundary, or simply a crystal edge). At the cathode, the pH will increase, while at the anode, hydrolysis of the metal

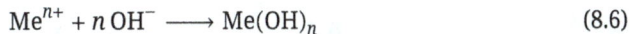

$$Me^{n+} + n\,OH^- \longrightarrow Me(OH)_n \tag{8.6}$$

may occur, consuming OH^- ions and causing a local pH decrease. The pH difference further manifests localized corrosion. Local acidification at the site of metal dissolution worsens the attack, especially in the presence of chloride ions.

Typically, the metal hydrolysis is not complete if the process exists at all, so that in a closed system with a small electrolyte volume (e. g., just a thin liquid layer), the overall pH increases over time.

8.1.2 Types of corrosion

Uniform corrosion
The most simple form of corrosion is the homogeneous dissolution or conversion of the metal. The anodic and cathodic sites of the local elements are randomly distributed and fluctuate with time. Depending on the metal and environmental conditions, the metal gets dissolved or converted to an oxide layer. The depletion of the metal measured by the corrosion current allows prediction of the corrosion rate CR, expressed, e. g., in $mm\,y^{-1}$ or the non-SI unit milli-inch per year (mpy).

The corrosion rate depends primarily on the metal and environment combination. A corrosion rate of $1\,mm\,y^{-1}$ is typical and acceptable for steel in seawater, while critical parts should show a rate below $0.1\,mm\,y^{-1}$.

If an oxide layer forms during the corrosion, it can passivate the metal and stop further oxidation. Alternatively, oxidation may continue due to the layer's porosity or tendency to peel off. The Pilling–Bedworth ratio (PBR) allows the classification of the different cases by relating the metal molar volume to the molar volume of the formed oxide:

$$\text{PBR} = \frac{M_{ox} \cdot \rho_{Me}}{n_{Me} \cdot M_{Me} \cdot \rho_{ox}} \tag{8.7}$$

M is the molecular weight and ρ the density with index $_{Me}$ denoting the metal, and $_{ox}$ the oxide. n_{Me} is the number of metal atoms in the molecular formula of the oxide. Three cases occur as illustrated in Figure 8.2:

– PBR < 1: The oxide layer is porous and cannot sufficiently protect the metal below. A typical example is magnesium (Mg/MgO: PBR = 0.81).
– 1 < PBR < 2: Full coverage of the metal by the oxide layer results in passivation. Prominent examples are aluminium (Al/Al_2O_3: PBR = 1.28) and titanium (Ti/TiO_2: PBR = 1.73).
– PBR > 2: High PBR causes stress in the oxide layer, which finally leads to chipping off. A typical representative is iron oxidized to iron(III)oxide, which is in its hydrated form major component of rust (Fe/Fe_2O_3: PBR = 2.14).

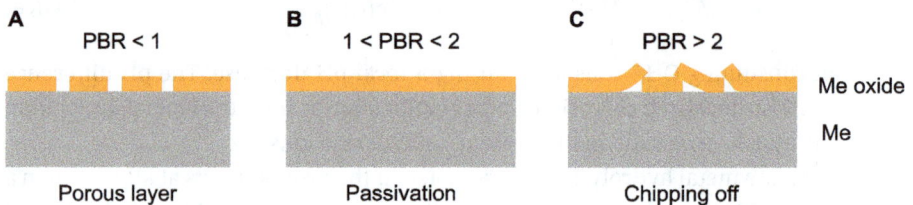

Figure 8.2: Different cases of passivation during uniform corrosion depending on the Pilling–Bedworth ratio (PBR). The oxide layer can be porous (A), lead to passivation (B), or peel off because of internal stress (C).

ℹ Rust is a mixture of different iron oxides formed in the presence of water or humidity. It is a mixture of the hydrated forms of iron(III) oxide $Fe_2O_3 \cdot n\,H_2O$ and of iron(III) oxide-hydroxide $FeO(OH) \cdot n\,H_2O$. Overall appearance is red or orange with a generally nonuniform appearance.

Pitting corrosion

In contrast to uniform corrosion, metal dissolution can occur in localized attacks forming deep pits. Pitting is often the case in the presence of halide anions (stainless steel with chloride ions in seawater or physiological solutions, fluoride as remains from microfabrication).

Once a localized attack of the passivating film started, and the pit overcame a critical size and initial metastability, pitting corrosion has an "autocatalytic nature" (Figure 8.3(A)). Oxygen depletion comes together with acidification because of the metal hydrolysis (equation (8.6)), and the formation of an anodic polarization attracts further halide ions by electromigration. A detailed discussion on localized corrosion and pit growth stability provides the articles series of Frankel, Li, and Scully [100–104].

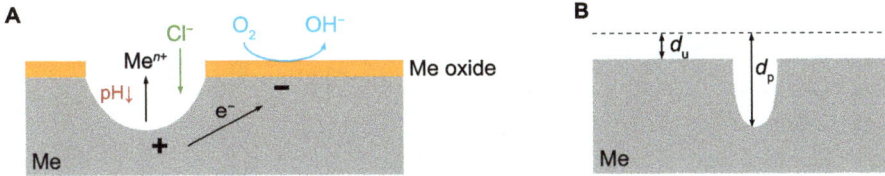

Figure 8.3: Pitting corrosion: localized attack of a passivating metal (A). The *pitting factor* is defined as ratio of the depth of deepest pit d_p to the depth of average penetration d_u (B).

On metals that do not show passivation, both uniform and pitting corrosion can cooccur. The *pitting factor* PF describes the degree of pitting in comparison to uniform corrosion:

$$PF = \frac{d_p}{d_u} \tag{8.8}$$

d_p is the depth of the deepest pit, d_u is the depth of uniform corrosion as illustrated in Figure 8.3(B). Typical pitting factors are in the single-digit range; uniform corrosion would correspond to PF = 1.

Crevice corrosion

Crevice corrosion occurs in confined spaces with different electrolyte compositions, such as the depletion of oxygen. Examples are cracks or gaps inside the connection of two parts. The inhomogeneous chemistry, e. g., lower oxygen content, is the primary driver of this attack. Often crevice corrosion comes along with galvanic corrosion.

Galvanic corrosion

Galvanic corrosion (bimetallic corrosion) occurs when two different metals come in contact. The position in the electrochemical series predicts the outcome: the less noble metal (lower E^0) gets dissolved (Figure 8.4(A)). Table 8.1 summarizes the standard reduction potentials for some metals.

Figure 8.4(B) and (C) illustrates galvanic corrosion with different coatings. Tin (Sn) on iron protects the iron only if it is defect-free. Any scratch causes a local element resulting in the dissolution of iron. In contrast, zink (Zn) shows lower E^0 (Table 8.1), thereby

Table 8.1: Standard reduction potentials (electrochemical series) of some metals.

Reaction	Potential vs. SHE (E^0)
$Au^+ + e^- \rightleftharpoons Au$	1.692 V
$Pt^{2+} + 2\,e^- \rightleftharpoons Pt$	1.188 V
$Ir^{3+} + 3\,e^- \rightleftharpoons Ir$	1.156 V
$Pd^{2+} + 2\,e^- \rightleftharpoons Pd$	0.951 V
$Ag^+ + e^- \rightleftharpoons Ag$	0.799 V
$Ru^{2+} + 2\,e^- \rightleftharpoons Ru$	0.455 V
$Cu^{2+} + 2\,e^- \rightleftharpoons Cu$	0.342 V
$Pb^{2+} + 2\,e^- \rightleftharpoons Pb$	−0.126 V
$Sn^{2+} + 2\,e^- \rightleftharpoons Sn$	−0.138 V
$Ni^{2+} + 2\,e^- \rightleftharpoons Ni$	−0.257 V
$Co^{2+} + 2\,e^- \rightleftharpoons Co$	−0.280 V
$Fe^{2+} + 2\,e^- \rightleftharpoons Fe$	−0.447 V
$Zn^{2+} + 2\,e^- \rightleftharpoons Zn$	−0.762 V
$Cr^{2+} + 2\,e^- \rightleftharpoons Cr$	−0.913 V
$Nb^{3+} + 3\,e^- \rightleftharpoons Nb$	−1.099 V
$Mn^{2+} + 2\,e^- \rightleftharpoons Mn$	−1.185 V
$Ti^{2+} + 2\,e^- \rightleftharpoons Ti$	−1.630 V
$Al^{3+} + 3\,e^- \rightleftharpoons Al$	−1.662 V
$Mg^{2+} + 2\,e^- \rightleftharpoons Mg$	−2.372 V

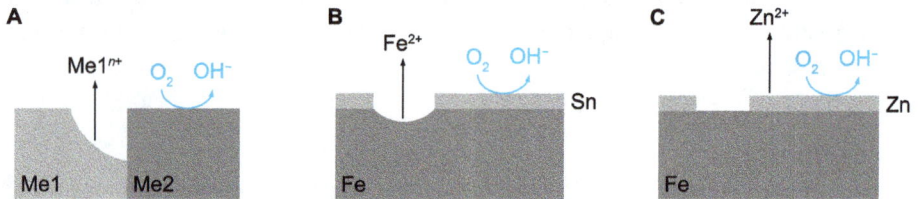

Figure 8.4: Galvanic corrosion: when two metals are in contact, the less noble metal dissolves (A). Coatings with a more noble metal (tin on iron) protect only when they are defect-free (B), while the other way round (zink on iron), the less noble metal gets "sacrificed" when a defect occurs (C).

protecting the iron fully. The zink dissolves and gets "sacrificed" instead of the iron (Figure 8.13).

ℹ️ Corrosion phenomena can be observed in the kitchen as well. In the humorously called "lasagne cell," a stainless steel container with some slightly acidic, salty food such as lasagne covered with an aluminum foil touching both food and container, galvanic corrosion occurs. The food acts as an electrolyte; the less noble metal (aluminum) is the anode and dissolves where it is in touch with the food. That said, you would prefer not to eat the food when this happened.

Waterline corrosion

A specific but highly relevant case is the *waterline corrosion*, a localized attack slightly below the liquid level, e. g., in a tank or on a ship's hull (Figure 8.5). At the top of the liquid, especially in the meniscus, the oxygen supply is higher. Accordingly, the cathode of the local elements gets localized, and OH- ions accumulate. Metal dissolution occurs below, and rust precipitates when the iron ions get hydrolyzed. The localization of the anode where the metal dissolves is, therefore, lower than the liquid level.

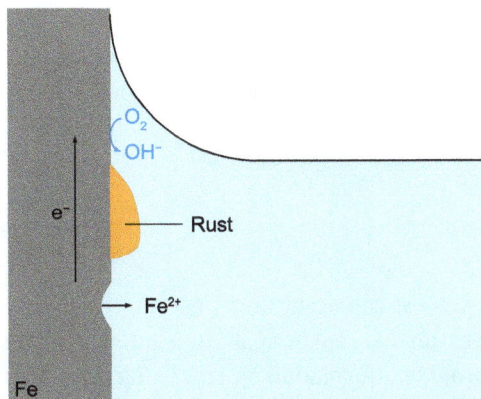

Figure 8.5: Waterline corrosion: localized attack slightly below the liquid level because of different oxygenation.

Seawater has a salt content of around 3.5 %, depending on the location and depth. Its pH varies between pH 7.5 and 8.4. While a NaCl solution is a first approximation, better equivalents should also contain salts such as $MgCl_2$, Na_2SO_4, $CaCl_2$, KCl, $NaHCO_3$, and KBr. The ASTM International suggests a "substitute ocean water" under the standard ASTM D1141-98 for corrosion experiments, considering the various salts, pH 8.2, and optionally heavy metals. However, it is essential to consider the lack of microbiological load when relying on artificial solutions only.

Microbiologically influenced corrosion

Microbiologically influenced corrosion (MIC) is a collective term for local corrosion caused by a microorganism, e. g., bacteria, fungi, or microalgae.

An example is sulfate-reducing bacteria, which can live in the absence of oxygen, e. g., in deep-sea. The bacteria can provide the reduction process to drive the anodic metal dissolution. A possible reaction is the reduction of sulfate ions to H_2S:

$$SO_4{}^{2-} + 10\,H^+ + 8\,e^- \longrightarrow H_2S + 4\,H_2O \qquad (8.9)$$

Additionally, the formed H_2S may attack iron chemically:

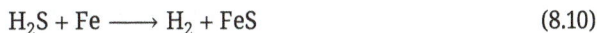

$$H_2S + Fe \longrightarrow H_2 + FeS \qquad (8.10)$$

Enning and Garrelfs provided an excellent overview of the evolution of understanding corrosion by sulfate-reducing bacteria, which may serve as a further introduction to the topic [105].

Especially critical is the formation of biofilms. Bacteria can aggregate together and produce a matrix to embed and protect themselves in a gel-like structure. Biofilms can accept harsh environmental conditions and may show much higher resistance against common disinfectants than individual bacteria. A biofilm can trigger corrosion by secreting substances to change the microenvironment, and some drive the metal dissolution directly, providing the cathode reaction.

The MIC research is of enormous economic interest because much corrosion damage in seawater or pipelines accounts for bacteria involved in the process. However, labs need the ability to grow bacteria and biofilms in addition to electrochemical expertise.

8.1.3 Thermodynamics: Pourbaix diagram

Thermodynamics allow the prediction if a process can occur under certain conditions. Still, it does not say if a reaction runs at a reasonable rate or at all. Here, kinetics come into the arena. Also, the diagrams do not provide information on other effects, such as the likelihood of pitting corrosion. However, knowing these limitations, looking at the data provides the first indication of corrosion or passivation and gives orientation about the species to expect.

The Pourbaix diagrams (Section 2.8.3) arrange this information in a potential-pH-landscape. The charts depend on the used database and which compounds the calculation considers. Usually, in corrosion research, the diagrams assume unit activity for solid and dissolved species with a concentration of $1\,\mu M$. Other values would change the position of the equilibrium lines but not the overall appearance. The slopes indicate the nature of the electrochemical reactions (ratio between H^+/OH^- and electrons in the reaction equation).

Figure 8.6 shows two well-known examples of Pourbaix diagrams, iron and copper, in pure water. Additional processes may occur in the presence of, e. g., halide ions. The book's webpage links software tools and databases to generate Pourbaix diagrams.

8.1.4 Kinetics: Evans diagram, Tafel plot

Kinetic plays a role as soon as the metal is not inert at the given potential-pH regime. Without external polarization, only the bulk pH is fixed while the open-circuit potential E_{corr} adjusts according to the reversible potentials and exchange current densities of the involved processes. Three cases can occur depending on the type of the metal: dissolution, passivation, and the breakthrough of the passivation.

A

B

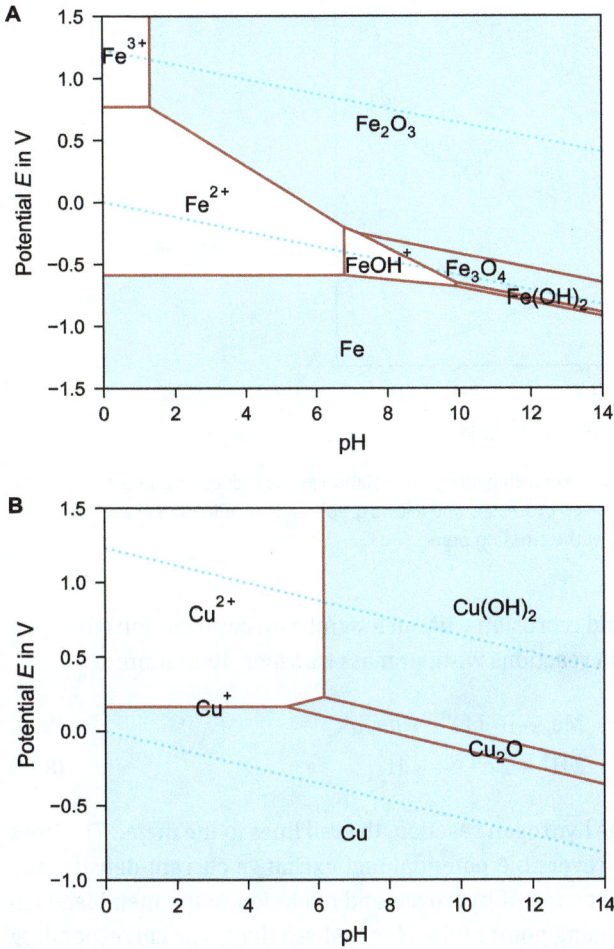

Figure 8.6: Pourbaix diagrams of iron (A) and copper (B) in pure water at 25 °C. The concentration of dissolved species is 1 µM.

Mixed potential

The mixed potential theory considers the influence of the different reactions and focuses on the case of no external current at E_{corr} with possibly a lot of action at the local elements of the metal surface. Here, the sum of all currents (anodic and cathodic current densities i_a and i_c) need to be zero:

$$\sum i_a + \sum i_c = 0 \tag{8.11}$$

In the *Evans diagram*, the Tafel lines of the involved reactions lead to a graphical derivation of E_{corr} and i_{corr} fulfilling the above condition (equation (8.11)). Figure 8.7

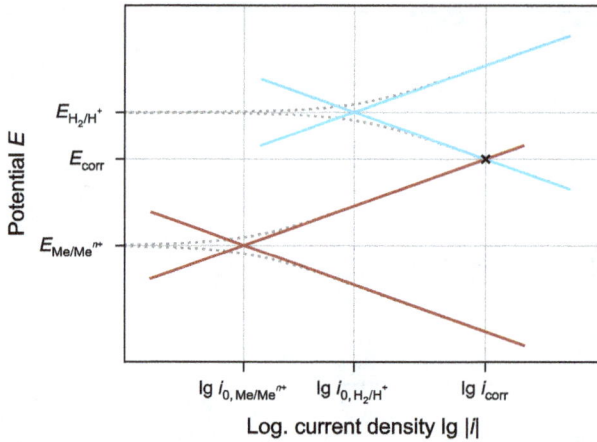

Figure 8.7: Evans diagram of metal corrosion with hydrogen formation as the reduction process. The chart combines the Tafel lines of the two involved processes and allows graphical determination of the corrosion potential E_{corr} and current density i_{corr} at the crossing point.

shows the Evans diagram for acid corrosion without a significant contribution from oxygen reduction involving the two reactions without mass transport limitation:

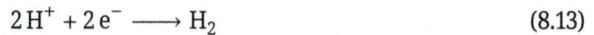

$$\text{Me} \longrightarrow \text{Me}^{n+} + n e^- \tag{8.12}$$

$$2\,\text{H}^+ + 2\,\text{e}^- \longrightarrow \text{H}_2 \tag{8.13}$$

The blue lines correspond to the hydrogen reaction, the red lines to the metal. The lines of each reaction cross at their reversible potential and exchange current density. The combination of the reduction reaction of hydrogen and oxidation of the metal leads to mixed potential E_{corr} at the crossing point of the blue and red lines. The corresponding x-value is the corrosion current density i_{corr}. The Evans diagram allows understanding the influencing factors on corrosion quickly. Inhibition of corrosion (lowering the corrosion current density) occurs when the reversible potential of the metal increases (anodic shift) or the counter process shifts more cathodically.

External polarization

External polarization around the corrosion potential favors either the oxidation or reduction reaction. Mass transport does not play a role in acid corrosion, so the current follows the Butler–Volmer equation (2.32) in or above example. Figure 8.8 shows the resulting voltammogram as a Tafel plot. In corrosion research, often the potentials are on the y-axis analog to the Evans diagram.

A

B

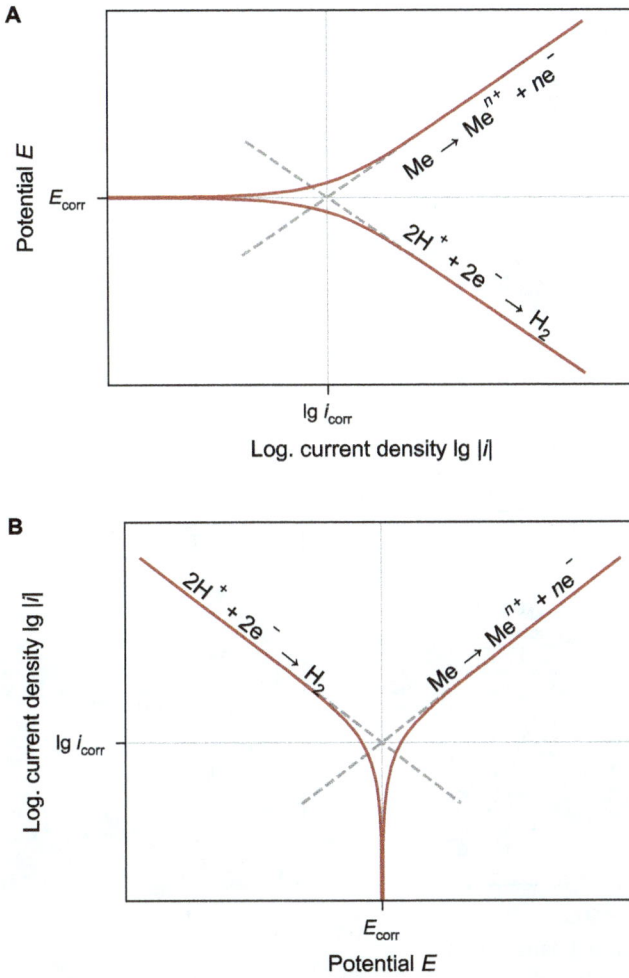

Figure 8.8: Tafel plot of metal corrosion with hydrogen formation as the reduction process. In corrosion research, the Tafel plot often is in the form E vs. $\lg(|i|)$ (A), directly matching the presentation in the Evans diagram. More consistent with cyclic voltammetry is the orientation with $\lg(|i|)$ versus E (B). The latter is also more rational in the understanding that what is on the x-axis is the excitation, and the y-axis shows the result.

Mass transport control

With oxygen reduction as the cathodic process, oxygen diffusion limitation dominates the corrosion. Figure 8.9(A) shows the current responses schematically with their Tafel representation (B). However, in practice, the exact geometry (e. g., open accessible surface vs. crevice) plays an essential role.

A

B

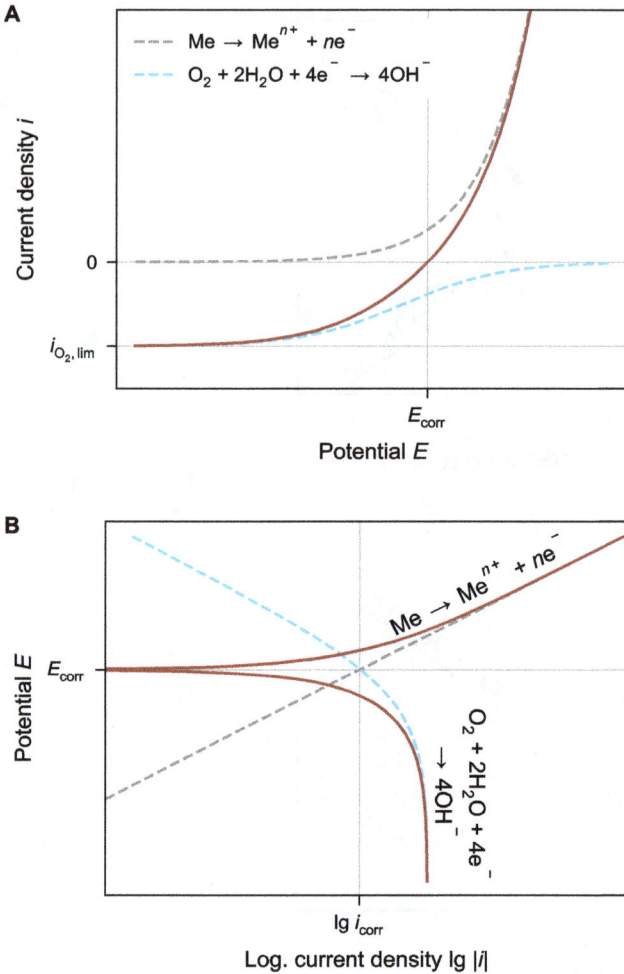

Figure 8.9: Metal corrosion with oxygen reaction as the counter process: the reduction process reaches a mass transport limited current (A). In the Tafel plot (B), the characteristic of a diffusion-controlled process is the deflection of the Tafel slope.

8.1.5 Passivation and transpassivity

The polarization of a metal over a large potential range (up to several volts) results in characteristic passivation curves. Figure 8.10 shows an idealized passivation curve with a clear separation of the three regions:

1. In the *active region* metal dissolution occurs. The current response follows the Butler–Volmer equation.
2. For higher potentials, the current suddenly drops once an oxide film on the metal *passivates* the surface. The associated value is the passivation potential E_p or *Flade*

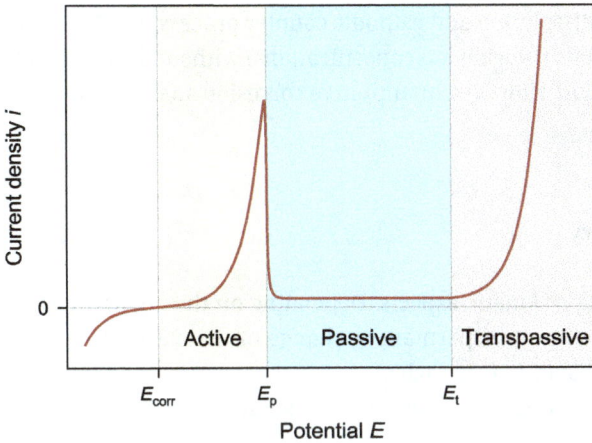

Figure 8.10: Idealized passivation curve with a clear separation between the active corrosion, passivation, and transpassive region once the oxide layer breaks through.

potential.[2] In real samples, there may be a larger transition region. The definition of E_p is inconsisent, according to ISO 8044 the passivation potential characterize the onset of the passivation [99]. For PBR < 1 or PBR > 2, a substantial current is visible in this region. Also in a dense passivation layer pitting can occur and metastable pits may cause flucations in the current.

3. In the *tranpassive region*, the passivation decomposes, or break-through of the layer occurs when molecular oxygen forms. The onset potential is the transpassivation potential E_t.

8.2 Methods to analyze corrosion

Electrochemical corrosion analysis is a quick method compared to the timescales of the derived predictions. Even if a measurement takes hours, it is still short compared to a corrosion rate CR in the units $\mathrm{mm\,y^{-1}}$ (assuming uniform corrosion). Accordingly, one should understand the predictive power more about the correct order of magnitude than a precise three-digit number.

8.2.1 Corrosion potential measurement

The primary parameter to investigate is the open-circuit potential (OCP), as it tells about the position of the corrosion potential E_{corr} of the local elements concerning the re-

---

2 Although the term Flade potential is often used synonymously for the passivation potential obtained by a voltammetric curve, initially, it was the transition potential in potentiometric measurements just before the breakdown of a dissolving passivation layer [106].

versible potentials of the metal dissolution and cathodic counter process. The OCP hints at influencing factors (e. g., various chloride ion concentrations) without disturbing the "natural" mixed potential situation. However, quantitative corrosion analysis is not possible.

8.2.2 Linear sweep voltammetry

Linear ramps allow the studying of kinetic aspects. Depending on the applied voltage range, the situation stays reversible, or a permanent change of the electrode surface occurs. The sample should be polished before each measurement to remove any possible natural oxide or other layer depending on the research question.

Polarization resistance

The polarization resistance or corrosion resistance R_p is the slope of the current-voltage diagram around the corrosion potential. Linearizing the Butler–Volmer equation is only possible for small numbers, e. g., $E_{corr} \pm 10$ mV or even less. A corrosion resistance measurement comprises two steps:

1. Monitoring of the open-circuit potential E_{corr} until it reaches steady-state (e. g., $dE/dt < 1\,\mu V\,s^{-1}$).
2. Polarizing the electrode to $E_{corr} - 10$ mV and slowly (e. g., $0.1\,mV\,s^{-1}$) ramping the potential to $E_{corr} + 10$ mV.

The slope in the current-voltage plot is $1/R_p$. The *Stern–Geary equation*

$$i_{corr} = \frac{1}{R_p} \cdot \frac{b_a \cdot |b_c|}{2.303(b_a + |b_c|)} \cdot \frac{1}{A} \tag{8.14}$$

enables the calculation of the current density (i_{corr}) from R_p and the Tafel slopes (b_a, b_b). The Tafel slopes should be known (Table 8.2) or require a separate Tafel analysis. A is the electrode area.

Table 8.2: Tafel slopes b_a, b_b (equation (2.40)) per decade change in current (density) depending on the number of electrons n, for $\alpha = 0.5$ at 25 °C.

n	Tafel slopes
1	±118 mV
2	±59.2 mV
3	±39.4 mV
4	±29.6 mV

Assuming uniform corrosion, the corrosion current density provides the corrosion rate:

$$CR = \frac{M}{nF\rho} \cdot i_{corr} \tag{8.15}$$

M is the molecular weight, and ρ is the density of the metal. Typical units of CR are $mm\,y^{-1}$ or the non-SI unit milli-inch per year (mpy).

While the polarization resistance measurement seems somewhat limited compared to a Tafel analysis, its simplicity facilitates continuous recording, and it is also applicable with material showing curves in the Tafel plot, which are difficult to interpret reliably. In contrast to measuring the polarization resistance by EIS (Section 8.2.3), the linear sweep voltammetry cannot separate R_p from the uncompensated resistance R_u.

Tafel analysis

The Tafel analysis is the interpretation of linear sweep voltammetry in the Tafel plot. The potential range around the open-circuit potential should be large enough to cover also nonlinear effects but still in the reversible range. Depending on the material system, relevant values are $E_{corr} \pm 100$ mV to ± 250 mV. A Tafel analysis measurement comprises three steps:

1. Monitoring of the open-circuit potential E_{corr} until it reaches steady (e. g., $dE/dt <$ $1\,\mu V\,s^{-1}$). For repetitive measurement on different samples, it might be helpful to check the open-circuit potential of a few electrodes and use a fixed value for all repetition of the experiments to ensure better comparability. Depending on the metal, reaching steady state requires an equilibration time in the hour range after immersion in the electrolyte.
2. Polarizing the electrode $E_{corr} - 100$ mV (or -250 mV) for a particular hold time (e. g., 10 s) to avoid any switching-on effect. This step is optional but avoids a steep decrease at the left branch.
3. Ramping the opential slowly (e. g., $0.1\,mV\,s^{-1}$) to $E_{corr} + 100$ mV (or $+250$ mV).

It is crucial to use the auto current range functionality of the potentiostat, as the measured current spans several orders of magnitude.

The analysis requires presenting the data in a Tafel plot ($lg\,|i|$ vs. E or vice versa). For a system limited by the electron transfer reactions, the resulting chart should resemble Figure 8.11; with mass-transport limitation, the cathodic branch should approach the diagram in Figure 8.9(B). In practice, the obtained Tafel plots are often in between. Sometimes lowering the scanrate can improve linearity. However, it is essential to keep all measurement parameters identically when comparing different materials or the influence of environmental conditions.

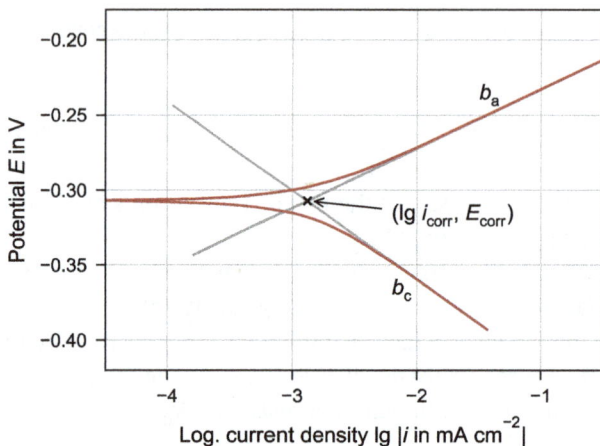

Figure 8.11: Tafel analysis of an ideal system provide E_{corr}, i_{corr}, b_a, and b_c. For a Tafel plot in the orientation $\lg(|i|)$ versus E the slopes are $1/b_a$ and $1/b_c$.

From the Tafel plot (Figure 8.11), the corrosion potential E_{corr}, the corrosion current density i_{corr}, and the Tafel slopes b_a, b_c are available. i_{corr} allows calculating the corrosion rate CR (equation (8.15)) and with b_a and b_c the polarization resistance R_p (equation (8.14)).

Ideally, the crossing of the two linear extrapolations should match the previously measured open-circuit potential. When unambiguous linear regions are available, a difference between the corrosion potential measured before and obtained from the intersection indicates that the system was not in steady state before or the scanrate was not low enough.

Large range linear sweep voltammetry

Large range linear sweep voltammetry leaves the potential range keeping the electrode in a reversible state. Depending on the material, the sample may get destroyed during the measurement or needs extensive polishing to recover the surface. A typical measurement comprises three steps:

1. Monitoring of the open-circuit potential E_{corr} until it reaches steady (e. g., $dE/dt < 1\,\mu V\,s^{-1}$) or using a guess value from previous experiments.
2. Applying a value below the open-circuit potential, e. g., $E_{corr} - 0.5\,V$, keep it for some hold time (e. g., 10 s). Otherwise, with the massive step from E_{corr} to 0.5 V below, an unwanted spike would occur.
3. Ramping the potential slowly (e. g., $0.1\,mV\,s^{-1}$) to $E_{corr} + 2\,V$ or more. The goal is to reach the transpassive region (compare Figure 8.10). If possible, trial measurements on a test sample help to find the optimal potential range. Depending on the metal and demand on the results, a larger scanrate (e. g., $5\,mV\,s^{-1}$) is acceptable, e. g., for quick testing or lab course experiments.

As before, it is crucial to use the auto-current range functionality of the potentiostat, as the measured current spans several orders of magnitude. Depending on the individual preference, the resulting plot can be current density versus potential or vice versa, with the current density shown directly or on a logarithmic axis ($\lg |i|$).

For large range linear sweep voltammetry with cyclic potential ramps it is helpful to compare the end of the passive region in the upward scan and the onset of passivation in the downward scan. A significant hysteresis indicates pitting corrosion instead of early transpassivity.

8.2.3 Electrochemical impedance spectroscopy

Electrochemical impedance spectroscopy (EIS) is a powerful method in corrosion analysis. Typically, EIS complements a Tafel analysis and large range linear sweep voltammetry. EIS is also well suited for long-term monitoring, especially with a multisine approach (7.1.3) or measuring at a single frequency as an indicator for changes in corrosion behavior. A typical EIS measurement comprises three steps:

1. Monitoring of the open-circuit potential E_{corr} until it reaches steady-state (e. g., $dE/dt < 1\,\mu V\,s^{-1}$). Alternatively, one can choose a fixed voltage, e.g, in the passivation region.
2. Applying a DC potential at the measured open-circuit potential or the chosen fixed voltage, keep it for some hold time (e. g., 10 s).
3. Measuring at different frequencies to obtain a spectrum. Typically, measurement of the polarization resistance asks for a lower frequency range, while characterization of a coating requires higher frequencies. Most important is to keep the AC amplitude reasonably small, ensuring the linearity of the system.

Polarization resistance

The metal as the sample, with the DC value of EIS at E_{corr}, allows measuring the polarization resistance. The impedance data for many systems resemble a Randles circuit (Section 7.2.3, Fgure 8.12(A)). The fit results provide the polarization resistance R_p separated from the uncompensated resistance R_u what is superior to the above-discussed methods to measure the polarization resistance by linear scan voltammetry. Also, the double layer capacity is available – a reasonable value indicates a proper setup and appropriate model.

Sometimes a CPE (Figure 7.7) instead of the capacitor improves the match. For systems with mass transport limitation, typically from oxygen reduction, the fit requires a Warburg element in the model to account for the deviation from the hemispherical Nyquist plot for lower frequencies (Figure 8.12(B)).

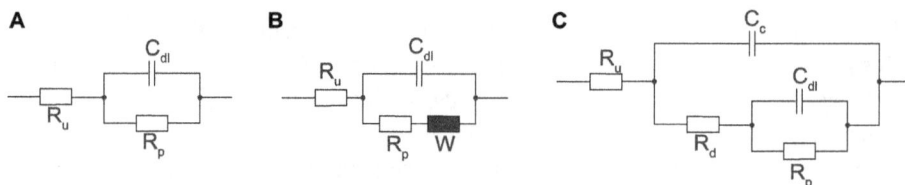

Figure 8.12: Different equivalent circuits relevant for corrosion analysis: measurement of the polarization resistance (A), optionally considering mass transport limitation in the reduction process (B). The model in (C) matches a coating with defects.

Coating

In EIS on samples with coatings, the corrosion occurs only at defects. Apart from those sites, the impedance would be purely capacitive. Here, EIS helps to observe the breakdown of a coating in a long-term experiment or with different substances in the electrolyte or allows analyzing the defects. Figure 8.12(C) shows an equivalent circuit matching such a situation. Ideally, only the coating's capacity (C_c) is present. When pores or similar flaws occur, the corrosion becomes visible as a Randles circuit with the polarization resistance R_p and a resistor R_d representing the ohmic drop in the defect. Measuring the polarization resistance of the bare metal separately in the specific electrolyte allows comparing the value to the polarization resistance of the pores and calculating an average value for the area of all defects together.

8.2.4 Short-circuit current measurement

Measurement of the short-circuit current between two metal samples has its application in the characterization of galvanic corrosion and for electrochemical noise (ECN) analysis (Section 6.3.3). A zero-resistance amperemeter (ZRA) connects the two samples. Depending on the investigation, the two electrodes should have the same area or resemble the situation in the final application while investigating galvanic corrosion.

8.2.5 Critical pitting temperature measurement

The critical pitting temperature (CPT) is when severe pitting sets in at an electrode polarized in the passivity range. This method requires a temperature-controlled electrochemical cell to increase by $1\,\mathrm{K\,min^{-1}}$, e. g., from room temperature to $100\,°\mathrm{C}$. The sample is polarized constantly; a substantial increase in current indicates the onset of pitting, and the respective temperature is the CPT.

The standard ASTM G150-99 defines the CPT when a current density of $0.1\,\mathrm{mA\,cm^{-2}}$ is exceeded and suggests a potential of 700 mV versus SCE in a sodium chloride solution. While the method is common to compare different materials, the overall approach is controversial as it does not consider the specific mechanism, especially concerning the pit growth stability and its temperature dependency.

8.2.6 Combined research methods

Research on corrosion and passivation, particularly considering alloys, benefits a lot from in situ analysis methods. Mass spectroscopy helps to identify corrosion products, e. g., by using flow cells linked to inductively coupled plasma mass spectroscopy (ICP-MS) or inductively coupled plasma atomic emission spectroscopy (ICP-AES), called atomic emission spectroelectrochemistry (AESEC) [107]. Atom probe tomography (APT) allows the study of structure composition due to corrosion in alloys [108]. Various scanning methods (Section 6.2) provide local information, especially relevant to study pitting. Gravimetry by EQCM provides direct data on the dissolution but can additionally benefit from combination with scanning methods [109]. Other approaches combine sensors to access the electrode's microenvironment or to measure the oxygen gas consumption and hydrogen gas evolution using respirometry (Section 6.3.2).

8.3 Methods to prevent corrosion

While the overall goal should be to select corrosion-resistant materials, many applications cannot avoid corroding materials for technical or economic reasons. Methods to prevent corrosion help overcome the materials' limitations by hindering the degradation or reducing the corrosion rate to an acceptable low value.

8.3.1 Cathodic protection

The operation principle of *cathodic protection* (CP) aims at lowering the metal's potential below E_{corr} (rendering the metal the cathode), preventing the dissolution process. Instead, oxygen reduction or hydrogen formation will occur. The latter is critical when the material is vulnerable to hydrogen embrittlement.

Two principles of cathodic protection are possible:
– Active protection by impressed current using an external power supply.
– Passive (galvanic) protection by connecting a less noble material as a sacrificial anode. The chemical energy of the anode drives the protecting current.

Driving a current through the metal to be protected and a second electrode, called *impressed current cathodic protection* (ICCP), lowers the potential of the metal into a safe zone but causes the reduction process (oxygen reduction, hydrogen formation) to occur and accordingly comes along with severe energy consumption. A typical example is an underground pipeline with a graphite anode next to the pipe in the soil.

Galvanic protection by a *sacrificial anode* does not consume electrical power; the anode dissolution provides the energy to keep the metal as the cathode. A classic example is a zink anode to protect metal Me (Figure 8.13). Here, the dissolution of zink prevents the metal from corroding but drives the oxygen reduction at the cathode.

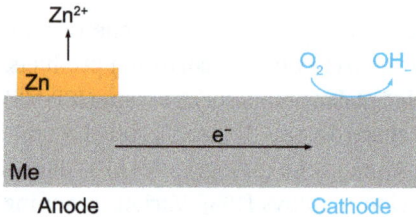

Figure 8.13: Cathodic protection by a sacrificial anode: the zinc anode on the metal Me, e. g., iron, dissolves, lowering the metal's potential below E_{corr} and driving the oxygen reduction instead of corrosion.

ⓘ Cathodic protection with a less noble material is widespread. Ship hulls in seawater use zinc, magnesium, or aluminum alloys as a sacrificial anode. When you are the next time at a harbor, check if you notice the several 10 cm up to meter-long metal bars attached to the ship's metal hull or the rudder blades, e. g., when a boat is on a dry deck for repair. You even do not have to go so far if you have a hot water boiler with a glass-lined steel container. Most likely, a magnesium anode does its job to keep your boiler tight, and you may be aware of it only when a warning signal shows up or once it was dissolved entirely for too long.

8.3.2 Anodic protection

Anodic protection (AP) works actively only. The metal becomes the anode and gets passivated. The prerequisite is that the metal has a pronounced passivation zone. Typically, a three-electrode setup with a potentiostat brings the metal's potential high enough to passivate but prevents transpassive decomposition. The important advantage of low power consumption comes along with the risk of causing severe corrosion when the potential is wrong.

8.3.3 Corrosion inhibitors

A *corrosion inhibitor* (CI) is a chemical added to the fluid, the electrolyte for the corrosion. Depending on the working mechanism, CIs are anodic, cathodic, or mixed inhibitors. Common to all inhibitors is the requirement of continuous addition to the fluid (e. g., oil in a pipeline). While it is an enormous market volume, recent developments focus not only on more economic CIs but consider environmental aspects and biodegradability ("green corrosion inhibitors").

Cathodic inhibitors prevent the oxygen decomposition. Anodic inhibitors poison the metal dissolution or catalyze the passivation reaction. Depending on the corrosion type, other mechanisms play a role. In microbiologically influenced corrosion, biocides act as an anodic inhibitor.

8.3.4 Protective barriers

Instead of the continuous addition of corrosion inhibitors, protective barriers on the metal are the better choice if feasible. Depending on the metal electrodeposited metal coatings (e. g., zinc, nickel, or chromium), the formation of an oxide layer (anodization of aluminum) or phosphating prevents corrosion. Furthermore, organic paints are widespread. A common disadvantage is the risk of local corrosion when a defect occurs, why protective barriers may require combination with, e. g., cathodic protection. Most recent developments include "smart coatings," multilayer systems with self-healing capabilities.

8.4 Platinum electrochemistry

Platinum is one of the most exciting electrode materials from both the electrochemical and the application perspectives. Here, the focus is on a refined look into the surface reactions, enhancing the general overview from the chapter on the classical methods (Section 5.3.3). The charge integration in the hydrogen area of the platinum cyclic voltammogram measures the surface roughness, which also links to deposition methods increasing and controlling surface roughness (Section 9.2.2).

8.4.1 Platinum surface reactions

The different surface reactions of platinum are best understood when discussing a cyclic voltammogram in more detail. From the point of didactics following the reactions along the curve of a CV is more straightforward. Of course, the same processes occur during chronoamperometry stepping to the specific potential regions.

We start our journey with bare platinum marked with the blue point in Figure 8.14:
1. At potentials around 0.8 V, chemisorption[3] of OH^- ions from physisorbed species (see region 9) sets in:

$$Pt + OH^- \longrightarrow Pt\text{--}OH + e^- \tag{8.16}$$

---

3 The notation Pt – OH instead of PtOH hints to the chemisorbed nature of the species.

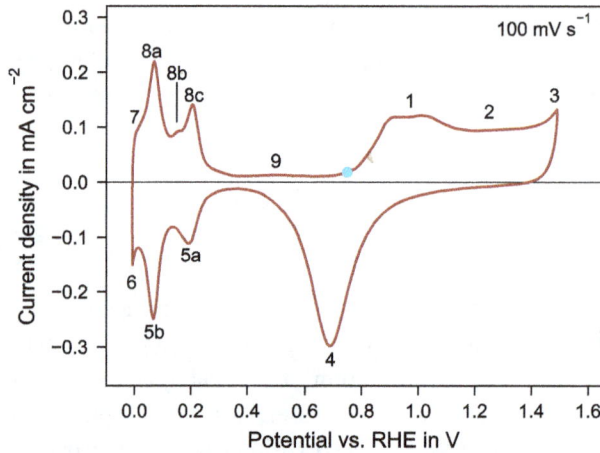

Figure 8.14: Cyclic voltammogram of a platinum electrode in oxygen-free 1 N H_2SO_4. See the text for the discussion of the different surface reactions.

The exact nature of the adsorbed species is under discussion. Jerkiewicz et al. suggested chemisorbed oxygen instead of $Pt-OH$ [110].

2. At higher potentials, the $Pt-OH$ gets oxidized to PtO:

$$Pt-OH \longrightarrow PtO + H^+ + e^- \tag{8.17}$$

The formed PtO layer is in the range of a monolayer. The charge required for a complete monolayer is $440\,\mu C\,cm^{-2}$ [111]. The PtO growth kinetics depends on the potential and how much PtO is there already. Alsabet et al. provided a comprehensive study on the PtO growth considering different hold times at a particular potential [112] with the exact results critically depending on the instrumentation, especially for chronoamperometry as demonstrated by McMath et al. [113]. The formation of PtO_2 does not occur within the water window.

3. The onset of the decomposition of water defines the water window. At the anodic side, molecular oxygen forms:

$$2\,H_2O \longrightarrow O_2 + 4\,H^+ + 4\,e^- \tag{8.18}$$

Within the potential range shown in Figure 8.14, the molecular oxygen is a dissolved gas. At higher potentials, gas bubbles become visible.

4. In the backward scan, the before formed PtO gets reduced:

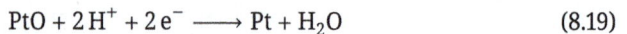

$$PtO + 2\,H^+ + 2\,e^- \longrightarrow Pt + H_2O \tag{8.19}$$

The formation of the PtO within two steps results in an overlap of the onset of the reaction in equation (8.16) with the beginning of the oxide reduction.

5. Protons adsorb at the electrode and form one monolayer:

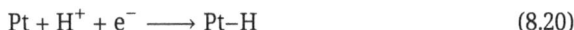

$$Pt + H^+ + e^- \longrightarrow Pt-H \tag{8.20}$$

The mechanism is an underpotential deposition (Section 9.2.4) with two different energy barriers causing two distinct peaks (5a and 5b). For a polycrystalline platinum electrode, the charge required is $210 \ \mu C \ cm^{-2}$ for one monolayer [114] (compare the infobox in Section 8.4.2).

6. The cathodic decomposition of water defines the lower end of the water window:

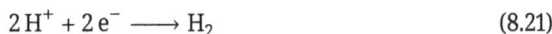

$$2H^+ + 2e^- \longrightarrow H_2 \tag{8.21}$$

Ideally, the onset of the hydrogen evolution is at 0 V versus RHE. Significant deviations from the zero-value with a planar electrode in acidic electrolyte hint at imprecise reference electrodes; for higher surface roughness of the electrode, the water window is broader, and the water decomposition becomes visible at potentials lower than 0 V versus RHE.

7. Here, a peak due to oxidation of H_2 can occur when the lower turning point is more into negative potentials and more H_2 forms at 6:

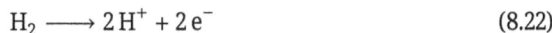

$$H_2 \longrightarrow 2H^+ + 2e^- \tag{8.22}$$

8. The monolayer $Pt-H$ formed in 5 oxidizes, showing very characteristic peaks:

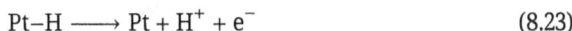

$$Pt-H \longrightarrow Pt + H^+ + e^- \tag{8.23}$$

The precise resolution of three peaks with a tiny 8b in the middle is characteristic of the acidic regime, contamination-free electrolytes, and pure electrode materials. At neutral pH, the peaks 8a and 8c are broader, hiding a possible peak at 8b.

9. In this region, the so-called *double layer region*, no redox process occurs. The only current contribution is from charging the double layer capacity. Additionally, the physisorption of OH^- (or anhydrous oxygen) occurs, which is not directly visible in the cyclic voltammogram. However, in the presence of other redox-active substances (e. g., glucose), the presence of the physisorbed species becomes evident (compare the discussion of peak 2 of Figure 5.25).

Adsorption is the adhesion of an *adsorbate* (atom, ion, or molecule) to a surface. Adsorption plays a role in many electrochemical phenomena, like forming the double layer, catalysis of redox reactions, or electrodeposition. Adsorption is an umbrella term for different mechanisms: in *physisorption*, the adsorbate links to the surface due to Van der Waals forces. The electronic structure of the adsorbate and the solid are unaffected. In contrast, in *chemisorption*, a chemical reaction between the adsorbate and the surface leads to a chemical bond and, accordingly, a change in the electronic structure of the involved partners. The major differences are much higher binding energies in chemisorption (in the range of 1 to 10 eV) than physisorption (10 to 100 meV) and the specificity due to the chemical reaction in contrast to the more unspecific physisorption. In material science, also *segregation*, the partitioning effect within solids causing local enrichment, counts as an adsorption process.

Overall the platinum cyclic voltammogram has a very distinct shape. With some experience, minor variations allow identification of instrumentation issues (compare Figure 3.7) or contaminations in the electrolyte or electrode material [115, 116]. H_2SO_4 solutions are still the defacto standard electrolyte for the acidic range. Many other acids have disadvantages, e. g., HCl because of halide ions or HF because of its toxicity and incompatibility with glass. $HClO_4$ would be an alternative but with substantially higher costs.

EQCM measurement

Seeing a platinum cyclic voltammogram for the first time and learning about the world of the surface reactions might feel strange. However, once with some practical experience, the features of the curve come alive and let you quickly identify clean electrode metal and electrolytes. A close link to an intuitive understanding provides the observation of mass change while the surface reactions occur.

Figure 8.15 shows a cyclic voltammogram in an electrochemical quartz-crystal microbalance (EQCM) measurement cell (Section 6.3.1). In the upward scan, at potentials above around 0.8 V, the mass steadily increases while Pt-OH adsorbs and PtO forms. Upon reversal, the EQCM signal remains roughly constant until the PtO reduction sets in below 0.8 V. The monolayer adsorption of protons is not visible in this curve as other contributions from physisorbed species dominate.

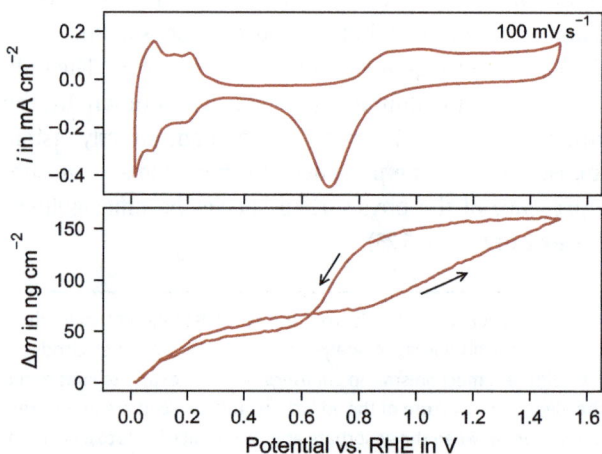

Figure 8.15: Cyclic voltammogram and EQCM signal of a platinum electrode in air-saturated 1 N H_2SO_4. The formation and reduction of platinum oxide dominate the mass signal.

8.4.2 Electrode roughness

The electrochemically active surface area A_{active} is an essential characteristic of electrodes, especially with micro- or nanostructures. Platinum provides an inherent measurement method by tracking the charge needed in the formation of the hydrogen monolayer. Knowing the electrochemically active surface area, the roughness factor Rf is the ration of the active to the geometrically (projected) electrode area:

$$Rf = \frac{A_{active}}{A_{geom}} \tag{8.24}$$

In contrast to optical or mechanical profilometry, the electrochemical method considers only active sites and better suits the demands when discussing, e. g., catalysis. The hydrogen adsorption and desorption processes are fast, resulting in a complete monolayer over a wide range of scanrates.

Figure 8.16 shows the principle to estimate the roughness factor from a cyclic voltammetry curve. Both the hydrogen adsorption and desorption process allow estimation of q_H, the charge required for one monolayer. However, for cyclic voltammograms showing no clear separation between the PtO reduction and the Pt–H formation, the approach with the Pt–H desorption as indicated in the figure provides more reliable results.

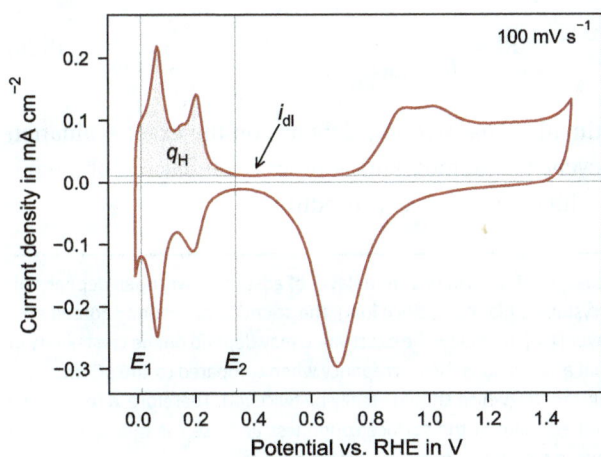

Figure 8.16: Cyclic voltammogram of a platinum thin-film electrode in oxygen-free 1 N H_2SO_4: integration of the hydrogen desorption area allows estimating the surface roughness. In this curve, the integration leads to $q_H = 261\ \mu C\ cm^{-2}$ corresponding to Rf = 1.2, a typical value for a thin-film electrode.

Prerequisites for meaningful data evaluation are high-quality cyclic voltammograms in an oxygen-free electrolyte. H_2SO_4 is the best choice to ensure comparability, but the method provides similar results in other electrolytes, e. g., PBS at neutral pH. Typical data evaluation comprises three steps:

1. The charging of the double layer capacity will add to the charge from the proton desorption when simply integrating the relevant part of the CV. Therefore, the double layer current density i_{dl} as visible in the double layer region is used as the lower current density boundary of the area. The assumption is that the double-layer capacity is the same in both the double layer region and hydrogen desorption area.
2. The onset of the hydrogen evolution E_1 is assumingly at the turning point in the downward scan between the hydrogen adsorption and H_2 formation. This procedure, instead of relying on zero-values in the RHE scale, ensures reliable integration boundaries also with high roughness.
3. Integration of the red area marked in the CV from E_1 to the beginning of the double layer region (end of proton desorption) at a potential E_2 and division by the scanrate v provides the measured charge per area:

$$q_{H} = \frac{1}{v} \cdot \int_{E_1}^{E_2} (i - i_{dl})\mathrm{d}E \tag{8.25}$$

Division of the measured charge by the reference value $q_{H,planar} = 210 \ \mu C \ cm^{-2}$ leads to the electrode roughness:

$$Rf = \frac{q_{H}}{q_{H,planar}} \tag{8.26}$$

Please note that the Rf is an estimation that strongly depends on the exact evaluation method (and the quality of the cyclic voltammogram). A precise comparison of the numeric value is meaningful only with the very same procedure.

---

i On single crystal electrodes, the charge required to form one monolayer of adsorbed hydrogen depends on the crystal orientation. In case of polycrystalline platinum, since long, the scientific community agreed on a value $q_{H} = 210 \ \mu C \ cm^{-2}$ for one monolayer [114]. In reality, the exact value may depend on the crystallinity of the polycrystalline electrode surface what also explains the discrepancy when compared to 440 $\mu C \ cm^{-2}$, the generally accepted value for one monolayer PtO (which should ideally be twice q_{H}). Therefore, a reasonable approach is to use q_{H} only for a general estimate of the surface roughness and use the ratio q/q_{H} when discussing the charge of different regions in the CV of a polycrystalline electrode.

An alternative procedure is the measurement of the active surface area by *CO stripping voltammetry*. The CO deposition forming an adsorbed monolayer occurs at low electrode potential with the electrolyte purged with CO gas. In a subsequent linear scan, analog to the classical anodic stripping voltammetry (Section 5.3.9), a peak shows up, providing the charge of the adsorbed monolayer.

8.4.3 Degradation of platinum

The Pourbaix diagram of platinum (Figure 8.17) suggests stability within the entire water window. However, potential cycling may cause degradation resulting in material loss and, accordingly, an imbalance in the cyclic voltammograms overall charge in the range of 10 $\mu C\,cm^{-2}$ [117]. The platinum dissolves when the cycling causes a substantial change of the surface [118, 119]. Also, the topology plays a role, and Cherevko et al. provides an overview comparing bulk and nanoscale platinum [120].

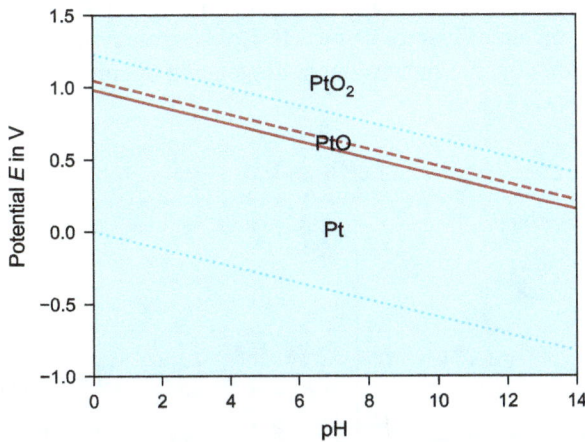

Figure 8.17: Pourbaix diagrams of platinum suggesting stability over the whole range. However, platinum can degrade during potential cycling, which a Pourbaix diagram does not cover.

Additionally, in the presence of chloride ions, platinum complexation occurs, which notably plays a role at lower pH:

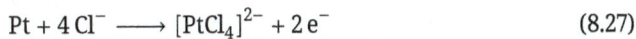

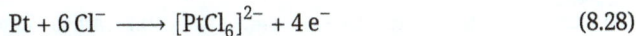

$$Pt + 4\,Cl^- \longrightarrow [PtCl_4]^{2-} + 2\,e^- \tag{8.27}$$
$$Pt + 6\,Cl^- \longrightarrow [PtCl_6]^{2-} + 4\,e^- \tag{8.28}$$

Complexation in acidic electrolyte is typically visible in the platinum cyclic voltammogram as an increase of the current before the onset of the oxygen evolution. Besides dissolution of the platinum complexes, chloride ions may further interfere with the surface reaction causing platinum degradation [121].

Platinum degradation plays a crucial role also in biomedical applications around pH 7. In brain stimulation and for cochlear implants, electrodes see cyclic polarization over many years. With explants of cochlear implants applied for years in humans, inconsistent findings occurred. In some cases, electrode degradation seems to cause a loss of electrode functionality. Shepherd et al. showed in animal models that platinum electrodes in the cochlear exposed to stimulation with high charge densities lead to the dissolution of platinum but did not observe functional changes [122]. The exact degradation mechanism in the human body and possible cofactors are still unclear and topics of current research. Doering et al. developed a framework for automated assessment of electrode degradation in vitro focusing on thin-film electrodes in neutral pH electrolytes [123].

8.4.4 Volcano plots

There are several rationales to explain the outstanding catalytic properties of platinum. Well known is Trasatti's volcano plot (Figure 8.18) describing the hydrogen formation reaction for various metals based on experimental data [124]. The diagram relates the exchange current density to the binding energy between hydrogen and the metal. The Sabatier principle explains the "volcano shape."

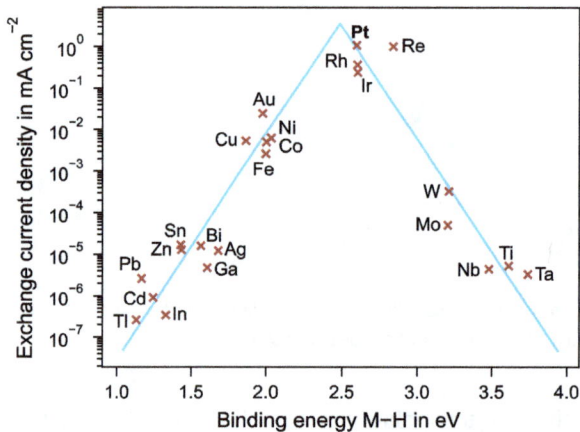

Figure 8.18: Volcano plot for the hydrogen formation reaction at various metals M. Data based on [124].

The Sabatier principle in heterogeneous catalysis relates the reaction rate to the binding energy between substrate and catalyst. There is an optimal binding energy leading to the highest reaction rate. For a substrate-catalyst combination with lower binding energy, the substrate does not reliably bind to the catalyst and cannot react. On the other side, too strong interaction hinders the product from dissociating. Presenting the situation in a diagram of reaction rate versus binding energy looks like an inverse parable or triangle, like a "volcano."

Today, the message of Trasatti's volcano plot is debated controversially, especially in the context of novel insights from DFT calculations [125, 126].

Nørskov et al. describe another arrangement leading to a "volcano shape" with platinum at the top for the oxygen reduction reaction [127].

9 Selected aspects: microfabrication and nanotechnology

Electrochemical methods play an essential role in the deposition of materials but also allows for subtractive techniques. This chapter focuses on examples in which electrochemical processes play a role in microfabrication and nanotechnology. Microsystems and nanosystems using electrochemical principles for their application are in Chapter 10.

The electroless and electrodeposition of material follow the same principles for conventional applications and micro-/nanostructures. Many electroplated coatings on macroscopic components are in the micrometer range. Most differences are the need for homogeneity over substantial surface areas versus the ability to coat evenly in narrow structures asking for different cell setups and electrode formulations. Nearly all chip fabrication for microelectronics and microsystems contain an electrodeposition step, while subtractive techniques are less common though nevertheless not less powerful.

Electroless and electrodeposition play a role in the wiring of classical microelectronics but are more prominent with microsystems and nanotechnology. In contrast to "dry" plasma processes in a cleanroom, the "wet" electrochemical methods offer distinct features. Apart from comparable low equipment costs and in situ control, often, maskless technologies are possible without the bias for planar substrates, as is usually the case with classical microfabrication technologies. Also, topology plays a minor role in electrochemical methods, while standard cleanroom processes require flat substrates. Connecting electrode structures allows further deposition or modification without masking while ensuring perfect alignment of the deposit. A great introduction to classical microfabrication technologies considering the electrochemical methods too are the textbooks of Madou [128, 129] or Menz et al. [130].

The last section in this chapter provides a brief overview of different nanoelectrodes with a strong focus on carbon nanoelectrodes.

Especially with electroless and electroplating baths, the warning at the beginning of Chapter 4 is essential. Many metal salts are highly toxic and often considered cancerogenic, mutagenic, and reproductive toxicants requiring careful handling and appropriate protection, especially when manipulating the salts in the dry state. Plating baths containing cyanide ions (CN^-) require very cautious handling. Additionally to the cyanide's toxicity itself, deadly fumes can evolve if the cyanide comes in contact with acids. If you are unfamiliar with chemistry and lab work, spend some time on a proper introduction from specialists.

9.1 Electroless metal deposition

Electroless deposition (electroless plating, autocatalytic plating) is the reverse case of corrosion understood as a local element with a reaction from the metal and a counter process (compare Figure 8.1). In corrosion, the metal dissolution happens at the anode

https://doi.org/10.1515/9783111488844-009

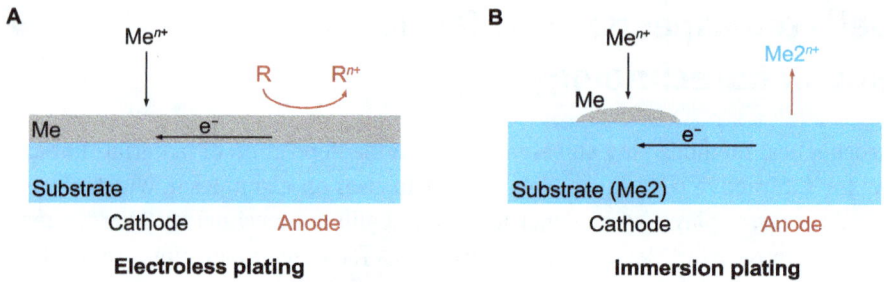

Figure 9.1: Electroless plating (A): metal (Me) deposits on a substrate with the help of a reducing agent (R). The deposited metal catalyzes the deposition reaction (autocatalytic plating). Immersion plating (B): metal (Me) deposits on a substrate with the metallic substrate (Me2) acting as the reducing agent. The process is self-limiting.

site with a cathodic counter process. For electroless deposition, the counter process is an oxidation reaction driving the deposition at cathode sites (Figure 9.1(A)).

9.1.1 Nickel-phosphorus plating

A typical electroless deposition is nickel-phosphorus plating. The reduction process leads to the metal deposition:

$$Ni^{2+} + 2\,e^- \longrightarrow Ni \tag{9.1}$$

The already deposited material catalyzes the cathodic reaction, therefore, the synonym autocatalytic plating. The anodic counter process is

$$H_2PO_2^- + H_2O \longrightarrow HPO_3^- + 3\,H^+ + 2\,e^- \tag{9.2}$$

with hypophosphite ($H_2PO_2^-$) as the reducing agent. The nickel source is nickel chloride ($NiCl_2$) or nickel sulfate ($NiSO_4$) in a concentration around 0.1 M. The reducing agent is typically sodium hypophosphite (NaH_2PO_2) in a concentration of the same order of magnitude. The ratio between the nickel salt and the reducing agent is a trade-off between fast deposition and bath stability. Reasonable plating rates of 10 to 20 $\mu m\,h^{-1}$ require a bath temperature of around 90 °C.

Typical plating baths contain other ingredients to optimize the deposition and suppress parasitic reactions or bath decomposition [131]:
-   *Complexants* act as gelating agents to stabilize the nickel ions and prevent precipitation. Examples are glycolic, lactic, malic, or citric acid.
-   Buffer substances control the pH of the deposition solution. The autocatalytic reaction benefits from a high pH and the anodic counter process from low pH, asking for tight control. Also, the phosphorus content in the deposit depends on the pH.

- The *stabilizer* maintains the bath composition by poisoning decomposition reactions. A typical stabilizer is Pb^{2+} ions in trace amounts or combination with thiourea.
- *Exaltants* increase the deposition rate, e. g., CN^- ions, by accelerating the anodic reaction.

Depending on the bath composition, the deposited layer has a different phosphorus content. Low phosphorus coatings (up to 5 wt% P) show a crystalline microstructure providing well solderable layers. High phosphorus coatings (more than 10 wt% P) are amorphous layers showing excellent corrosion resistance. Medium phosphorus coatings have a mixed crystalline and amorphous structure, leading to high tensile strength.

The substrate material for the electroless plating can be conductive or non-conductive. Conductive metal substrates, which are more electropositive than nickel, can catalyze the deposition reaction without further treatment. For other conductive substrates, a seed layer of nickel formed by electrodeposition starts the electroless deposition. Nonconductive substrates require surface activation. A typical procedure is the formation of palladium seeds from the dipping of the substrate in an HCl solution with $PdCl_2$.

Many coatings on technical and household products rely on nickel-phosphorus plating for surface protection. Typical layer thicknesses are in the range of several 10 μm. In microfabrication, electroless nickel-phosphorus electrically connects different layers or thickens the metal layer at a bond pad to enable, e. g., wire-bonding.

The silver mirror reaction is another example of electroless deposition. This method was applied to prepare mirrors for optical applications for a long time. The Tollens' reagent ($[Ag(NH_3)_2]OH$ in NaOH solution) helps in organic analysis to distinguish ketones and aldehydes. In the presence of aldehyde, the metallic silver indicates the positive test result. In silver mirroring, glucose is a typical reducing agent causing the release of metallic silver. The Tollens' reagent requires fresh preparation from $AgNO_3$, NaOH, and NH_3. Warning: wrong composition or aged solutions can result in silver nitride (Ag_3N), which is explosive!

9.1.2 Other electroless plating processes

Similar processes exist, e. g., for electroless nickel-boron coating with sodium borohydride ($NaBH_4$) instead of the sodium hypophosphite. Copper is another metal commonly deposited electroless:

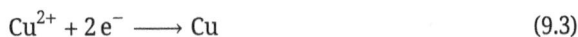

$$Cu^{2+} + 2\,e^- \longrightarrow Cu \tag{9.3}$$

The anodic counter reaction classically relies on formaldehyde (HCOH):

$$2\,HCOH + OH^- \longrightarrow 2\,HCOO^- + 2\,H_2O + H_2 + e^- \tag{9.4}$$

Also, formaldehyde-free electroless plating baths are possible with, e. g., glyoxylic acid $(C_2H_2O_3)$ as reducing agent.

Münch compiled in an excellent review the state-of-art of electroless plating for nanotechnology [132].

9.1.3 Immersion plating

In immersion plating, no autocatalysis occurs, and the bath contains no reducing agent. Instead, the metallic substrate material drives the reaction, as illustrated in Figure 9.1(B). The reaction is a replacement reaction and self-limiting once a dense layer of the depositing metal covers the substrate.

Typical examples are immersion plating silver onto copper or gold onto nickel to improve solderability. The latter is also well suited to coat pads for wire bonding.

The overall reaction in immersion plating of gold onto nickel is a simple replacement reaction:

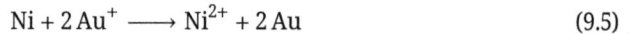

$$Ni + 2\,Au^+ \longrightarrow Ni^{2+} + 2\,Au \tag{9.5}$$

The immersion plating bath uses $KAu(CN)_2$ salt as the source of the gold ions in an aqueous solution of ethylenediaminetetraacetic acid (EDTA) as the complexing agent for the nickel ions and citric acid.

The Electroless Nickel Immersion Gold (ENIG) process combines electroless nickel-phosphorus deposition followed by immersion plating of gold. Typical nickel-phosphorus layer thickness is 5 µm with a gold layer up around 50 to 100 nm. A further improvement is the Electroless Nickel Electroless Palladium Immersion Gold (ENEPIG) plating with an electroless deposited palladium layer between the nickel and gold. Both processes are common in printed-circuit board (PCB) fabrication.

Another common example is copper on steel. A typical recipe contains $CuSO_4$ as the copper ion source in an aqueous solution of sulfuric and optionally tartaric acid.

9.2 Electrodeposition

Electrodeposition refers to all deposition processes comprising an external voltage or current to drive the reaction. The electroplating of metals is most common among the different techniques and is therefore often synonymous with electrodeposition.

In contrast to decorative or corrosion protection applications, in microtechnology, electrodeposition typically requires localized deposition. Two strategies are possible: when the chip layout defines electrodes as opening in a passivation layer on metal lines, the electrode opening is the plating area (Figure 9.2(A)). Using a masking layer is the alternative when the deposited structures should be smaller than the conducting area

A

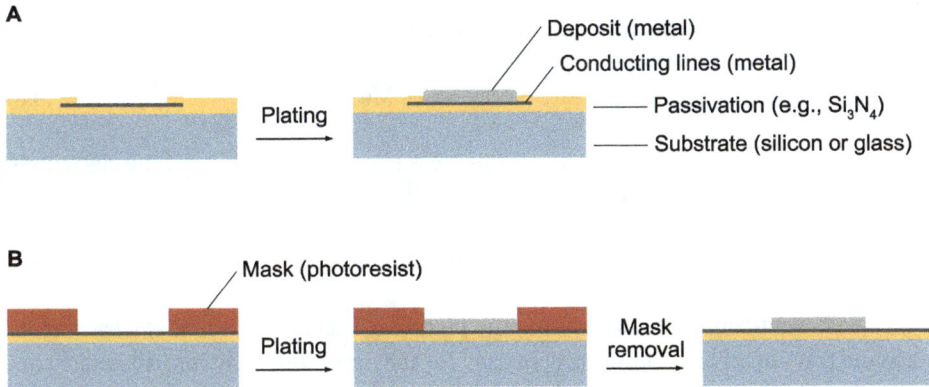

Deposit (metal)

Conducting lines (metal)

Plating

Passivation (e.g., Si_3N_4)

Substrate (silicon or glass)

B

Mask (photoresist)

Plating

Mask removal

Figure 9.2: Electrodeposition in microfabrication: metal plating in openings of the passivation layer (A); the conducting lines should be accessible outside the plating bath to connect the current source. Alternatively or in combination, a masking layer (photoresist) defines the areas of electrodeposition (B). The aspect ratio of the drawing is not to scale.

of the substrate (Figure 9.2(B)). The masking layer can be a photoresist removed after the electrodeposition.

For electrodeposition on wafer-level, a combination of both approaches is often the best choice: electrode openings on individual chips define the deposition area, masking material covers structures like connection pads. All electrodes that should see electrode-position must be connected with connection lines separated when dicing the wafer. An alternative to the liquid photoresist is a dry-film resists providing more design freedom with multilayer masking and more complex structures [133].

It is essential to check the stability of the masking layer in the electroplating bath; e. g., positive photoresists based on novolak resin and diazonaphthoquinone ("AZ photoresists") are critical in harshly alkaline elec-trolytes. Negative photoresists involving cross-linking may be a more stable alternative. The resist profile matter, hanging edges may cause local bath depletion and unreproducible deposition. Also, good adhesion is essential to avoid underplating. Using ultrasonication, generally preferred to remove gas bubbles from edges and achieve homogenous deposition, may conflict with the adhesion of the resits.

9.2.1 Electroplating of metals

Electroplating of metals is the deposition of material from an electrolyte driven by ap-plied voltage or current. Using a current allows simple control of the deposited mass (according to Faraday's law, equation (2.3)) and works fine with a two-electrode arrange-ment. Voltage-controlled deposition may work in a two-electrode cell, too, but often re-quires a reference electrode and a potentiostat. The advantage of using a voltage is the more accurate control of which processes will occur, e. g., to avoid gas evolution.

The deposition occurs at the cathode, inversely to the metal dissolution in corrosion:

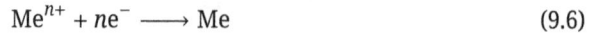

$$Me^{n+} + ne^- \longrightarrow Me \tag{9.6}$$

Using the reverse reaction as the counter process allows replenishing of the electrolyte by dissolving from an anode of the same material as deposited:

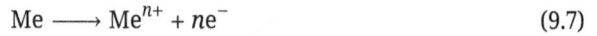

$$Me \longrightarrow Me^{n+} + ne^- \tag{9.7}$$

In a current-controlled deposition, the *plating efficiency* or *current efficiency* measures the fraction of the overall current used for the deposition reaction. Adequate control of the plating bath conditions and deposition parameters yields typical plating efficiencies higher than 95 %. When the bath formulation requires the anode to dissolve, one needs to consider both anodic current efficiency (how well the counter replenishes the metal ions in the electrolyte) and cathodic current efficiency (the plating efficiency). Typically, the anodic current efficiency is more critical.

Nickel plating

Classical nickel plating often uses the "Watts bath," going back to the works of Oliver Patterson Watts in the 1920s (Table 9.1). The bath operates at around pH 4 at elevated temperatures (e. g., 50 °C). Nickel ions get reduced in a two-electron process and deposits onto the cathode:

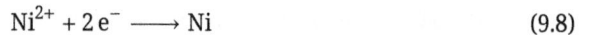

$$Ni^{2+} + 2e^- \longrightarrow Ni \tag{9.8}$$

Typical current densities are in the range of several $10\,\mathrm{mA\,cm^{-2}}$, resulting in deposition rates of around $12\,\mathrm{\mu m\,h^{-1}}$ per $10\,\mathrm{mA\,cm^{-2}}$ depending on the plating efficiency. Apart from the deposition rate, the current density affects the mechanical properties of the deposits, such as internal stress and hardness. The chloride ions from the $NiCl_2$ in the Watts bath facilitate the dissolution of the anode.

Table 9.1: Watts bath for nickel plating, the nickel content is between 1 and 1.5 M (60 to 90 g l^{-1}). Typical operation conditions: pH 3.5 to 4.5, temperature 40 to 60 °C, and current density 20 to 70 mA cm^{-2} [134].

Substance	Content	Concentration
$NiSO_4 \cdot 6\,H_2O$	240 to 300 g l^{-1}	0.91 to 1.14 M
$NiCl_2 \cdot 6\,H_2O$	30 to 90 g l^{-1}	0.13 to 0.38 M
H_3BO_3	30 to 40 g l^{-1}	0.49 to 0.73 M

Other plating baths contain nickel sulfamate instead of sulfate (Table 9.2). For decorative applications, commercial bath compositions contain several additives to control,

Table 9.2: Sulfamate bath for nickel plating, the nickel content is between 0.9 and 1.5 M (55 to 90 g l^{-1}). Typical operation conditions: pH 3.5 to 4.5, temperature 40 to 60 °C, and current density 20 to 150 mA cm^{-2}. A chloride-free composition can be a good choice for microstructures [134].

Substance	Content	Concentration
$Ni(NH_2SO_3)_2 \cdot 4\,H_2O$	300 to 450 g l^{-1}	0.93 to 1.39 M
$NiCl_2 \cdot 6\,H_2O$	0 to 30 g l^{-1}	0 to 0.13 M
H_3BO_3	30 g l^{-1}	0.49 M

e. g., the layer's brightness. Common to all plating bath formulations is the high salt content compared to other electrolytes as discussed in Section 4.5.1. Apart from providing enough metal ions, the salt ensures sufficient conductivity to achieve a uniform coating.

In microtechnology, electrolyte formulations with nickel sulfamate as the only nickel source and high nickel content (up to 90 g l^{-1}) are optimal for high aspect ratio structures [130]. Skipping nickel chloride reduces film stress of the deposit at the cost of less effective anode dissolution. Additionally, an anionic wetting agent helps filling narrow trenches.

The anode material (the "nickel target") needs to be pure metal to avoid contamination of the bath leading to impurities in the deposits. While less critical for decorative applications, in technology, especially in microfabrication, the quality of the anode material is crucial. Also, the placement of the anode relative to the substrate onto which the metal gets deposited is essential. The current paths from the anode to the substrate should be homogeneous. If a substantial distance from the substrate is not possible, several anodes could improve the deposition. The Nickel Institute, a global association of primary nickel producers, provides helpful information on practical aspects of nickel plating [134].

Throwing power is a term expressing the uniformity of the coating. Two cases are distinguished: micro- and macrothrowing power. Microthrowing power is the ability to fill small structures with high aspect ratio such as trenches, holes, or crevices. Macrothrowing power indicates the even coating of large surfaces. High microthrowing power is also essential for macroscopic deposition to obtain defect-free layers.

Copper plating

Since the beginning of the 19th century, acidic copper plating baths were in use. Most simple acidic copper solutions contain copper sulfate as the metal source and sulfuric acid (Table 9.3).

At the cathode, copper deposits following a two-electron process (in details discussed as a sequence of two one-electron reactions):

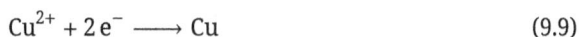

$$Cu^{2+} + 2\,e^- \longrightarrow Cu \qquad (9.9)$$

Table 9.3: Classical acid copper bath, the addition of Cl$^-$ ions from HCl or NaCl up to 1 mM helps dissolve the anode and maintain the bath composition.

Substance	Content	Concentration
$CuSO_4 \cdot 5 H_2O$	250 g l^{-1}	1 M
H_2SO_4	50 g l^{-1}	0.5 M

Typically, the anode is copper, providing new metal ions to keep the bath composition stable. Also, inert counter electrodes are possible if the copper ion concentration is maintained independently. Typical current densities are in the range of 10 to 50 mA cm^{-2}.

Other plating baths are alkaline (both cyanide and noncyanide systems are available) or based on pyrophosphate or fluoborate solutions. However, the acid copper bath is the far most commonly used electrolyte.

Damascene process

Copper interconnects in integrated circuits were an essential driver of what we today see as state of the art in microprocessors. Before, the wiring consisted of aluminum showing lower conductivity and higher electromigration. In 1997, IBM introduced a CMOS chip with copper interconnects. Therefore, the chip manufacturer developed a microfabrication process replacing the vapor deposition of aluminum with an electroplating method [135]. They coined the name "damascene copper electroplating," assumingly in acknowledging the great work of the creators of the Damascus steel, a comparable enabling technology for the medieval production of swords.

The damascene process allows filling trenches or holes (vias) in a passivation layer (e. g., SiO2 or a low-κ dielectric) to form the interconnects (Figure 9.3(A)). Before the plating a seed-layer formed by physical vapor deposition (PVD) covers all areas, including the sidewalls of the structures. The trenches and holes get filled with copper in the plating step, and a top layer of metal that covers everything forms. Finally, chemical mechanical polishing (CMP) planarizes the surface and interrupts the short-cuts caused by the continuous top layer.

The "dual-damascene" process further improves the formation of interconnects. Here two passivation layers are plated at once. The trenches and holes can span only the upper or both layers [136].

A critical aspect when filling narrow gaps or vias is the ability to deposit the metal without voids. Conformal and especially subconformal coating (Figure 9.3(B) and (C)) risk void formation in the deposit. Preferably, the plating bath and deposition methods favor superconformal coating, referred to as "superfilling" [136], which means filling the structure from the bottom (Figure 9.3(D)). Superfilling is possible even with the standard $CuSO_4/H_2SO_4$ bath by appropriate additives [137, 138].

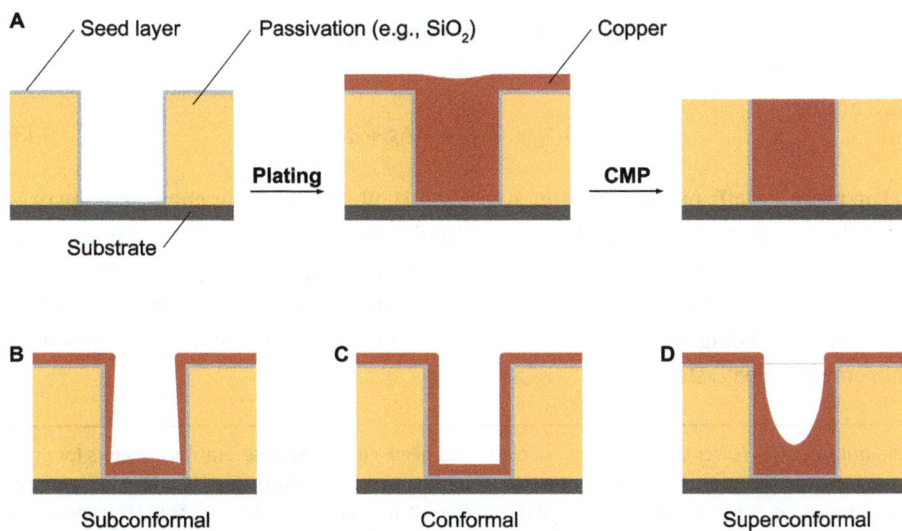

Figure 9.3: Damascene copper electroplating to form interconnects in integrated circuits (A). Subconformal (B) or conformal coating (C) may lead to void formation. Especially for high-aspect ratio structures, the plating bath should enable superconformal deposition (D).

■□□ Task 9.1: Assume a nickel, a silver, and a copper plating bath. For each metal, you run a deposition for 10 minutes with a current density of $10\,\text{mA cm}^{-2}$ on a separate sample. What is the height of the metal layers you obtain? You can assume 100 % current efficiency and that the density of the deposited layers is the same as the bulk density (Table 4.4). The molar masses are Ni: $58.69\,\text{g mol}^{-1}$, Ag: $107.9\,\text{g mol}^{-1}$ and Cu: $63.55\,\text{g mol}^{-1}$.

Silver plating

Silver is a widely used material in electronics because of its superior conductivity and stability. It is also used in catalysis, for mirrors, and of course, for decorative applications. Silver plating is of high importance for electrochemical microsensors to form an Ag/AgCl reference electrode.

The most common silver plating bath formulations are alkaline cyanide solutions. The main constituents are silver cyanide (AgCN) and potassium cyanide (KCN), with typically K_2CO_3 to set the pH and increase the electrolyte conductivity. In contrast to the above-discussed deposition examples, the electrolyte contains the metal ions as complexes. Depending on the concentration of the Ag^+ and CN^- ions different complexation of the silver occurs:

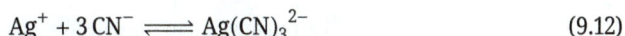

$$Ag^+ + CN^- \rightleftharpoons AgCN \tag{9.10}$$

$$Ag^+ + 2\,CN^- \rightleftharpoons Ag(CN)_2^- \tag{9.11}$$

$$Ag^+ + 3\,CN^- \rightleftharpoons Ag(CN)_3^{2-} \tag{9.12}$$

Typically, the $Ag(CN)_3{}^{2-}$ complex is the predominant species, resulting in the following overall deposition reaction:

$$Ag(CN)_3{}^{2-} + e^- \longrightarrow Ag + 3\,CN^- \tag{9.13}$$

For more details on the reaction steps, see [139]. The cyanide chemistry ensures bath stability; an excess of CN^- facilitates the dissolution of silver from the anode to maintain the bath composition. Variants of the cyanide plating bath work without free cyanide ions and an inert counter electrode (e. g., platinum). Also, cyanide-free formulations came up, being a possible replacement for the very toxic cyanide chemistry with compromises in bath stability.

---

i Microfabricated silver/silver chloride reference electrodes need a thickness of several micrometers for good performance. Therefore, electroplating is the best choice to obtain a silver layer, e. g., 10 μm Ag in a cyanide-based plating bath. Subsequently, a fraction of the Ag (e. g., 2 μm) gets oxidized in a 0.1 M KCl solution at low current density resulting in an AgCl layer with approximately double the thickness[1] of the depleted Ag (e. g., 4 μm):

$$Ag + Cl^- \longrightarrow AgCl + e^- \tag{9.14}$$

For details of the process to fabricate Ag/AgCl electrodes in microtechnology, see, e. g., [140].

The process of silver oxidation in a Cl^--containing electrolyte also allows the conversion of silver wires into Ag/AgCl electrodes. The wire surface must be oxid-free, e. g., by mechanical aberration just before the process. Typical conditions are 2 to 5 mA cm^{-2} for 10 to 20 minutes in 0.1 M KCl with a platinum counter electrode. The silver wire connects as the anode (positive terminal of the current source). For quick experiments, applying a voltage in the range of 2 to 3 V versus a substantially larger Pt counter electrode provides quite good results (process duration in the minute scale until a darker color on the silver wire becomes visible).

9.2.2 Platinum black and hierarchical platinum structures

While electrodeposition of smooth platinum is common too, microrough platinum electrodes are of high interest. Since the 19th century, scientists have generated platinum overcoats with grayish or black appearance, e. g., for applications in bolometers. The dark color coined the deposit's name "platinum black." A similar coating helps to improve the equilibrium process at the hydrogen electrode, typically named "platinized platinum."

Aqueous solutions of chloroplatinic acid (H_2PtCl_6) are the most common electrolytes for platinization. Typical concentrations are in the range of 10 to 50 g l^{-1} (25 to 120 mM). Deposition follows a direct pathway with a four-electron step or an indirect route with two consecutive two-electron steps:

---

1 The density of AgCl (5.6 g cm^{-3}) is roughly half of the density of silver (10.5 g cm^{-3}).

$$PtCl_6^{2-} + 4\,e^- \longrightarrow Pt + 6\,Cl^- \tag{9.15}$$

$$PtCl_6^{2-} + 2\,e^- \longrightarrow PtCl_4^{2-} + 2\,Cl^- \tag{9.16}$$

$$PtCl_4^{2-} + 2\,e^- \longrightarrow Pt + 4\,Cl^- \tag{9.17}$$

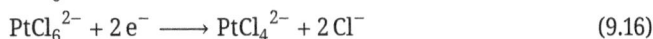

At the counter electrode, the most likely reaction is the formation of chlorine gas:

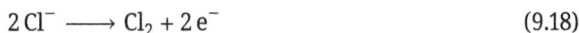

$$2\,Cl^- \longrightarrow Cl_2 + 2\,e^- \tag{9.18}$$

The electrolyte may optionally contain HCl or $NaClO_4$. The most critical control over the final appearance of the deposit is by the addition of lead. The lead's exact role is not clear but seems to inhibit crystal growth, leading to new crystals and attributing the cauliflower-like appearance of the deposits. Additionally, the deposition reaction shows a lower overpotential and prevents the parasitic H_2 formation increasing the current efficiency. Recipes suggest lead acetate ($Pb(CH_3COO)_2$) as the source for the Pb_2^+ ions, added in a concentration in the range of 30 to 2000 mg l^{-1} (0.1 to 6 mM).

The literature describes various current densities from 10 to several 100 mA cm^{-2}. Feltham and Spiro compiled a comprehensive overview of the traditional processes and the influence of Pb_2^+ ions [141]. In general, lower current densities are associated with better adhesion and compact greyish coatings, while higher values lead to black and sometimes even powdery deposits with excellent roughness factors. A possibility is to start the deposition with a lower current density for good adhesion and increase to a value leading to black coating with huge roughness.

Figure 9.4(A) shows a cyclic voltammogram of a typical platinum black deposit onto a thin-film platinum electrode. While the appearance of the CV curve resembles a planar electrode, for platinum black, the current densities are much higher (compare, e. g., Figure 8.16). Looking at the platinized electrode (Figure 9.4(B)) and an unmodified thin-film electrode of the same size (Figure 9.4(C)), one can clearly understand the origin of the term platinum black.

Increased current densities for a prolonged duration result in higher roughness, but overall, the platinum black process is typically limited to Rf values up to several hundred before adhesion gets poor. Much greater roughness factors are possible with current-less platinum nanostructures formation or combining the current-less process with a platinum black deposition leading to hierarchical micro/nanostructures.

Platinum nanostructures

Chloroplatinic acid solutions allow for electroless deposition, too. Sun et al. proposed a recipe with formic acid (HCOOH) as the reducing agent [142]:

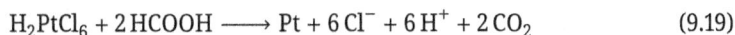

$$H_2PtCl_6 + 2\,HCOOH \longrightarrow Pt + 6\,Cl^- + 6\,H^+ + 2\,CO_2 \tag{9.19}$$

The aqueous electrolyte contains just 3 mM H_2PtCl_6 compared to 1.3 M HCOOH. Depending on temperature and deposition time, nanowires or nanoflowers form. Pro-

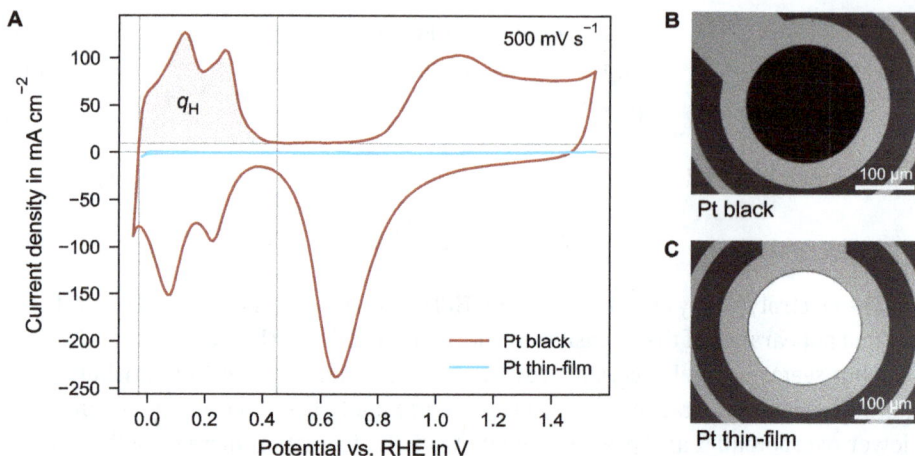

Figure 9.4: Platinum black electrodeposited onto a thin-film platinum electrode with 200 µm diameter. The cyclic voltammogram (A) of platinum black is much larger than the thin-film electrode itself but shows the typical features of platinum. Evaluating the roughness factor according to Section 8.4.2 leads to Rf = 280. Microscopy images with the black appearance of a platinized electrode (B) in comparison to a bright unmodified platinum thin-film electrode (C).

longed deposition duration up to 7 days in stagnant solutions provides nanostructured decoration in the form of a nanowire mesh well-adhering to platinum thin-film electrodes (Figure 9.5). Such electrodes come along with a notable roughness factor Rf of around 1000 [18].

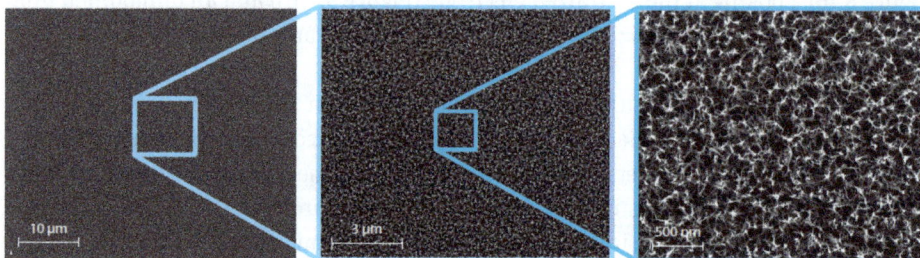

Figure 9.5: Scanning electron microscopy images of a platinum nanowire mesh deposited on thin-film electrodes at different magnifications [18]. Reproduced from Daubinger et al., Electrochemical characteristics of nanostructured platinum electrodes – a cyclic voltammetry study, 2014, DOI 10.1039/C4CP00342J, under CC BY 3.0.

Chen and Holt-Hindle provide an excellent overview of platinum-based nanostructured materials and their applications [143]. Their review is more than a decade later, still one of the most comprehensive resources.

Hierarchical platinum micro/nanostructures

The combination of platinum black electrodeposition and electroless deposition of nanostructures results in hierarchical micro/nanostructures (Figure 9.6). The platinum black deposit enhances the adhesion of the nanostructure, forming a stable structure with roughness factors of several thousand. Apart from catalytic applications, such structures provide a selectivity mechanism toward slow reactions, making them good candidates for nonenzymatic glucose sensors [144].

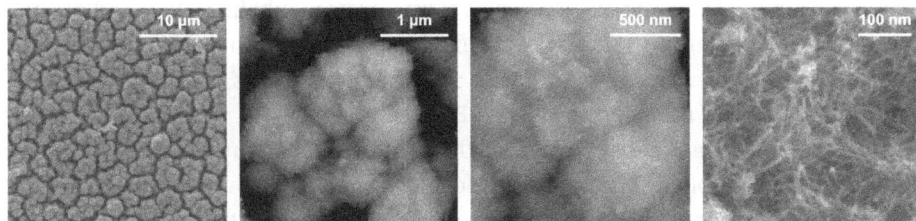

Figure 9.6: Scanning electron microscopy images of hierarchical platinum micro/nanostructures deposited on thin-film electrodes at different magnifications [144]. Reprinted from Journal of Electroanalytical Chemistry, 816, Unmüssig et al., Nonenzymatic glucose sensing based on hierarchical platinum micro/nanostructures, 215–222, Copyright 2018, with permission from Elsevier.

Al-Halhouli et al. investigated the impact of hierarchical arrangements as internal (porous) versus external structures on mass transport using 3D simulation [145, 146]. The choice of the optimal design strongly depends on the application.

9.2.3 Anodically electrodeposited iridium oxide films

Iridium oxide is an exciting electrode material for sensors, neurostimulation, and previously also because of its electrochromism. Yamanaka proposed a deposition solution for *anodically electrodeposited iridium oxide films* (AEIROF) on ITO electrodes targeting electrochromic applications [33]. The deposition solution (Table 9.4) uses $IrCl_4$ as the iridium source, forming an iridium oxide in an alkaline environment. Oxalic acid $(COOH)_2$ is the complexing agent to avoid precipitation of the developed iridium oxide. Additionally, the solution contains H_2O_2 allowing deposition at lower current densities but with

Table 9.4: AEIROF electrodeposition solution according to Yamanaka [33]. The concentrations are for reference only.

Substance	Content	Concentration
$IrCl_4 \cdot H_2O$	$1.5\,g\,l^{-1}$	4 mM
H_2O_2 (30 %)	$10\,ml\,l^{-1}$	0.1 M
$(COOH)_2 \cdot 2\,H_2O$	$5\,g\,l^{-1}$	40 mM
K_2CO_3	adjust to pH 10.5	approx. 0.5 M

an unclear role in forming the complex. After preparing, a rest time in the range of days in which the solution changes its color helps to obtain reproducible depositions.

Yamanaka suggested $[Ir(COO)_2(OH)_4]^{2-}$ as the iridium oxide complex in the deposition solution and for the anodic electrodeposition a reaction releasing CO_2:

$$[Ir(COO)_2(OH)_4]^{2-} \longrightarrow IrO_2 + 2\,CO_2 + 2\,H_2O + 2\,e^- \tag{9.20}$$

For ITO substrates, Yamanaka recommended anodic deposition with $0.16\,mA\,cm^{-2}$, stirring the plating bath, and oxygen removal by nitrogen bubbling. He found a maximum deposition rate at $15\,°C$. The deposited films show a brownish to intense blue color. Also, a reduction process exists, causing less color-intense deposits, the so-called *cathodically electrodeposited iridium oxide films* (CEIROF).

Over the years, many researchers used the recipe without improving the deposition solution much but applying different electrochemical methods on various substrate materials. Most works focus on anodic deposition with current densities in the range of 0.1 to $2\,mA\,cm^{-2}$ [147–149] and even below $0.5\,\mu A\,cm^{-2}$ for UME [150]. Other approaches consider fixed potential [151, 152], pulsed deposition, or cyclic voltammetry [149, 152–155]. In general, constant current or voltage results in faster deposition, while cyclic methods seem to improve film quality.

Figure 9.7(A) shows a cyclic voltammogram on a platinum EQCM in the Yamanaka's deposition solution. Permanent mass change occurs mainly at potentials above $1.3\,V$, the region where AEIROFs get formed. The lower turning point is too high to observe a significant formation of a CEIROF. AEIROF electrodes appear in different shades of blue (the layer in Figure 9.7(B) is a typical example). Also, brownish layers are possible.

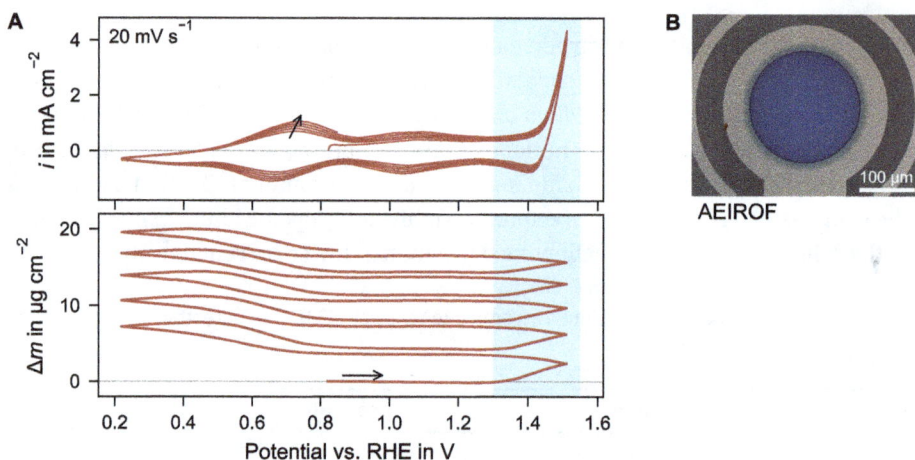

Figure 9.7: Cyclic voltammogram on a platinum EQCM in the iridium oxide deposition solution according to Yamanaka [33]. The AEIROF forms in the marked region (A). Typical appearance of iridium oxide deposited onto a platinum thin-film electrode (B).

There are various methods to obtain iridium oxide electrodes (and all with abbreviations following the same pattern). Excessive potential cycling or applying potential pulses to an iridium electrode forms an *activated iridium oxide film* (AIROF). Such films can reach a thickness of several µm. While AEIROFs are typically thinner and preferred for pH sensing applications, AIROF better suits the demands as a neural stimulation electrode. However, for both statements, you find examples to prove the opposite. Another possibility well matching a typical microfabrication process flow is reactive sputtering; the layer's name is *sputtered iridium oxide film* (SIROF) accordingly. Also, thermal decomposition of iridium salts providing a *thermal iridium oxide film* (TIROF) is a possibility but less common because of the required temperature above $300\,°C$.

9.2.4 Unterpotential deposition

Metal deposition happens by reduction of metal cations (or a complex, which even may have an opposite charge such as $Ag(CN)_3{}^{2-}$ or $PtCl_6{}^{2-}$). This process occurs at a potential lower than the equilibrium potential. In contrast, *underpotential deposition* (UPD) works at potentials above the equilibrium, "under" the required negative potential to drive the regular metal plating (bulk deposition).

UPD occurs in systems where the energy for a metal ion to adsorb on another metal, the substrate, is higher compared to the interaction of the metal ions with each other. A typical example is the UPD of lead ions on a gold substrate:

$$Pb^{2+} + Au + 2\,e^- \longrightarrow Au\text{–}Pb_{UPD} \tag{9.21}$$

The process ends after the formation of one monolayer. Figure 9.8 illustrates the relation of UPD and bulk deposition to the equilibrium potential. However, two or more

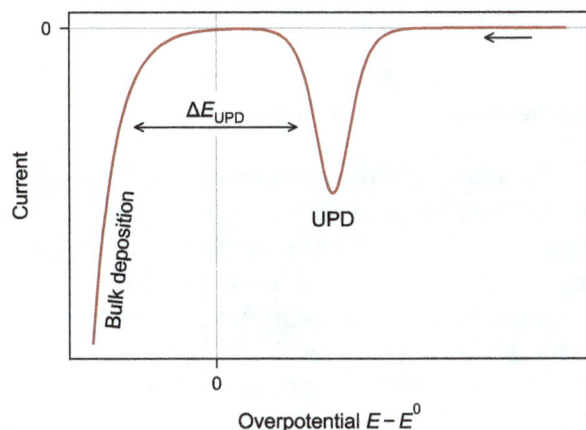

Figure 9.8: Idealized voltammogram of underpotential deposition: metal deposition up to one monolayer occurs more positively than the equilibrium potential E_0. At potentials below E_0, bulk deposition sets in. Practical systems typically comprise more than one and often less pronounced peaks.

distinct UPD peaks are visible in practical systems [156], which might be rather diffuse and overlapping on a polycrystalline substrate material. The potential difference between the UPD and bulk deposition is the UPD shift ΔE_{UPD} with values in the range of 0.2 to 0.4 V. Sudha and Sangaranarayanan provide an exact definition of ΔE_{UPD} and relate experimental values with derivation from thermodynamics [157]. Table 9.5 provides some typical material combination for UPD of metals.

Table 9.5: Typical material combinations for underpotential deposition of metals. The ΔE_{UPD} values are calculated [157]; the same source also provides experimental data.

Substrate	UPD metal ion	UPD shift ΔE_{UPD}
Au	Cu^{2+}	0.22 V
Au	Pb^{2+}	0.40 V
Pt	Pb^{2+}	0.44 V
Ag	Pb^{2+}	0.16 V

Practical applications of metal UPD are in an electrochemical atomic layer deposition (next section) or calibration of an EQCM. UPD also allows the investigation of the surface of noble metals [158].

Similarly, the proton adsorption (Section 8.4.1) on platinum is an underpotential deposition:

$$H^+ + Pt + e^- \longrightarrow Pt\text{–}H_{UPD} \tag{9.22}$$

In this system, two characteristic peaks are visible on a polycrystalline platinum electrode (5a and 5b in Figure 8.14). The accompanying bulk process occurring at potentials more negative than the equilibrium is the formation of hydrogen gas.

9.2.5 Nanofilm deposition: electrochemical atomic layer deposition

In microfabrication, physical vapor deposition (PVD) enables metal thin-films with a thickness of a few 100 nm. Intending a layer height in the micrometer range asks for other methods, preferable electroplating as discussed before. The lower end is a layer with a few 10 nm, quickly revealing any inhomogeneity as a defect. Here is the domain of atomic layer deposition (ALD) building up a film atomic monolayer by monolayer. Classical gas-phase ALD typically provides passivating layer, the most popular is Al_2O_3. Further improvement also allow the deposition of elemental metals, including platinum [159].

UPD of metals provides a similar tool to form layers bottom-up but without the cost of an ALD machine. Additionally, the electrochemical approach implicitly includes a patterning possibility as the layers only form on connected electrode sites. Stickney coined the name electrochemical atomic layer deposition (E-ALD) for the processes exploiting UPD and optionally replacement reactions to grow nanofilms layer-by-layer and popularized this approach [160].

Both classical ALD and the E-ALD were known by atomic layer epitaxy (ALE). Now, the more general term ALD/E-ALD became more common. Strictly, epitaxy refers to crystalline growth only, while ALD considers a layer-by-layer deposition generally, including amorphous structures.

Combination of UPDs

Film formation with UPD alone is limited to a monolayer. A possibility to achieve layer-by-layer growth is the combination of two different reactions. One example is the formation of cadmium telluride (CdTe) bilayers where the Te_2^- ions undergo UPD in an oxidation and the Cd_2^+ ion in an reduction reaction:[2]

$$Te^{2-} \longrightarrow Te_{UPD} + 2\,e^- \tag{9.23}$$
$$Cd^{2+} + 2\,e^- \longrightarrow Cd_{UPD} \tag{9.24}$$

The first reaction provides a monolayer of tellurium, followed by the second reaction generating a cadmium layer on top [161–163]. Repetitive cycles allow the layer-by-layer growth of a CdTe nanofilm (Figure 9.9(A)). Practically, the UPDs run in different baths at different potentials with rinsing in between the two deposition steps. Automatization is possible, e. g., by using a flow cell [164].

Combination of UPD and SLRR

The approach to combine two UPD as described above provides bilayers. Elemental deposits are also possible by using cycles with UPD and a replacement reaction. One example is the deposition of copper nanofilms on gold substrates with lead as the intermediate UPD species (Figure 9.9(B)) [165–167].

UPD of lead is possible on both the substrate (gold) and the nanofilm material (copper) when polarizing to the optimal UPD deposition potential E_{UPD}:

$$Pb^{2+} + 2\,e^- \longrightarrow Pb_{UPD} \tag{9.25}$$

E_{UPD} needs to be negative enough to ensure high UPD coverage but strictly avoids the onset of bulk deposition. Once a monolayer has formed, the electrode is kept open-circuit in a bath comprising copper ions. Spontaneously, a surface-limited redox replacement (SLRR) reaction sets in, replacing the lead with a copper monolayer (ML):

$$Pb_{UPD} + Cu^{2+} \longrightarrow Pb^{2+} + Cu_{ML} \tag{9.26}$$

---

2 The reaction equations follow the convention, not to mention the substrate, although it is essential for the UPD reaction.

A

Te UPD on Au   Cd UPD on Te   Te UPD on Cd

Cycles

B

Pb UPD on Au   SLRR   Pb UPD on Cu   SLRR

Cycles

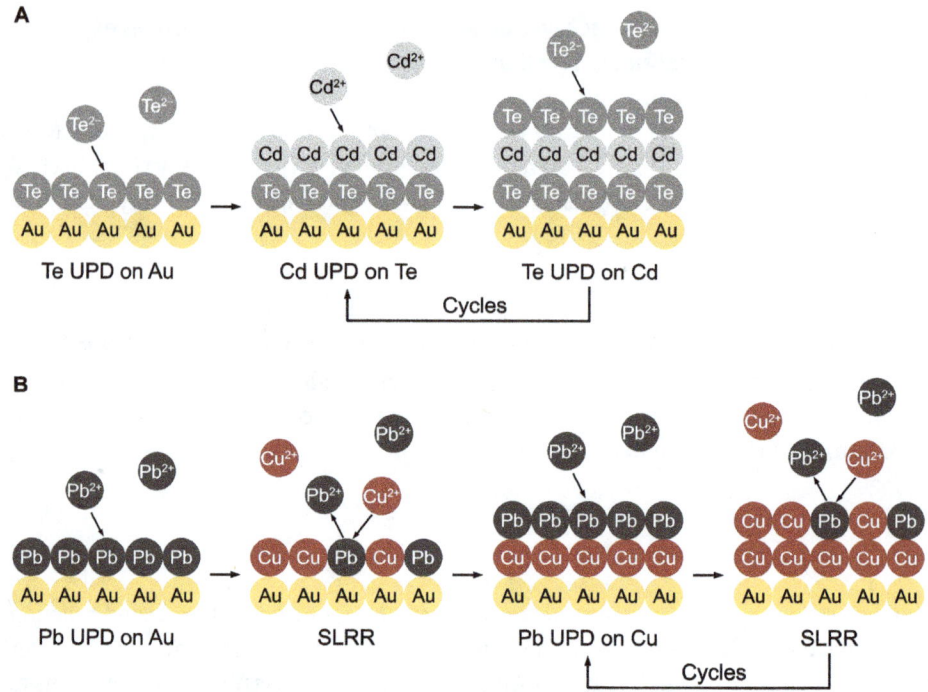

Figure 9.9: Electrochemical atomic layer deposition (E-ALD): CdTe nanofilm formation by alternating underpotential deposition (UPD) with oxidation and reduction reactions (A). Copper nanofilm formation by alternating UPD of lead as intermediate substance and surface-limited redox replacement (SLRR) to copper (B). Please note that the sketch oversimplifies the resulting structure by ignoring the size differences of the involved atoms and possible lattice mismatch.

During the SLRR, the open-circuit potential rises to the equilibrium potential of copper (Figure 9.10). An essential prerequisite and driver for the SLRR is a higher standard potential of the metal, which replaces the UPD layer. From this perspective, the UPD metal should be selected with a low standard potential, but also a too low E_{UPD} might harm because of interfering with the hydrogen evolution.

Appropriate parameters ensure layer-by-layer growth resulting in defect-free films replicating the surface structure and roughness of the substrate without incorporation of lead in the copper deposit.

The replacement efficiency relates the charge obtained when anodically stripping the obtained copper film to the charge needed for the lead UPD summed up over all cycles [166]:

$$\text{Replacement efficiency} = \frac{Q_{\text{Cu,strip}}}{\Sigma \, Q_{\text{Pb,UPD}}} \tag{9.27}$$

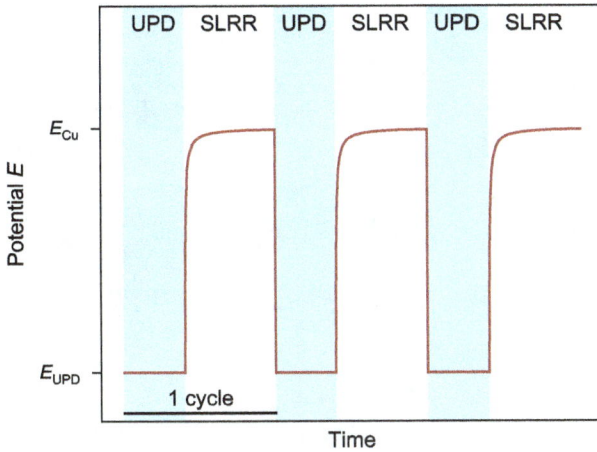

Figure 9.10: Electrochemical atomic layer deposition (E-ALD): in each cycle, the potentiostat applies E_{UPD} to the electrode, followed by an open-circuit phase, in which the SLRR happens.

Typical values are above 90 %, essentially depending on traces of oxygen in nominal oxygen-free electrolytes.

Cooper nanofilm deposition alternating between UPD and SLRR is possible without solution exchange. Using a "one-pot" bath with a much higher concentration of lead ions than copper ions (10 mM vs. 0.3 mM) allows to select the process by either polarization to E_{UPD} for lead or keep the system open-circuit, enabling the SLRR. By minimizing the UPD duration and the concentration difference between lead and copper ions in the solution, parasitic copper deposition during the UPD phase is acceptable low. We presented a robust process compatible with standard microfabrication technology to form copper nanofilms between a few monolayers and more than 30 nm on polycrystalline gold [168].

Also, other metal nanofilms are possible, including platinum [169–171], palladium [172–174], ruthenium [175], and silver [176]. Lead-free alternatives are the usage of zink as the intermediate UPD species for copper deposition [177, 178] or even UPD of hydrogen atoms to obtain platinum films [179, 180].

9.2.6 Electrophoretic deposition

In *electrophoretic deposition* (EPD), the deposition solution contains charged particles. The particles move by electrophoresis in an electrical field and deposit onto a conductive substrate of opposite charge. Both anodic and cathodic EPD exist, referred to by anaphoretic and cataphoretic deposition (Figure 9.11). In contrast to electroplating, the coating does not depend on charge transfer. A current flow is a parasitic effect only; that is why the bath has preferably high resistance. Nonaqueous media are preferred to avoid the gas formation occurring in aqueous solutions. Typical EPD materials are poly-

A

B

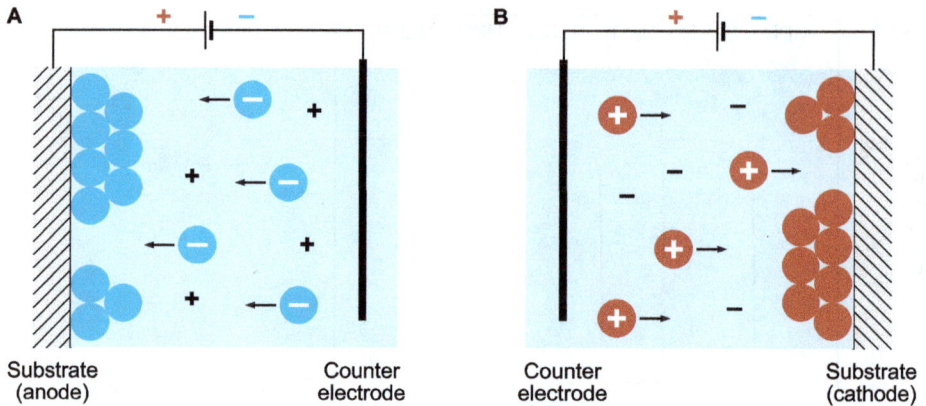

Substrate   Counter   Counter   Substrate
(anode)   electrode   electrode   (cathode)

Figure 9.11: Electrophoretic deposition: charged particles in suspension move toward the substrate in an electrical field. Depending on the substrate's polarity, the process is also called anaphoretic (A) and cataphoretic deposition (B).

mers, ceramics (e. g., Al_2O_3), or metals. Apart from the latter, the process is self-limiting. The applied electrical field strength $\mathcal{E}$ is in the range of up to $100\ \text{V cm}^{-1}$, strongly dependent on the solution type.

The drift velocity v_d of the charged particles depends on their electrophoretic mobility μ_{ep} and the electrical field:

$$v_d = \mu_{ep} \cdot \mathcal{E} \tag{9.28}$$

A better relation to describe the deposition itself is the *Hamaker equation*:

$$m = \mu_{ep} \cdot C_s \cdot A \cdot \mathcal{E} \cdot t \tag{9.29}$$

m is the mass deposited in a period t, and C_s (e. g., in g cm^{-3}) is the solid content of the solution. The relation assumes a parallel arrangement of flat electrodes with the area A.

An important parameter is the zeta potential ζ of the charged particle as it describes the stability of the suspension but also relates to the electrophoretic mobility by *Henry's equation*:

$$\mu_{ep} = \frac{2\epsilon_r \epsilon_0 \zeta}{3\eta} \cdot F(\lambda_D^{-1} r) \tag{9.30}$$

The Henry function $F(\lambda_D^{-1} r)$ is between 1 and 1.5 depending on the ratio between the colloid's radius r and the Debye length λ_D. In an aqueous solution, mostly the Smoluchowski approximation applies ($r \gg \lambda_D$) leading to $F(\lambda_D^{-1} r) = 1.5$. In non-polar systems, the colloids are assumed as a point charge ($r \ll \lambda_D$), the Hückel–Onsager approximation, with $F(\lambda_D^{-1} r) = 1$. For example, Lowry et al. provide an easy to follow, but more detailed description of the relation between zeta potential and the electrophoretic mobility [181].

EPD has many applications like organic and ceramic protective layers, e. g., in the automotive industry, or electrophoretic paints for decorative applications. Besra and Liu give a broad overview of fundamentals and applications of EPD (but an inconsistent description of Henry's function) [182].

There are also photoresists with charged micelles in an aqueous emulsion, allowing electrophoretic deposition. After coating, the deposits form a smooth film in a subsequent baking step. Depending on the chemistry, both positive and negative tone photoresists are available. Such resists enable conformal coating also on noneven microstructures.

9.2.7 Electropolymerization

Electropolymerization or electrochemical polymerization forms a polymeric film on the electrode by electrochemically triggering the polymerization reaction. The obtained films may be conductive or passivating depending on the polymer but also on possible dopants. Commonly, the polymerization is anodic with the monomers oxidized at the electrode, the formed cationic radical causing the polymerization. A prominent example known since the 19th century is the oxidation of aniline forming the conductive polymer polyaniline (PANI).

Also, cathodic polymerization exists, e. g., the deposition of polyphenylene vinylene (PPV), another conducting polymer suggested for applications in organic solar cells or light-emitting diodes.

Apart from electropolymerization with constant current, often cyclic voltammetry is used. Figure 9.12 shows the electrodeposition of a polymeric film from 1,3-diaminobenzene (1,3-DAB) or m-phenylenediamine (mPD) by cyclic voltammetry. Such films play an essential role in enzymatic biosensors (Section 10.1.4) as a semipermeable membrane blocking substances like ascorbic acid, acetaminophen, or uric acid while allowing smaller molecules like H_2O_2 to pass [183]. Additionally, such layers with a thickness around 10 nm allow immobilization of enzymes directly at the electrode [184] or fixate a dip-coated enzyme layer by electropolymerization [185].

Another electropolymerized material gaining a lot of attention in the field of bioelectronics is poly-3,4-ethylene dioxythiophene (PEDOT), which shows mixed ionic and electronic conductivity adjustable by doping [186].

9.3 Subtractive electrochemical techniques

This section, opposing the previous two, focuses on subtractive techniques with processes roughly analog to corrosion. In detail, especially at the edge between micro- and nanofabrication, careful selection of electrolytes and parameters are key to success-

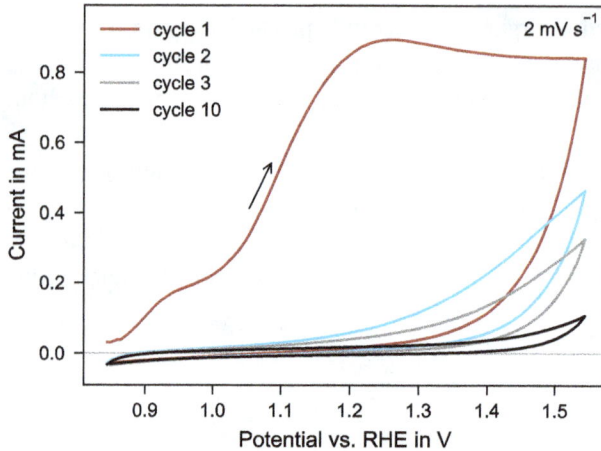

Figure 9.12: Electropolymerization of a nanofilm from 3 mM 1,3-DAB in phosphate buffer by cyclic voltammetry for interference blocking in enzymatic biosensors. The process is self-limiting as visible from the decaying current with increasing cycle number.

ful technology. While the fundamentals become clear, processing on your own requires recipes and tips from the specific literature beyond this textbook.

9.3.1 Electrochemical machining

In conventional *electrochemical machining* (ECM), a tool as cathode with the negative shape of the desired result approaches the metal substrate, which is the anode. The anode dissolves in a die-sinking process by relatively high current density (10 to 100 A cm^{-2}). Typical electrolytes are highly concentrated aqueous solutions of KCl, NaCl, or KNO$_3$. The tool is lowered into the substrate, maintaining a distance between anode and cathode in the range of 150 µm. Strong convection, also in the gap, ensure removal of dissolved metal and stable bath conditions.

Apart from the installment costs, expensive tools, and efforts to maintain the electrolyte, ECM offers advantages over other methods such as electrical discharge machining or milling, making it attractive for critical applications. The abrasion of the material occurs at low temperature and without shear forces resulting in stress-free parts. The surface roughness is in the micrometer range or slightly below.

Electrochemical micromachining
In electrochemical micromachining (ECMM or EMM), the cathode does not have the workpiece's shape. Typically, a tiny tool structures the surface like a milling machine would do. In contrast to conventional ECM, the gap is less (around 10 µm), and short pulses are applied instead of constant current. Spatial resolution in the 10 µm range is

possible with surface roughness around 100 nm. Bhattacharyya et al. compared in their still relevant review on ECMM the differences to conventional ECM in more detail [187].

9.3.2 Electrochemical etching

Closely related to ECM is *electrochemical etching*. Similarly, anodic metal dissolution is the mechanism of structuring:

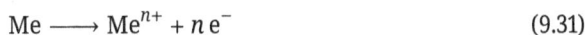

$$\text{Me} \longrightarrow \text{Me}^{n+} + n\,\text{e}^- \tag{9.31}$$

In an aqueous electrolyte, at the counter electrode, primarily hydrogen formation occurs:

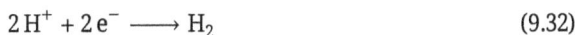

$$2\,\text{H}^+ + 2\,\text{e}^- \longrightarrow \text{H}_2 \tag{9.32}$$

In microtechnology, the preferred method is the through-mask etching, e. g., with a structured dry-film resists on thin metal foils. Typical electrolytes are neutral salt solutions or of acids like HCl and HNO_3, depending on the material.

9.3.3 Electropolishing

In electrochemical polishing, the surface of the substrate surface gets etched, too. The etching conditions in the acidic electrolyte ensure that predominantly spikes and protrude dissolve, resulting in reduced surface roughness but not much bulk attack.

9.4 Nanoelectrodes, nanomaterials

The nanoelectrode is an ambiguous term referring to either an electrode with a characteristic dimension below one micrometer or an electrode assembly comprising a nanomaterial. Nanoelectrodes of the first meaning are essential tools to access biological structures, e. g., to approach biological cells or even penetrate, accessing the intracellular space electrochemically (but also accepting that the cell sees not at all physiological conditions). In this section, the focus is on nanomaterials and their use in electrode assemblies.

A length below one micrometer at least in one dimension is the main characteristic of a nanomaterial. Often, the requirement of different properties compared to the same material as bulk is part of the definition. One-dimensional nanomaterials are nanotubes or nanowires; two-dimensional objects are nanosheets. 3D arrangement of nanowires can lead to brushes or even hierarchical structures. Nanoparticles often account as zero-dimensional, although their diameter defines their properties.

Common to all nanomaterials is the challenge of bringing them in contact with a substantially larger electrode, with dimensions from 10 µm to more than 1 cm. One strategy is directly synthesizing the nanomaterial onto a carbon or metal electrode connecting the nanostructures by adsorption or covalent bonding. Another common approach is the incorporation into a polymeric binder. Membrane formation of nanomaterials is possible, too, like the "buckypaper" consisting of carbon nanotubes [188] or a "graphene oxide paper" consisting of graphene oxide flakes [189].

A convenient strategy is hierarchical arrangements benefiting from the nanostructures while ensuring sufficient conductivity through a micrometer-sized backbone. One example is the hierarchical micro/nanostructures of platinum (Figure 9.6).

⚡ Nanomaterial safety is an essential field growing along with the development of novel materials. The unique properties because of their small size also suggest much easier interaction with the human body through different exposure pathways, most critically inhalation and dermal contact. While there is no conclusive picture on nanomaterial toxicity, especially lacking long-term data, the WHO formulated generally accepted guidelines to protect workers from the potential risks of nanomaterials.[3] Based on the experience with asbestos and how long it took to realize the severe health hazards, many scientists suggest overcareful handling, especially with novel nanomaterials.

9.4.1 Carbon nanomaterials

Carbon materials differ entirely depending on their bonds. When carbon atoms bind to four neighbors (sp^3 hybridization), the carbon allotrope is a diamond, with large hardness and very low conductivity. In contrast, sp^2 hybridization, each carbon atom having three neighbors, provide graphitic layers showing completely different properties such as high conductivity (Section 4.3.3). Here, two import nanomaterials are introduced: carbon nanotubes and carbon nanofibers. Sharma provides a broader overview of the different carbon nanomaterials and their application in micro and nano devices [29].

Carbon nanotubes

Carbon nanotubes (CNTs) are hollow cylinders composed of graphitic sheets. Single-walled carbon nanotubes (SWCNTs) consist of a single, rolled-up graphene sheet. The orientation of the graphene's crystal axis relative to the direction of the tube defines the electrical properties of the SWCNT between metallic conduction and a semiconductor. The typical diameter is between 1 and 2 nm with a tube length of 100 nm up to the mm range in extreme cases. In contrast, multiwalled carbon nanotubes (MWCNTs) have larger diameters (10 to 100 nm) and always show metallic conduction.

---

3 WHO guidelines on protecting workers from potential risks of manufactured nanomaterials. Geneva: World Health Organization; 2017.

CNT synthesis is mainly by a chemical vapor deposition (CVD) process. Plasma-enhanced CVD allows defining a growth direction and optionally synthesize the CNTs directly onto a substrate material, e. g., as a "nano-forest." The ends of the CNTs are closed (like a long stretched fullerene) when they come from synthesis; therefore, a cap opening process, e. g., by plasma etching or chemical oxidation in aqueous HNO_3, is standard.

The electrochemical response of CNTs is still under discussion [190–192]. In general, nanostructures favor faradaic reactions in the sense of maximal accessibility with minimal double layer capacity. A common interpretation is that the redox activity is mainly an edge effect at the tip or defects of a SWCNT. In more detail, the predominance of the tube's end depends on the redox-active substance [193].

CNTs alone are not much redox-active toward oxygen, hydrogen, or methanol. Therefore, many approaches use modifications to functionalize the nanotubes with noble metal particles, mainly platinum ("decorated CNTs").

Carbon nanofibers

Carbon nanofibers (CNFs) are cylindrical structures based on stacked graphene layers. Diameters are in the range of 5 to 500 nm with a length in the μm range. CNFs are typically larger than CNTs and come along with more defects. An interesting capability of both CNFs and CNTs is the direct electron transfer from an enzyme to the electrode in biosensors.

A risk of popular topics is that a few works are driven more by wishful thinking than sound experiments and data interpretation. Such trend was for long an issue with CNTs harming the reputation of excellent contributions from other researches and seems more recently coming along with graphene. Wang et al. wrote an ironic paper on "guano-doped graphene" [194], which might entertain but also should us all be a warning to question our experiments and their interpretation continuously, especially in the context of a highly competitive scientific landscape generating a pressure to succeed.

!

---

10 Selected aspects: microsystems and nanosystems

Introductory courses often describe microsystems as the extension of microelectronics to other physical domains. Optics and electrochemistry are complementary fields that provide access to the information gained in the microsystems and nanosystems. Observing the physiology of our body, we can understand the power (and complexity) of electrochemistry as the default approach converting between the (bio)chemical information and information processing in the nervous system and brain. Accordingly, many microsystems and nanosystems rely on an electrochemical method in their operation, either using electroanalysis directly to measure a parameter or use electrochemistry to read out the results of a biological receptor element in biosensors. Consequently, a large part of this chapter deals with sensors. Also, in actuators, electrochemical approaches play a role and are briefly discussed.

The chapter closes with a look at the contribution of electrochemistry in neurotechnology, starting from the classical procedures in neurostimulation, briefly touching the recording in brain implants, and coming to sensing in neuroscience, including innovative approaches using the implant's electrodes.

---

i *Microsystems* or *micromachines* are systems with components with a least one dimension in the size of 1 to 100 μm. Microelectromechanical systems (MEMS) evolved from microelectronics considering other physical domains, like mechanics. Pragmatically also microsystems without mechanical functionality but optics, electrochemistry, or microfluidic account to MEMS. Especially microsystem incorporating microfluidics often have large dimensions but still accounts for the microworld, when the characteristic length scale is in the microrange. On the other end, nanosystems or nanoelectromechanical systems (NEMS) connect, where the essential dimension is between 1 to 100 nm. The transition is smooth, also driven by the preference for "nano" instead of "micro," hunting for a more fancy-sounding description.

Feynman envisioned in his speech "There's Plenty of Room at the Bottom," dating back to 1959, the route from our directly experiential world down to ultimate small systems on the atomic length scale. Decades later, we arrived at his wish to "arrange the atoms the way we want; the very atoms, all the way down" [195] – with the help of electrochemical methods. E-ALD (Section 9.2.5) allows building films monolayers by monolayer; SECM (Section 6.2.1) enables the placement of individual atoms.

10.1 Sensors

Sensors are elements that convert physical, chemical, or biological quantities into another quantity, usually electrical. That said, sensors are a particular case of a *transducer*, an element converting one form of energy into another. In chemical and biosensors, the input quantity is nearly exclusively concentration (or partial pressure of a gaseous analyte). In general, the term sensors may be limited to self-contained devices; however, with electrochemical sensors, this aspect is often considered not as strict. In many cases, there is a smooth transition from electroanalytical procedures in the lab to sensor devices.

https://doi.org/10.1515/9783111488844-010

10.1.1 Potentiometric ion-selective sensors

Miniaturization of ion-selective electrodes needs deviation from the classical setup with an internal liquid electrolyte (Figure 5.4). Instead, a solid ion-to-electron-transducer translates between the ionophore membrane and the electron-conducting connection (Figure 10.1). Typical materials are conducting polymers or carbon nanostructures.

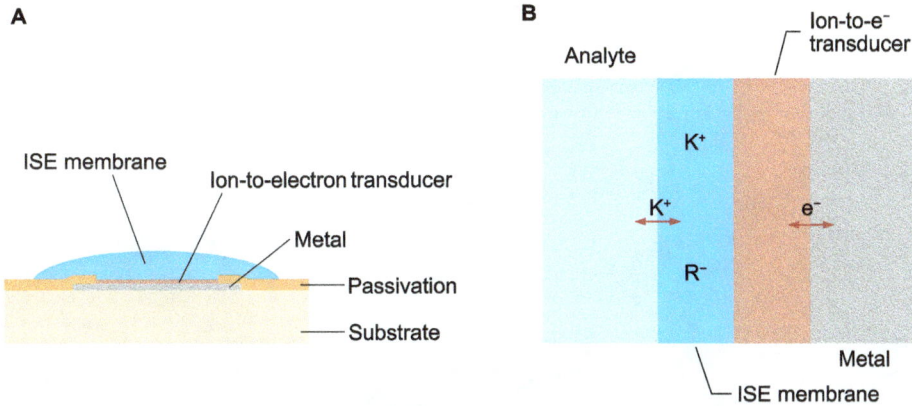

Figure 10.1: Miniaturized ion-selective electrode (ISE) fabricated in microtechnology (A). Instead of the liquid inner filling in a classical ISE, a solid ion-to-electron transducer connects to the electrode (B).

Similarly, a solid-contact reference electrode with a solid internal electrolyte contributes to an all-solid-state ion-selective sensor. Typical membrane materials are polyvinyl chloride (PVC), polyurethane (PU), or acrylate polymers. As with the liquid analog, the membrane must contain chloride ions, e.g., from KCl or an ionic liquid. Bieg et al. provide an introductory overview of solid-contact ion-selective and reference electrodes [196].

Metaloxide pH sensors

Metal oxide pH sensors are a particular class of ion-selective sensors. Instead of using a miniaturized pH glass membrane combined with a solid inner electrolyte, metal oxide electrodes are simpler to produce by microfabrication and provide more reliable results. Among the different metal oxides, iridium oxide is the most used material for pH sensing.

In general, the pH-dependent equilibrium of iridium oxide observed in potentiometric measurements is at the transition between Ir(III) and Ir(IV):

$$Ir(IV)oxide + q\,H^+ + n\,e^- \rightleftharpoons Ir(III)oxide + r\,H_2O \qquad (10.1)$$

q, n, and r are stoichiometric coefficients; the ratio q/n determines the pH sensitivity:

$$q/n \cdot \frac{2.303 \cdot RT}{F} \quad \text{per pH} \tag{10.2}$$

In the cyclic voltammogram of iridium oxide (Figure 10.2), a prominent peak around 0.9 V versus RHE represents the oxidation of Ir(III)oxide to Ir(IV)oxide with the reverse reaction occurring in the downward scan. Overlead as the vertical blue line is the open-circuit potential observed at different pH values. Its position at the transition between Ir(III) and Ir(IV) agrees with equation (10.1).

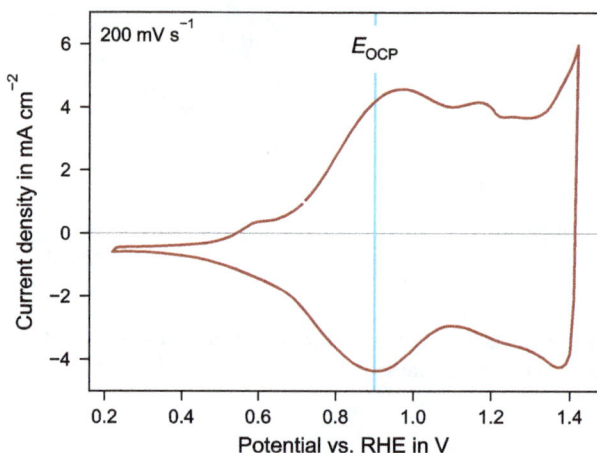

Figure 10.2: Cyclic voltammogram of an AEIROF electrode in air-saturated 1 N H_2SO_4. The blue vertical line indicates the open-circuit potential E_{OCP} found from measurements in electrolytes with different pH, which is at the transition between Ir(III) and Ir(IV). Because of the RHE potential scale's pH correction, the E_{OCP} appears approximately independent of pH.

Anhydrous iridium oxide films, as obtained by reactive sputtering or thermal decomposition of an iridium salt (see the infobox in Section 9.2.3), show a q/n ratio of 1 and, therefore, a sensitivity of $-59\,\text{mV}\,\text{pH}^{-1}$ at room temperature. Two equilibrium reactions have been suggested differing in the Ir(III)oxide:

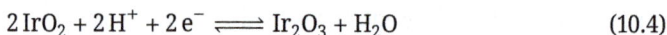

$$\text{IrO}_2 + \text{H}^+ + \text{e}^- \rightleftharpoons \text{IrO} \cdot \text{OH} \tag{10.3}$$

$$2\,\text{IrO}_2 + 2\,\text{H}^+ + 2\,\text{e}^- \rightleftharpoons \text{Ir}_2\text{O}_3 + \text{H}_2\text{O} \tag{10.4}$$

Several mechanisms were postulated for hydrated iridium oxide films; most works agree on the formulation by Burke and Whelan [197]:

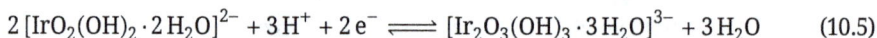

$$2\,[\text{IrO}_2(\text{OH})_2 \cdot 2\,\text{H}_2\text{O}]^{2-} + 3\,\text{H}^+ + 2\,\text{e}^- \rightleftharpoons [\text{Ir}_2\text{O}_3(\text{OH})_3 \cdot 3\,\text{H}_2\text{O}]^{3-} + 3\,\text{H}_2\text{O} \tag{10.5}$$

The q/n ratio is 1.5 and, therefore, the sensitivity is $-89\,\mathrm{mV\,pH^{-1}}$ at room temperature. More refined models [198] consider partial hydration leading to q/n ratios in between 1 and 1.5 matching typical experimental findings for AEIROF. Figure 10.3 shows a calibration of an AEIROF electrode around neutral pH. The sensitivity depends strongly on the deposition procedure and subsequent thermal treatment.

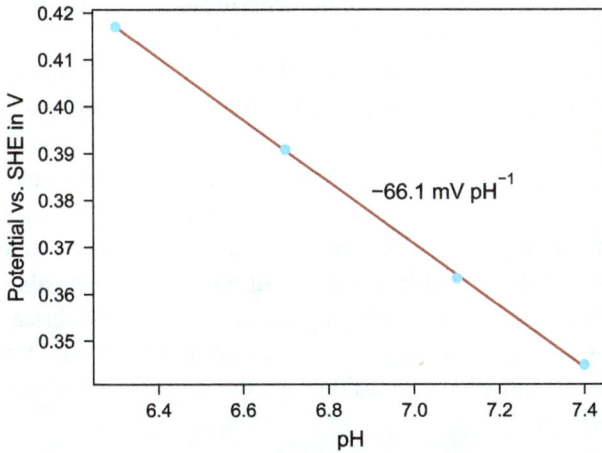

Figure 10.3: Calibration curve of an iridium oxide microelectrode in phosphate buffer solution with a different pH at room temperature. The pH-sensitive layer was an anodically electrodeposited iridium oxide film (AEIROF) using Yamanaka's recipe [33] with the process described in [140].

Olthuis et al. provided another interpretation [199]. The sensing mechanism is the uptake or release of protons from hydroxyl groups in the iridium oxide without electron transfer. Instead, the valency of the iridium changes. Prepolarization of the iridium oxide to above or below the equilibrium changes the valency of the iridium in the oxide. The sensitivities were between $-60\,\mathrm{mV\,pH^{-1}}$ (reduced state of the iridium oxide film) and $-80\,\mathrm{mV\,pH^{-1}}$ (fully oxidized). The authors also explained the contradictory results reported in the literature regarding cross-sensitivity to dissolved oxygen by the different states of the oxide. Electrodes preconditioned to their reduced form showed an oxygen-dependent potential, while fully oxidized films did not. Assumingly, reduced iridium oxide electrodes get oxidized in the presence of dissolved oxygen and, therefore, have different pH sensitivity.

10.1.2 Microsensors with gas-permeable membrane

Miniaturized liquid inner electrolytes are more common in combination with a gas-permeable membrane. The principle goes back to the Clark-type oxygen sensor with gas-permeable membrane separating the sensor electrolyte from the measurement medium as characteristic the characteristic feature.

Clark-type oxygen sensor

Clark's invention back in the late 1950s was to separate the sensor electrolyte from the measurement medium by a gas-permeable membrane to eliminate cross-sensitivity issues [200]. He used a two-electrode setup with a platinum working and silver counter electrode. However, the breakthrough came from the separated sensor electrolyte. Therefore, I recommend using the term "Clark-type" to emphasize the separation with a gas-permeable membrane rather than for devices exactly resembling the electrode materials of Clark's historical approach.

Miniaturized Clark-type oxygen sensors are attractive for application with electrolytes prone to contaminate noble metal electrodes, e. g., with blood or cell culture medium. Figure 10.4(A) shows a typical planar realization. The inner electrolyte is often a hydrogel brought onto a planar electrode chip and covered with a gas-permeable membrane, e. g., silicone. The hydrogel contains salts but can dry out during fabrication to allow deposition of the silicone. Upon usage, the sensor needs immersion into an aqueous measurement medium, and vapor passes through the thin gas-permeable membrane to rehydrate the hydrogel.

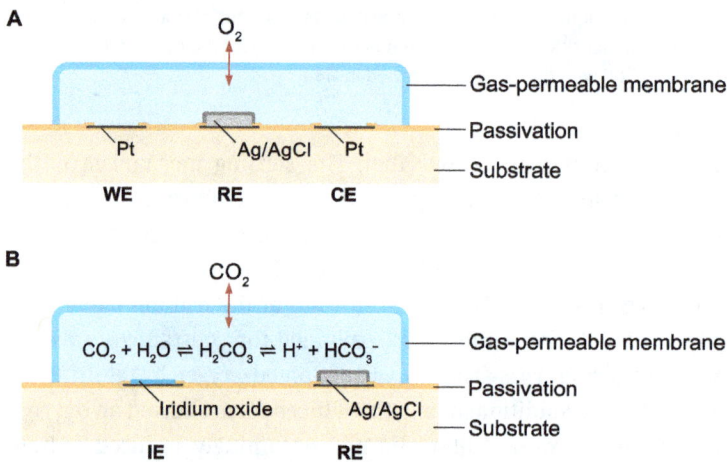

Figure 10.4: Microsensors with gas-permeable membrane: the Clark-type sensor measures oxygen by amperometry (A) in a three-electrode setup with working (WE), reference (RE), and counter electrode (CE). The Severinghaus carbon dioxide sensor relies on a carbonate buffer and measures the pH by potentiometry (B) between an indicator electrode (IE) and a RE.

An elegant concept to maintain the electrolyte composition and achieve net zero-oxygen consumption is the *Ross principle* which was proposed in 1966 [201] and named as such later by Jobst et al. [202]. This principle compensates for the oxygen consumption at the working electrode by oxygen evolution at the counter electrode while maintaining the pH:

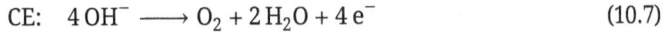

$$\text{WE:} \quad O_2 + 2\,H_2O + 4\,e^- \longrightarrow 4\,OH^- \tag{10.6}$$

$$\text{CE:} \quad 4\,OH^- \longrightarrow O_2 + 2\,H_2O + 4\,e^- \tag{10.7}$$

Liebisch et al. realized a miniaturized Clark-type sensor for cell culture application and used measurements of the counter electrode potential to verify the Ross principles [22]. This work also contains an overview of different Clark-type microsensors and their performance data.

Severinghaus carbon dioxide sensor
While the Clark electrode uses amperometry as the measurement method, potentiometric approaches allow, e. g., to measure the internal electrolyte's pH, depending on the analyte concentration. The Severinghaus electrode, described first in 1958, employs this approach to measure CO_2 [203].

CO_2 pass through the gas-permeable membrane, dissolves in the aqueous sensor electrolyte, and dissociates:

$$CO_2 + H_2O \rightleftharpoons H_2CO_3 \rightleftharpoons HCO_3^- + H^+ \tag{10.8}$$

Figure 10.4(B) illustrates a miniaturized Severinghaus sensor with an iridium oxide indicator electrode to measure the pH change. Depending on the desired measurement range, a carbonate solution with appropriate concentration is the inner electrolytes.

Severinghaus-type potentiometric gas sensors for other analytes exist, but not all have been miniaturized so far. Table 10.1 summarizes a few examples.

Table 10.1: Severinghaus-type potentiometric gas sensors with their typical inner electrolyte.

Analyte	Inner electrolyte	Equilibrium reaction
CO_2	$NaHCO_3$	$CO_2 + H_2O \rightleftharpoons HCO_3^- + H^+$
NH_3	NH_4Cl	$NH_3 + H_2O \rightleftharpoons NH_4^+ + OH^-$
NO_2	$NaNO_2$	$2\,NO_2 + H_2O \rightleftharpoons NO_3^- + NO_2^- + 2\,H^+$
SO_2	$NaHSO_3$	$SO_2 + H_2O \rightleftharpoons HSO_3^- + H^+$

10.1.3 Biosensors

A *biosensor* is a sensor with a biological recognition element selectively measuring an analyte concentration. Figure 10.5 illustrates a biosensor in its most general understanding. Biosensors comprise different transduction principles, including optics or mass detection, but electrochemical methods are the most common leading to an electrochemical biosensor.

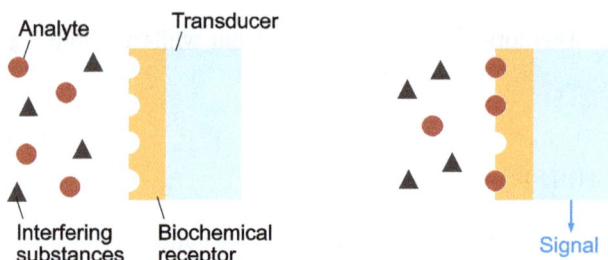

Figure 10.5: A biosensor combines a biological recognition element (biochemical receptor) with a transducer, often based on electrochemical methods. The sensor shows selectivity on the analyte against the background of interfering substances.

Most commonly is the IUPAC definition: "An *electrochemical biosensor* is a self-contained integrated device, which is capable of providing specific quantitative or semi-quantitative analytical information using a biological recognition element (biochemical receptor), which is retained in direct spatial contact with an electrochemical transduction element" [204].

Possible biological recognition elements are enzymes, ionophores, antibodies, oligonucleotides, cell organelles, cell membrane parts, whole cells, tissue slices, or controversially even multicellular organisms. With the advent of biomimetic approaches such as "artificial enzymes" and molecularly imprinted polymers (MIPs), the term biosensor got diluted and is sometimes in use here, too, also strictly there is no biological recognition element.

10.1.4 Enzymatic biosensors

A substantial class of biosensors relies on enzymes as biological recognition elements. Enzymes are proteins that act as biochemical catalysts. Typically, but not necessarily, enzymes have high selectivity towards a substrate, the analyte, in the context of a sensor. The most well-known enzymatic biosensor is the glucose sensor, first introduced by Clark and Lyons in 1962 [205]. The enzyme glucose oxidase converted the analyte to gluconolactone, which reacts further to gluconic acid:

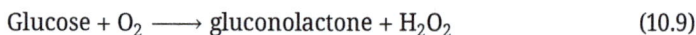

$$\text{Glucose} + O_2 \longrightarrow \text{gluconolactone} + H_2O_2 \tag{10.9}$$

The accompanying pH drop was a measure for the glucose concentration.

Most enzymatic biosensors still use enzymes from the group of oxidases, mainly with O_2 as the electron acceptor like the glucose oxidase. In general, the enzyme catalyzes the reaction of a substrate (S) to a product (P):

$$S \longrightarrow P \tag{10.10}$$

Common to those oxidases is the cofactor *flavin adenine dinucleotide* (FAD), which accepts the electrons needed for the substrate reaction:

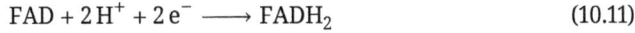

$$FAD + 2H^+ + 2e^- \longrightarrow FADH_2 \tag{10.11}$$

The enzyme reaction completes with molecular oxygen as the electron acceptor to reverse the above reaction and keep electroneutrality:

$$O_2 + 2H^+ + 2e^- \longrightarrow H_2O_2 \tag{10.12}$$

Altogether, the reaction catalyzed by an oxidase with O_2 as the cosubstrate leads to H_2O_2, which is today the preferred species for the electrochemical transducer in the biosensor:

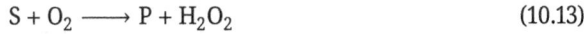

$$S + O_2 \longrightarrow P + H_2O_2 \tag{10.13}$$

Many different oxidases are available, but only some of them are stable enough for application in biosensors. Table 10.2 lists some common oxidases with their EC number[1] and the catalyzed reaction in the form of equation (10.13).

Table 10.2: Common oxidases used in enzymatic biosensors with the catalyzed reactions. For lactate oxidase, the share between the two listed reactions depends on the enzyme's environment, with only the first relevant for sensor application.

Enzyme	EC number	Catalyzed reactions
Glucose oxidase	EC 1.1.3.4	Glucose + O_2 $\longrightarrow$ gluconolactone + H_2O_2
Lactate oxidase	EC 1.13.12.4	Lactate + O_2 $\longrightarrow$ pyruvate + H_2O_2
		Lactate + O_2 $\longrightarrow$ acetate + CO_2 + H_2O
Glutamate oxidase	EC 1.4.3.11	Glutamate + O_2 + H_2O $\longrightarrow$ oxoglutarate + NH_3 + H_2O_2
Galactose oxidase	EC 1.1.3.9	Galactose + O_2 $\longrightarrow$ galacto – hexodialdose + H_2O_2

Figure 10.6 illustrates possible variations of the electrochemical transduction principle. First-generation biosensors rely on the natural cosubstrate oxygen and the detection of H_2O_2 produced by the enzyme reaction. In low oxygen concentration applications, the dependency on oxygen may become dominant. A possible solution is the involvement of a redox mediator (e. g., $[Fe(CN)_6]^{3-}$) instead of relying on oxygen, often referred to as a second-generation biosensor. Third-generation biosensors use direct electron transfer from the enzyme.

---

[1] The ENZYME repository is a nomenclature database providing additional information on enzymes, accessibly by the enzyme's name or the Enzyme Commission number (EC number), a classification system for enzymes. The database is a part of the bioinformatics resource portal of the SIB Swiss Institute of Bioinformatics: https://enzyme.expasy.org

O$_2$ as cosubstrate
"1st generation"

Redox mediator
"2nd generation"

Direct electron transfer
"3rd generation"

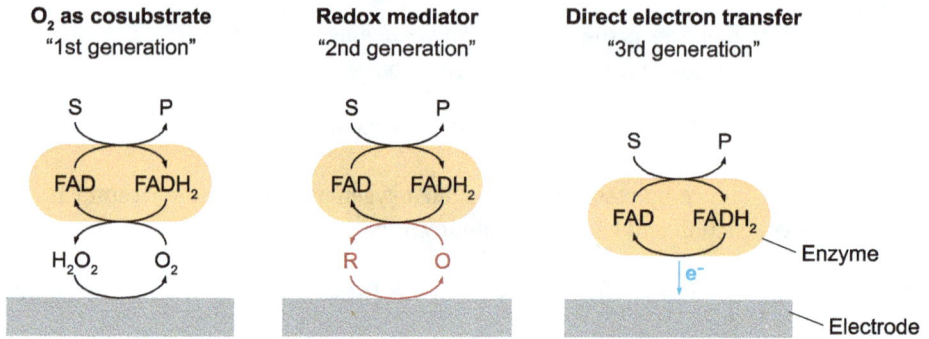

Figure 10.6: Enzymatic biosensors: Different principles to electrically link to the enzyme, illustrated for a generic oxidase as the enzyme.

> **i** New is not always better: classifying enzymatic biosensors by generation reflects historical development but not general improvements. Both first- and second-generation biosensors allow a high enzyme concentration resulting in stable signals even when some enzymes become inactive. This "overloading" possibility makes them the best candidates for long-term applications. Third-generation sensors with direct electron transfer have fewer interference issues and easier fabrication, but the overall sensor performance depends more on enzyme stability.

Michaelis–Menten kinetics

The *Michaelis–Menten kinetics* is the commonly accepted theory to describe enzyme kinetics. It considers the reaction of the enzyme E with the substrate S forming the enzyme-substrate complex ES. The associated rate constants are k_1 for the forward and k_{-1} for the backward reaction. The catalyzes step with rate constant k_2 is the reaction of the ES complex to a product P and freeing the enzyme:

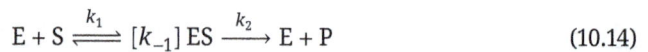

$$E + S \underset{k_{-1}}{\overset{k_1}{\rightleftharpoons}} ES \overset{k_2}{\longrightarrow} E + P \tag{10.14}$$

The Michaelis–Menten equation relates the reaction rate v to the substrate concentration [S] with the Michaelis–Menten constant K_M:

$$v = \frac{v_{max} \cdot [S]}{K_M + [S]} \tag{10.15}$$

The maximal reaction rate v_{max}, when all enzymes are continuously active, depends on the total number of enzymes $[E]_0$ and the rate constant of the catalyzes step:

$$v_{max} = k_2 \cdot [E]_0 \tag{10.16}$$

The Michaelis–Menten constant K_M depends on the enzyme as such but not the amount of enzyme:

$$K_M = \frac{k_2 + k_{-1}}{k_1} \tag{10.17}$$

At a substrate concentration equal to K_M, the reaction is $v_{max}/2$. For $[S] \gg K_M$, the reaction rate approaches v_{max}, and the reaction turns zero-order ("all enzymes are working all the time"). In contrast, for $[S] \ll K_M$ the reaction rate depends approximately linearly on the substrate concentration (see Figure 10.7).

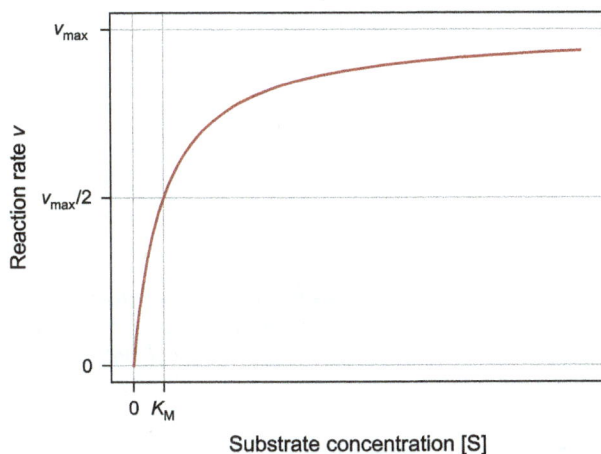

Figure 10.7: Michaelis–Menten kinetics: the dependency on the reaction rate v on the substrate concentration $[S]$ is approximately linear for low concentrations and saturates at v_{max} when all enzymes "continuously work." The Michaelis–Menten constant K_M characterizes the curve shape.

In an amperometric enzymatic biosensor with the enzyme reaction being the rate determinating step, the output signal follows the Michaelis–Menten kinetics. Figure 10.8 shows a fictive example comparing enzyme kinetic and diffusion limitation. Within the optimal working range, a dependency on the enzyme kinetics yields the highest current. However, for higher analyte concentration, the current approaches a maximum, and the sensor loses sensitivity (change in current per change in analyte concentration). Forcing the biosensor into diffusion limitation regime, e. g., by introducing a diffusion limiting membrane, the output current is much lower but linear. More severe diffusion limitation increases the linear range at the cost of lower sensitivity.

Another essential difference between biosensors operated under enzyme kinetic and diffusion limitation is when signal stability over longer measurement time matters. In a kinetically limited sensor, the inactivation of some enzymes causes instant sensitivity loss (v_{max} depends on $[E]_0$). In contrast, for a diffusion-limited sensor, a substantial fraction of the enzymes may turn inactive without affecting the sensitivity as long as enough enzymes operate to keep the mass transport limitation. More severe diffusion limitation causes a lower sensor signal but with the benefit of extended signal stability, assuming the inactivation of enzymes is the dominating degradation mechanism.

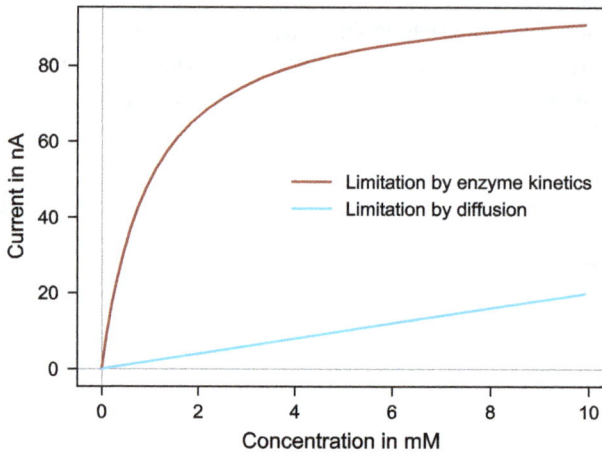

Figure 10.8: Amperometric enzymatic biosensor: limitation by enzyme kinetics results in a transfer function with a shape following the Michaelis–Menten equation. Forcing the sensor into the diffusion limiting regime leads to a linear transfer function independent of the enzyme kinetics at the cost of a lower signal.

Biosensors with oxygen as cosubstrate ("first generation")

Classical enzymatic biosensors with oxidases rely on molecular oxygen as the cosubstrate. Early works considered measuring oxygen depletion to deduce the analyte concentration but with accuracy issues for low analyte concentration and dependency on the oxygen concentration. Therefore today, nearly exclusively hydrogen peroxide is the intermediate substance in this type of biosensors.

Figure 10.9 shows a typical membrane stack at the working electrode of advanced glucose biosensors. Other analytes would look similar but with different enzymes and accordingly different reactions. The essential layer is the enzyme membrane hosting the

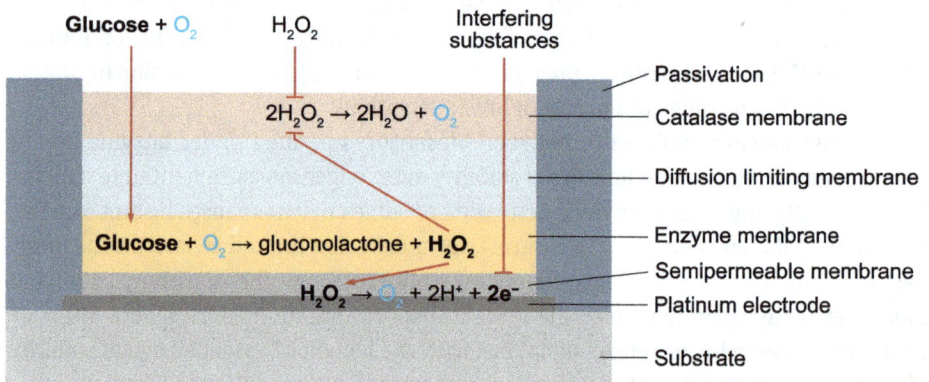

Figure 10.9: Working electrode of an advanced glucose biosensor using oxygen as cosubstrate ("first-generation biosensor").

glucose oxidase, e. g., in a pHEMA hydrogel. The exact membrane composition is vital to immobilize the enzyme in a sufficiently high concentration without losing functionality. While glucose oxidase is a pretty stable enzyme, other oxidases are more critical.

The other indispensable component is the electrode acting as the electrochemical transducer to measure the H_2O_2. Platinum allows both oxidation and reduction of H_2O_2. Still, oxidation is the better choice because of the recycling of O_2 molecules and the independence of the oxygen, which would also contribute to the current at the reduction potential. An alternative electrode material is Prussian blue (see the infobox), allowing H_2O_2 reduction at a potential less prone to oxidation of interferents. However, O_2 is not recycled as with the H_2O_2 oxidation at platinum, and overall electrode stability is much lower, limiting its application, e. g., for cost-effective single-use sensors.

Another scheme to detect hydrogen peroxide from the enzyme reaction is to combine glucose oxidase with horseradish peroxidase (HRP). HRP catalyzes the reduction of H_2O_2 while its structure facilitates direct electron transfer to the electrode.

Prussian blue (PB) is a ferric hexacyanoferrate ($Fe_4^{III}[Fe^{II}(CN)_6]_3$), presented in aqueous solution as a colloidal dispersion, typically in the form of $KFe^{III}[Fe^{II}(CN)_6]$. As a deposit, it keeps its characteristic blue color allowing both oxidation and reduction. The reduced form turns colorless (Prussian white), while the oxidized form is Berlin green (Prussian yellow).

PB deposits on gold and platinum but also on various carbon electrodes. Different coating methods are possible: the electrochemical route is cyclic voltammetry in an electrolyte containing $K_3[Fe(CN)_6]$ and $Fe(Cl)_3$. Chemical deposition occurs by spontaneous reaction between $K_3Fe(CN)_6$ and $Fe(Cl)_3$ in the presence of KCl and HCl. For both approaches, a heat treatment at 100 to 120 °C increases stability. Ricci et al. provide a comprehensive review on PB and its application in biosensors [206]. However, PB films degrade over time, especially in alkaline environments. Bilayers of ferric and nickel hexacyanoferrate may present a solution to overcome the stability issues [207, 208].

The other layers shown in Figure 10.9 are optional but improve the sensor performance depending on the desired application. The diffusion limiting membrane can be the same hydrogel as the enzyme membrane limiting the mass transport of the glucose to the enzyme. The layer thickness allows controlling the linear range of the sensors.

A semipermeable membrane prevents huger interfering substances from reaching the electrode and helps to avoid electrode poisoning. The retention mechanism is mainly by size-exclusion allowing small molecules like H_2O_2 and O_2 to pass but retains, e. g., ascorbic acid. The electropolymerized layer from 1,3-diaminobenzene shown in Figure 9.12 is a typical example. Another commonly used material is Nafion, a sulfonated tetrafluoroethylene-based fluoropolymer-copolymer.

Hydrogen peroxide generated in the enzyme reaction may leave the membrane stack. Apart from causing a flow rate dependency when such electrodes are part of a microfluidic system, the released H_2O_2 can adversely affect the environment, e. g., in a cell culture monitoring system (Section 10.1.8). Additionally, the sensor shows cross-sensitivity to hydrogen peroxide as the electrode cannot distinguish between H_2O_2 generated by the enzyme reaction and coming from the outside. Another hydrogel membrane containing the enzyme catalase (EC 1.11.1.6) on top of the diffusion limiting membrane solves both issues. The catalase catalyzes the decomposition of hydrogen peroxide:

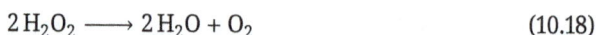

$$2\,H_2O_2 \longrightarrow 2\,H_2O + O_2 \qquad (10.18)$$

A positive side effect is the recycling to O_2, turning the sensor signal less dependent on available oxygen in applications with high glucose concentration but low oxygen background.

A complete sensor requires a working electrode with the described membrane stack together with counter and reference. Especially for application in the life sciences, it is routine to use a second working electrode as a blank electrode having the same membrane stack as the first working electrode but without the enzyme. Subtraction of the two currents leads to further interference elimination as long as the background is not too high compared to the signal from the enzyme reaction.

Great examples of first-generation biosensors are the works of Jobst et al. and Moser et al. on microfabricated biosensor arrays for glucose, lactate, glutamate, and glutamine [209–211].

---

i ■■□ Task 10.1: Sketch and label with the correct reactions a possible membrane stack for a lactate biosensor's working electrode. Consider the need for mass-transport limitation and an application in an environment containing hydrogen peroxide.

Biosensors with mediator ("second generation")

In enzymatic biosensors based on oxidases, mediators may take over the molecular oxygen's natural role as the electron acceptor for the oxidation of $FADH_2$ back to FAD (the reverse reaction to equation (10.11)). A mediator is the oxidized form of a redox couple; a straightforward candidate is $[Fe(CN)_6]^{3-}$ added to the solution. In general, mediators immobilized together with the enzyme in a membrane covering the electrode are the better choice. Both inorganic and organic mediators are possible. An alternative is the entrapment of the enzyme in a redox polymer.

Avoiding the route via hydrogen peroxide has several advantages: most prominently is the independence of oxygen, but also the hydrogen peroxide limits the enzyme lifetime in some cases. Additionally, mediators avoid or shift the risk of influence by other electroactive substances interfering with hydrogen peroxide.

Biosensors with direct electron transfer ("third generation")

Direct electron transfer (DET) from the enzyme seems the most elegant approach. However, it turns out that existing solutions still cannot compete with enzymatic biosensors of the above-discussed principles regarding long-term stability. Possible materials enabling direct electron transfer are organic salts such as TTF–TCNQ based on tetrathiafulvalene (TTF) and tetracyanoquinodimethane (TCNQ) or conducting polymers such as polyethylenedioxythiophene (PEDOT). Other strategies involve nanomaterials such as carbon nanomaterials, especially SWCNT (Section 9.4.1), for easier access to the enzyme's active sites. With the latter approach, even enzymes like glucose oxidase generally not thought to allow DET may be accessible. Recommended review articles to deepen this field are, e. g., the works from Luong et al. [212] and Schachinger et al. [213].

10.1.5 Glucose meter

Having learned a lot about instrumentation, electrochemical, and now biosensors, you may wonder what is inside one of the most widely spread biosensor systems with electrochemical readout: the self-monitoring of blood glucose (SMBG) devices with disposable sensor strips.

In SMBG, the patient collects a droplet of capillary blood by finger pricking and places it at the entrance of the sensor strip. Typically, less than 1 µl of blood is enough to fill the single-use strip by capillary forces. Inside is a disposable measurement cell with a glucose biosensor. In current commercial systems, the enzyme is often FAD-dependent glucose dehydrogenase (EC 1.1.5.9). This enzyme provides high selectivity toward glucose against other monosaccharides without the need for oxygen as a cosubstrate. Instead, a quinone gets reduced to a quinol:

$$\text{Glucose} + \text{quinone} \longrightarrow \text{Gluconolactone} + \text{quinol} \qquad (10.19)$$

Brazg et al. provide some insight into the electrochemical methods and instrumentation of the Accu-Chek Guide from Roche Diabetes Care [214]. The system uses multifrequency impedance measurement to obtain correction factors, e. g., for the hematocrit followed by pulsed-amperometric detection for the glucose measurement. Interestingly, the manufacturer did not develop an application-specific integrated circuit but relies on the ADuCM350 potentiostat and impedance analyzer circuit from Analog Devices (Section 3.2.7).

10.1.6 Immunoassay, immunosensor

Antibodies are another attractive detection element offered by nature, sometimes even more selective than enzymes and much more versatile in the targets. In the organism, antibodies help detect unwanted molecules called the antigen. The high selectivity of the antibody-antigen binding reaction also allows technical application in *immunoassays* (a procedure to measure the presence or concentration of an antigen) or *immunosensors* (continuous measurement of an antigen). The term immunosensor, in current literature, also can refer to systems for single-shot measurements.

In biochemical laboratories, immunoassays are standard procedures applied in microtiter plates with optical readout. Among different formats, the enzyme-linked immunosorbent assay (ELISA) is the common one. Here, an enzyme bound to an antibody or antigen catalyzes a reaction enabling the detection by optical or electrochemical means. For the latter, the enzyme could be glucose oxidase generating H_2O_2 from a glucose solution allowing electrochemical detection.

Figure 10.10 illustrates two possible ELISA formats. First, a capture antibody gets immobilized to the surface of a microtiter plater or a microfluidic channel. Incubation with proteins like bovine serum albumin (BSA) helps to block the surface to prevent non-specific adsorption of the antigen to the surface next to the antibodies. Often the preparation (I) takes place immediately before the measurement, but also strategies to store prepared surfaces exist. The different steps during preparation and measurement typically require washing in between. Depending on the assay format, the labeling enzyme comes attached to a second antibody or an antigen added to the sample.

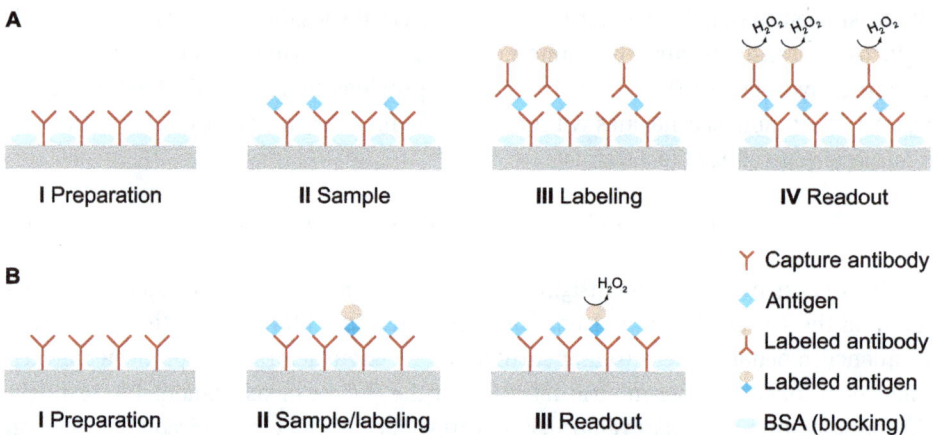

Figure 10.10: Enzyme-linked immunosorbent assay (ELISA): in the sandwich assay (A), the label is at the capture antibody; in a competitive assay (B), a labeled antigen competes with the sample. Glucose oxidase catalyzing the formation of H_2O_2 in the presence of glucose is a common enzyme used as the label.

In a *sandwich ELISA* (Figure 10.10(A)), the antigens from the samples bind to the capture antibody (II). A second antibody linked with the enzyme label binds in a subsequent step to the immobilized antigens (III). Finally, the enzyme reaction helps to read out the assay (IV). In the case of glucose oxidase as the label, the enzyme catalyzes the decomposition of glucose to hydrogen peroxide (equation (10.9)), which gets oxidized at a platinum electrode.

In a competitive assay (B), the sample with the antigen comes to the surface, together with a labeled antigen (II). Both antigens compete for accessible binding sites of the capture antibodies. Upon reading the labels (e. g., by flushing with a glucose solution in case of glucose oxidase), the signal is inverse to the antigen concentration in the sample (III).

Horak et al. describe implementing different immunoassays on a microfluidic platform with an on-chip electrochemical cell to detect hydrogen peroxide [215–217]. Apart from enzymes, other labels are possible, too, e. g., quantum-nanodots (QDs) with different metals, allowing multiplex detection of several antigens in parallel by stripping voltammetry. Kokkinos et al. give a comprehensive overview on different electrochemical immunosensors [218].

An interesting electrochemical method to amplify signals is using a label generating a redox couple and combing the immunosensor with redox cycling.

10.1.7 Redox cycling

Redox cycling as an electrochemical method is a powerful tool for signal amplification. Figure 10.11(A) illustrates the concepts: a pair of interdigitated electrodes form two working electrodes polarized to potential driving either oxidation or reduction. The species of interest as the reduced form R of a redox couple

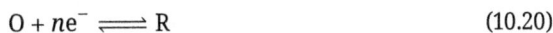

$$O + ne^- \rightleftharpoons R \qquad (10.20)$$

gets oxidized at one electrode ("generator") and reduced at the other ("collector"). The close distance in the micrometer range or below results in feedback between the species generated at one electrode and consumed at the other. The amplified signal, therefore, depends on the mass transport between the electrodes but not on the reaction kinetics. Depending on the initial and bulk concentration of the redox couple, a steady-state value establishes. The principle is equivalent to SECM in feedback mode (Section 6.2.1).

Two parameters evolved to describe the performance of different systems: The ratio between the current with redox cycling ("dual-mode") compared to the value with one active electrode only ("single-mode") is the *amplification factor*:

$$\text{Amplification factor} = \frac{I_{\text{dual-mode}}}{I_{\text{single-mode}}} \qquad (10.21)$$

The *collection efficiency* measures how much of the generated species can react at the collector electrode:

$$\text{Collection efficiency} = \left| \frac{I_{\text{collector}}}{I_{\text{generator}}} \right| \qquad (10.22)$$

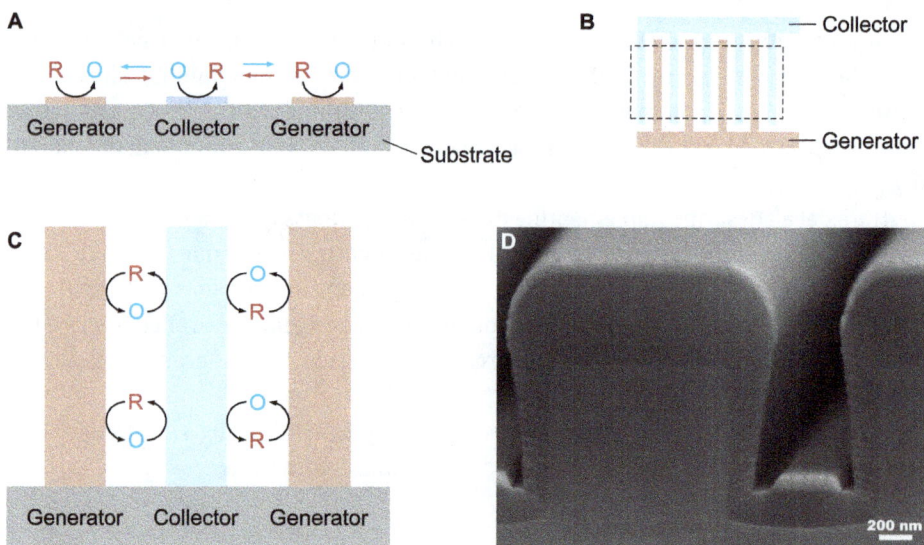

Figure 10.11: Redox cycling: a set of generator and collector electrodes allow oxidation and reduction of a redox couple (A). Top view on an interdigitated electrode array (IDA); typically, a passivation layer covers the whole structure but a rectangular opening indicated with dashed lines at the electrode fingers (B). 3D electrode structures enable redox cycling at the electrode sides leading to flow-rate independence (C). IDA with one electrode covering the entire ridge, the other electrode is at the bottom of the nanocavity (D) [219]. D: Reprinted with permission from Partel et al., ACS Nano, 10, 1086–1092 (2016). Copyright 2015 American Chemical Society.

Sanderson and Andersson described redox cycling with interdigitated electrodes back in 1985 [220]. Afterward, many works improved both electrode fabrication, the theory of the steady-state current, and the electrochemical application [221–231].

Typical electrode materials are carbon or noble metals, depending on the desired redox couple and possible interferents. The planar arrangement of the interdigitated electrode array (IDA) Figures 10.11(A) and (B) has the disadvantage of strong dependency on convection in the solution. Three-dimensional structures (Figure 10.11(C)) result in higher amplification factors and largely flow-independent signals [232]. A compromise between 2D and 3D is the work of Partel et al., allowing the efficient fabrication of interdigitated electrode arrays in a lift-off-free process resulting in nanocavities [219]. Figure 10.11(D) shows a cross-section of the electrode arrangement, and Figure 10.12 the characterization of the IDA with ferrocene methanol as the redox couple.

10.1.8 Sensortechnology for different applications

Many microsensor systems are to some extend fused with the measurement area rather than being a self-contained device, like a macroscopic pH glass electrode immersed in

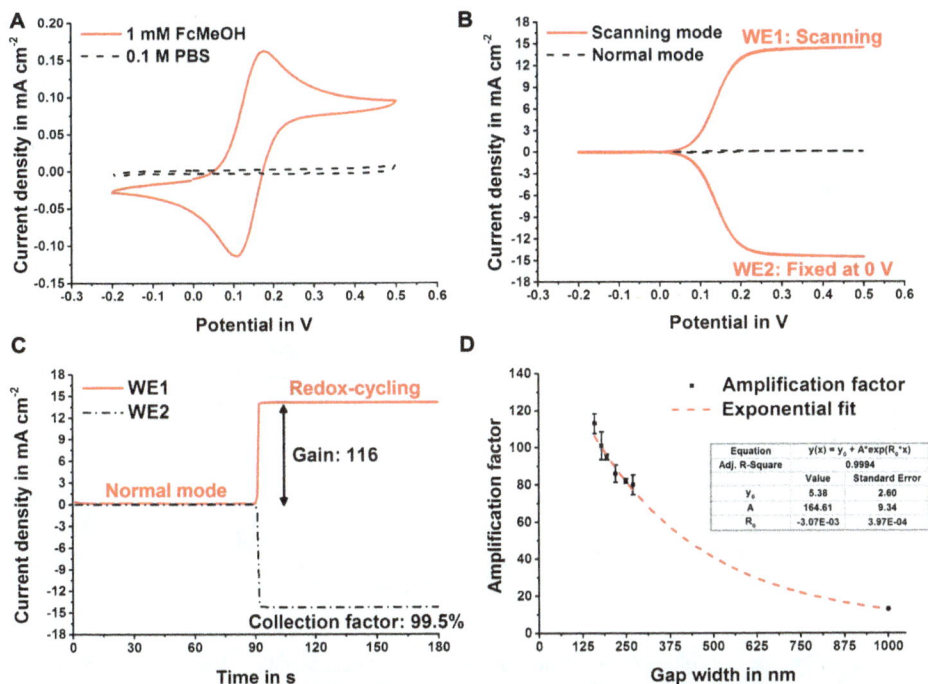

Figure 10.12: Redox cycling using gold IDAs as shown in Figure 10.11(D): cyclic voltammogram of the used redox couple in single-mode (A). Comparison of single-mode ("normal") and dual-mode ("scanning") with WE2 fixed at a potential driving the reduction (B). Amperometric measurements providing amplification factor and collection efficiency (C). Comparison of amplification factors for different geometries (D). [219]. Reprinted with permission from Partel et al., ACS Nano, 10, 1086-1092 (2016). Copyright 2015 American Chemical Society.

the solution to be measured. The sensor technology strongly depends on the realization of the overall system. While the outcome of such an approach benefits the application, the commercialization of a stand-alone sensor device would be, of course, more straight. This section does not aim for completeness but presents microsensor use cases from different domains.

Sensing Sell Culture Flask

Cell culturing is a routine procedure, e. g., in fundamental research and the selection process for drug candidates. Especially in cancer research, investigation of cancer cells in different oxygen environments and their ability to survive even low oxygen concentrations (hypoxia) is of tremendous interest as therapy resistance and risk of metastasis comes along with hypoxia [233, 234]. One possibility is the cell culturing in a flat tissue culturing flask with a monolayer of cells growing attached to the bottom. The placement of sensors must be at a fixed position relative to the cell layer to capture reliably concentration gradients formed to and from the cells what makes approaches with dip-in

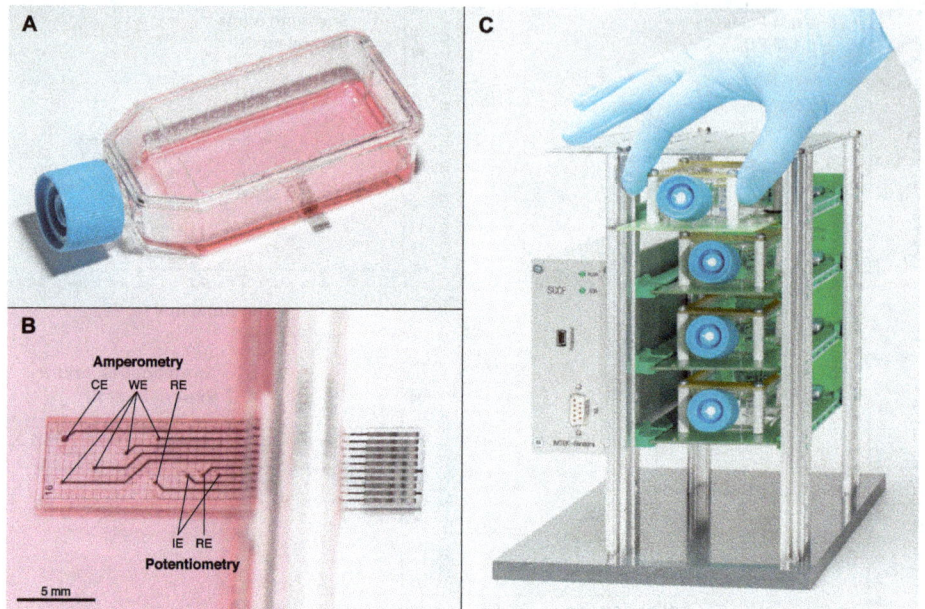

Figure 10.13: Sensing Cell Culture Flask: an electrochemical cell culture monitoring system with a transparent microsensor chip embedded in the bottom of a conventional tissue culture flask (A). The sensor chip features different electrodes for amperometry and potentiometry (B). The system comprises a rack system including a potentiostat with a multiplexer to operate the system, e. g., in an incubator (C). [140]. Reproduced from Kieninger et al., 2018, DOI 10.3390/bios8020044, under CC BY 4.0.

sensors cumbersome. The Sensing Cell Culture Flask (SCCF) is a monitoring system comprising sensor chips at the bottom of conventional tissue culture flasks (Figure 10.13).

The SCCF platform allows the integration of different chemo- and biosensors with amperometric or potentiometric readout [140]. With such as system, it is possible to track the cell's oxygenation and respiration with direct amperometric oxygen sensors and help to standardize cell culture experiments [235]. The close placement of the sensors to the cells also allows the measurement of short-lived species such as superoxide [236].

A variant of this technology platform is the Sensing Cell Culture Well (SCCW), providing similar functionality in the format of a well plate. The transparency of the sensor chip gives optical access to the cell layer, e. g., to research with electrochemical microsensors the cell response from photodynamic therapy [237].

In such platforms, there is no clear separation between the microsensor and the measurement domain. Working and indicator electrodes are placed close by the cells, while reference and counter electrodes may be at some distance. The cell culture medium is the sensor's electrolyte simplifying the technology tremendously also concerning the requirement of disposable systems. For short-lived species, a separate sensor electrolyte would not be possible at all. Cell culture medium offers sufficient conductivity, a known chloride ion concentration to use an Ag/AgCl reference electrode without an inner electrolyte, and a buffer keeps the pH around neutrality. However, several additives such as amino acids or antibiotics require careful testing in the specific medium and self-cleaning sensor protocols to avoid poisoning of the sensor electrodes. Often, cell culture medium contains serum with unknown composition requiring thorough characterization, as it is a natural product for some applications, even for each batch.

We provided a general overview on microsensor systems for metabolic monitoring in cell culture and organ-on-chip in [5].

Our body, with its 10^{14} cells, is a highly complex system that we often may see its functionality as a sheer miracle. Studying the physiology and pathophysiology of the human body is a big challenge, especially as soon it gets to the cellular level. Therefore, isolating cells as the fundamental bricks of life is an essential tool to learn about the underlying mechanism and allows us to do tests we want not to do with humans, such as probing different compounds for optimal treatment. Personalized medicine promises to find the optimal therapy for the individual patient, e. g., by testing the response of various cancer treatments on cells taken from the patient's tumor before administrating the best matching drug to the patient. Since Wilhelm Roux kept the embryonic chick cells alive outside the animal body in 1885, many scientists improved how to keep cells in culture. A new era in cancer research started when George Otto Gey isolated human cervical carcinoma cells from Henriette Lacks in 1952, still available in many labs worldwide known as HeLa cells.[2] For a long time, growing the cells as a monolayer at the bottom of a dish, a well or tissue culture flask was standard and is still done in many cases. However, it became evident that two-dimensional cultured cells resemble the in vivo situation much less than 3D approaches. Combining different cells with the promise of organ-like functionality further bridged the gap between cell culture and functional units in the human body. Such so-called *organ-on-chip* (OOC) systems showed superior modeling of the in vivo situation but also implied novel challenging on the sensor integration [5, 239].

Microphysiometry in a lab-on-chip

The above-discussed SCCF system fits very well the static nature of conventional cell culture. However, gaining data about a compound's effect on cells should take only a short time. Ideally, one would measure the cell's physiological response to an environment that might be refreshed frequently or exchanged with a compound-containing solution. Lab-on-chip systems allow such microphysiometry, combining a cell culture area, microfluidics, and microsensors (Figure 10.14). The microfluidic cyclically refresh the medium or provides compounds. Sensors are within the cell culture area (e. g., oxygen and pH sensors to monitor the cellular respiration and acidification) or downstream (e. g., biosensors to monitor glucose consumption or release of lactate).

Weltin et al. described the fabrication and application of such a microphysiometry system to monitor the metabolism of human cancer cells [240]. Different methods read

---

2 While the HeLa cells helped in research for decades, the patient died without knowing what will happen to her cells, nor have the family been informed appropriately. It took more than five decades before communication started between the patient's family and scientists [238].

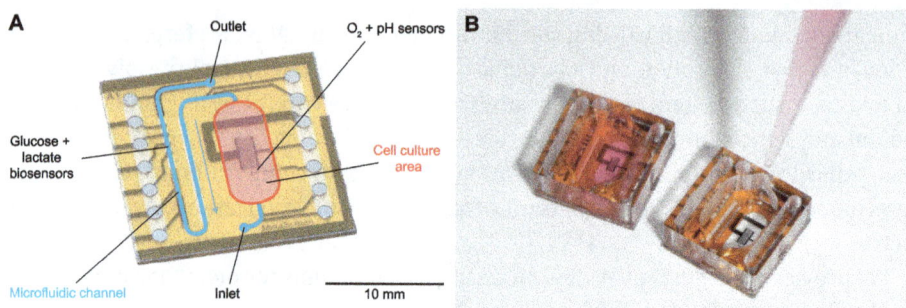

Figure 10.14: Lab-on-chip for microphysiometry: microsensors observe the metabolism of cells cultured in the chip, cyclic microfluidic actuation refreshes the medium in the cell culture or provides stimuli such as drugs to the cells (A). Photography of the microphysiometry chip with the cell culture well filled for cell culturing phase (B) [234, 240]. Reprinted from Pettersen et al., Targeting Tumour Hypoxia to Prevent Cancer Metastasis. From Biology, Biosensing and Technology to Drug Development: The METOXIA Consortium. J Enzym Inhib Med Ch, 2014, 30 (5), 689–721, by permission of Taylor & Francis Ltd, http://www.tandfonline.com.

the sensors in this lab-on-chip: biosensors use amperometry, the pH sensors are potentiometric, and the oxygen sensors are direct amperometric sensors with a chronoamperometric protocol to ensure long-term stability.

Metabolic monitoring in organ-on-chip systems
Organ-on-chip systems (see the info box) enable much better modeling of the in vivo situation than simple cell culture systems. At the same time, the progress in 3D cell culturing, often in different but connected compartments, poses new challenges for accessing cellular metabolism with electrochemical sensors. Figure 10.15 shows an organ-on-chip system with two compartments for cell growth and three microchannels to supply the cells with fresh medium and remove waste products [241]. The middle channel also allows for communication by signal molecules between the cell compartments when the medium is stagnant, or the communication can be interrupted when the flow in this channel is switched on.

The system comprises eleven oxygen sensor working electrodes and biosensors for lactate and glucose at different locations. All sensors share a common counter electrode and up to three reference electrodes. The electrochemical instrumentation needs to consider the joint counter and reference electrodes with various working electrodes in the same electrolyte. Additionally, parallel operation of constant amperometry for the biosensors with chronoamperometric oxygen sensor protocols, including open-circuit phases, must be possible. Preferred instrumentation is a multi-channel potentiostat. However, multiplexing in sync with the microfluidic actuation is also possible.

Another challenge of such systems is the interpretation of the results. In the example of Figure 10.15, patient-derived breast cancer cell lines were grown to spheroids in the

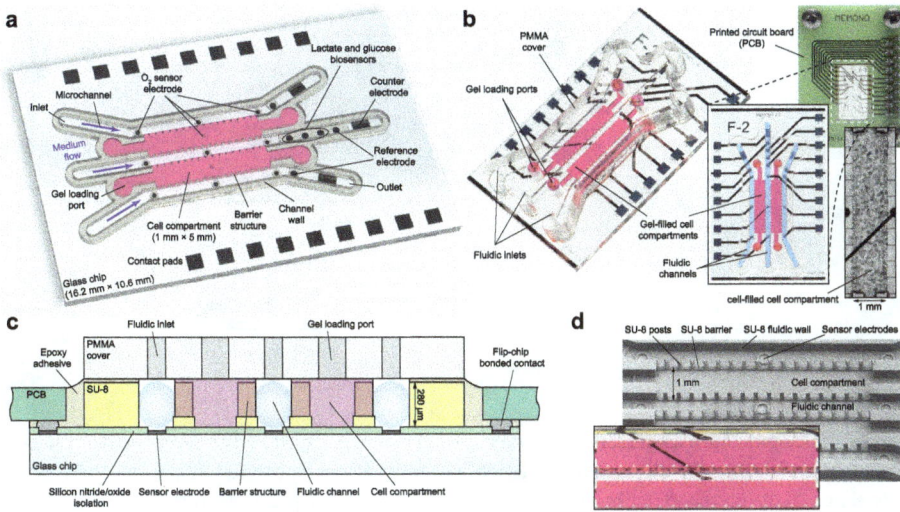

Figure 10.15: Organ-on-chip-system for metabolic monitoring. The chip features two cell compartments (purple in a and b) with three microchannels in between. Different oxygen sensor and biosensor electrodes are spread in the systems (a). Along with the photographs, a phase-contrast micrograph shows breast cancer spheroids in the cell compartments (b). Schematic cross-sectional view (c) and illustration of micro-fabricated barrier structures to confine the cell compartments (d) [241]. Reproduced from Dornhof et al., Microfluidic organ-on-chip system for multi-analyte monitoring of metabolites in 3D cell cultures, 2022, DOI: 10.1039/D1LC00689D, under CC BY 3.0 (https://creativecommons.org/licenses/by/3.0/).

cell compartments. While each spheroid replicates a patient's tumor, the sensor access is limited to integral quantities over many spheroids. Bioprinting is a possible solution, bringing a single spheroid next to the sensor. Figure 10.16 shows an example with single breast cancer spheroids deposited into an oxygen sensor microelectrode well [242]. It is essential to understand that the tiny drop in dissolved oxygen concentration can only be resolved if the concept enables the confinement of the analyte volume, which is realized in this system by a glass slide as a lid to close the microwell.

The term *in vivo* addresses the situation in an organism, while *in vitro* is the culturing of cells outside the body. Please note the different definitions in biochemistry, with in vivo describing the situation in a cell, while in vitro addresses the subcellular elements outside the cells.

Flexible microsensor for in vivo applications

Flexible microsensor strips allow the placement of electrochemical sensors in vivo. Such systems focus on acute experiments in anesthetized animals rather than aiming for permanent implantation or application in humans. Figure 10.17 shows polyimide sensor strips equipped with different membrane-covered platinum electrodes to operate biosensors or oxygen sensors [243–245].

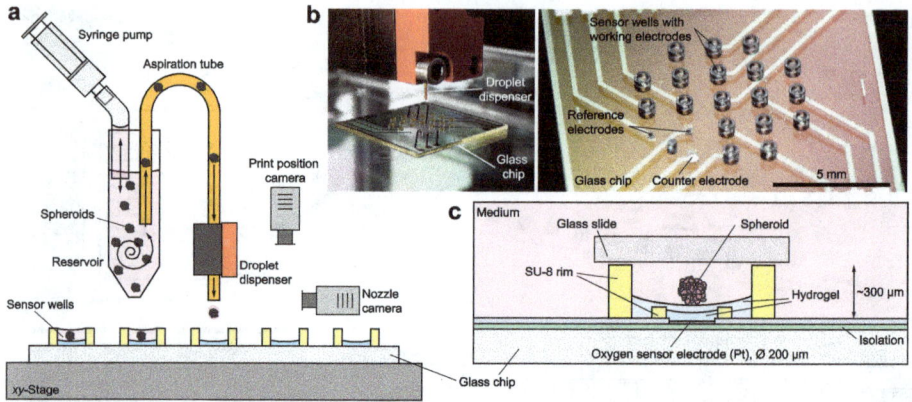

Figure 10.16: Bioprinting of single spheroids into microwells comprising an oxygen sensor working electrode. Scheme of the bioprinting process (a), photograph of the dispenser and sensor chip (b), and cross-sectional view of the microwell for the oxygen measurement (c) [242]. Reproduced from Dornhof et al., Bioprinting-based automated deposition of single cancer cell spheroids into oxygen sensor microelectrode wells, 2022, DOI: 10.1039/D2LC00705C, under CC BY 3.0 (https://creativecommons.org/licenses/by/3.0/).

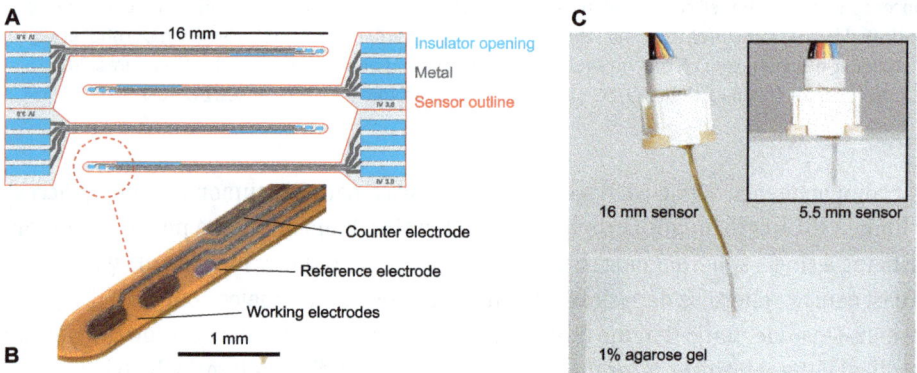

Figure 10.17: Flexible microsensor strips for in vivo applications: layout of the masks for sensor fabrication (A); image of the sensor tip at a fabrication stage after metal patterning and insulator opening (B); electrically connected sensors inserted into a gel emulating brain tissue (C) [243]. Reprinted from Biosensors and Bioelectronics, 61, Weltin et al., Polymer-Based, Flexible Glutamate and Lactate Microsensors for In Vivo Applications, 192–199, Copyright 2014, with permission from Elsevier.

A typical configuration to apply such a microsensor strip in neuroscience comprises a glutamate and lactate biosensor (Section 10.1.3). Glutamate is the major excitatory neurotransmitter in the central nervous system, and lactate is a marker for the anaerobic energy metabolism. Another application is the research on wound infections [246]. Such flexible microsensor strips also allow measurements outside an organism, e. g., in 3D cell culture of hepatocyte spheroids [245].

Weltin et al. provided a comprehensive review on microfabricated enzyme-based biosensors for in vivo applications [247].

Process monitoring in microreactors

The last use case discussed in this section is independent of biomedical applications. Hydrogen peroxide is a liquid oxidizing agent relevant to many industries. Its decomposition resulting in water, makes it attractive for "green chemistry" if the production process is eco-friendly. Heterogeneously catalyzed direct synthesis of hydrogen peroxide in microreactors is an attractive option but requires tight process control to let hydrogen and oxygen react without reaching the explosive range. Figure 10.18 shows sensor plugs to measure O_2, H_2, and H_2O_2 at the same platinum working electrode.

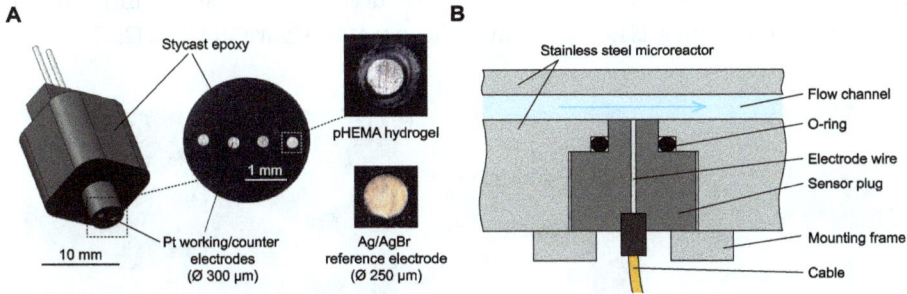

Figure 10.18: Process monitoring for hydrogen peroxide direct synthesis: sensor plugs (A) to measure O_2, H_2, and H_2O_2 at the same working electrode in a stainless steel microreactor (B) [248]. Reprinted from Sensors and Actuators B: Chemical, 273, Urban et al., Electrochemical Multisensor System for Monitoring Hydrogen Peroxide, Hydrogen and Oxygen in Direct Synthesis Microreactors, 973–982, Copyright 2018, with permission from Elsevier.

The sensors are as simple as possible, with polished wires embedded in an epoxy housing to withstand harsh conditions such as high pressure up to 100 bar. Hydrogel covers the working electrodes to ensure diffusion limitation; the reference electrodes are Ag/AgBr relying on the known bromide ion concentration in the electrolyte. A chronoamperometric protocol changing the platinum surface to either bare or oxide covered and careful data evaluation allows separation of the O_2, H_2, and H_2O_2 concentrations [248–250].

10.2 Actuators

Electrochemical actuators are the logical opposite to sensors, however, not as widespread and multifaceted. Still, especially in the upcoming field of soft robotics, electrochemical actuators play a prominent role as artificial muscles. Also, electrochemical pumps were proposed, e. g., for drug delivery from implants.

10.2.1 Release due to membrane corrosion

The release of substances from cavities by membrane corrosion is a method exploiting the scalability of microsystems. While each compartment open only once, integration of a multitude allows the release of controllable amounts or at different time points. This concept uses a thin gold layer sealing a microcavity and allows one-time opening by causing dissolution of the gold membrane (Figure 10.19(A)). In a chloride-containing solution, the gold forms a soluble complex with chloride ions before the onset of passivation. Fabrication of an array with many individual addressable cavities allows for release at different time points. Santini et al. introduced the principle by fabricating an array of 34 cavities with a 25 nl volume in a silicon wafer (Figure 10.19(B)) [251].

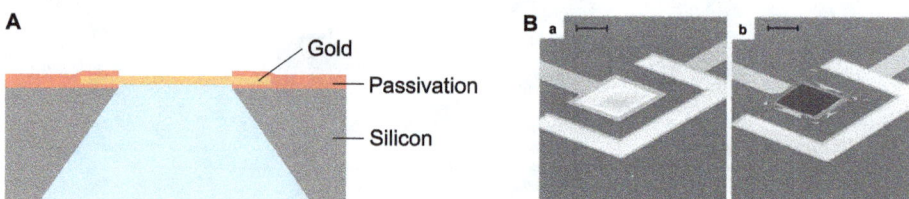

Figure 10.19: Release of substances from a microcavity: a thin gold layer seals the cavity (A), which permanently opens upon membrane corrosion at an appropriate potential. Practical realization (B) with a closed (a) and already opened (b) cavity [251]. The scale bar is 50 µm. B: Reprinted by permission from Springer Nature: A controlled-release microchip, Santini et al., Nature, vol. 397, 335–338 (1999). Copyright 1999.

Separating the release mechanism from the stored substances promises flexible application. However, the technological effort for such membrane arrays is high. In between, electroactive polymers got more attention. Here, a redox reaction of the polymer triggers the release. Typical electroactive polymers are polyaniline or polypyrrole; also, the usage of biodegradable electroactive polymers is feasible [252–254].

10.2.2 Electrolysis actuators and pumps

Electrolysis of water forming gas bubbles generates pressure in situ. The mechanism enables driving a liquid out of a reservoir, e. g., for drug delivery, as illustrated in Figure 10.20. The reaction at the anode is oxygen formation:

$$2\,H_2O \longrightarrow O_2 + 4\,H^+ + 4\,e^- \tag{10.23}$$

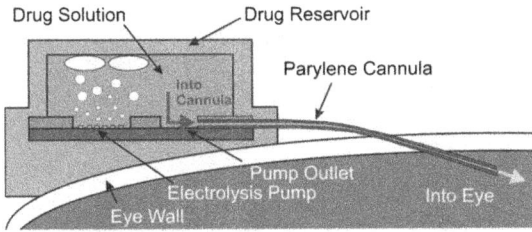

Figure 10.20: Electrolysis pumps proposed for intraocular drug delivery: Pressured generated by gas formation drives a solution out of the cannula [256]. Reprinted from Sensors and Actuators A: Physical, 143, Li et al., An electrochemical intraocular drug delivery device, 41–48, Copyright 2008, with permission from Elsevier.

Hydrogen formation occurs at the cathode:

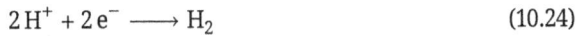

$$2\,H^+ + 2\,e^- \longrightarrow H_2 \tag{10.24}$$

A refinement of such an electrolysis pump is separating the electrolysis chamber from the pumped medium by a membrane or bellow structure [255].

Also, reversible actuation is possible. Here, e. g., oxygen formation/reduction at platinum is paired with copper as the opposite electrode in an electrolyte contains copper sulfate. The desired reactions are

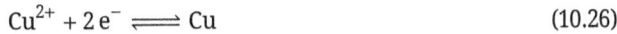

$$O_2 + 4\,H^+ + 4\,e^- \rightleftharpoons 2\,H_2O \tag{10.25}$$
$$Cu^{2+} + 2\,e^- \rightleftharpoons Cu \tag{10.26}$$

Parasitic oxygen reduction at copper during electrolysis needs to be prevented by a membrane [257].

Although academic literature has described electrolysis actuators and pumps for more than two decades, no larger commercial application emerged, assumingly because in many fields the effort to encapsulate a liquid electrolyte does not justify the benefits.

10.2.3 Bending beam actuators

The scientific literature describes electrochemical bending beam actuators for long. The demand for soft actuators as "artificial muscles" brought a new drive into the field, with many advances in the last decade. Bending occurs due to stress caused by the different densities in the films. Practical realizations are a bilayer with an inert and an electroactive film or a homogenous material in which the redox reaction causes differing domains.

An example of the latter is an actuator based on graphdiyne described by Lu et al.; see Figure 10.21. Graphyne, not to be confused with graphene, is a monolayer thick pla-

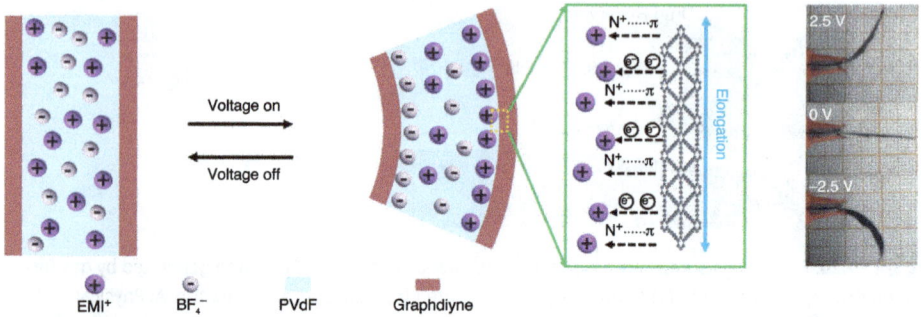

Figure 10.21: Bending beam actuator based on graphdiyne [258]. Reproduced from Lu et al., 2018, DOI 10.1016/j.bios.2005.10.020, under CC BY 4.0.

nar carbon allotrope comprising sp and sp^2-hybridized atoms. Graphyne-2 is a specific variety called graphdiyne.

Apart from carbon-based materials, electroactive polymers are typical materials. Well known is polypyrrole (PPy), proposed since long as actuator material [259]. In PPy doped with dodecyl benzenesulfonate (DBS), the redox reaction

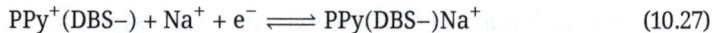

$$PPy^+(DBS-) + Na^+ + e^- \rightleftharpoons PPy(DBS-)Na^+ \tag{10.27}$$

causes different densities with shrinkage in the oxidized state. Xu et al. demonstrated the application of a bilayer made of PPy(DBS) and gold as a valve for a drug delivery device [260].

10.3 Neurotechnology

Neuroprosthesises allow restoration of functions for patients with neurological impairment. The devices aim at both the sensory system, e. g., cochlear or retina implants, and the motor system, e. g., to control the emptying of the bladder. It is reasonable that the field associates with the biblical promise that the blind see, the lame walk, and the deaf hear. More profane is the mechanism of all the neuroprosthesis: the stimulation of the nervous system by electrical activation.[3]

10.3.1 Neural activation: action potential

All biological cells show some potential difference between intra- and extracellular space. The voltage between inside and outside, the *membrane potential* or transmem-

---

[3] By principle, a neuroprosthesis does not restrict to electrical activation but may incorporate other methods like stimulation using light. In optogenetics, the control of neurons by light pulses is possible. However, as the technique requires genetic modification of the cells to express sensitivity to light, application in humans is not yet realized.

brane potential, originate in the concentration gradient of ions, controlled by ion pumps and ion channels in the cell membrane. In excitable cells (neurons, muscle cells), a change of the extracellular electrical field can trigger a spike in the membrane potential, called an action potential, which might propagate further to transport information.

Figure 10.22 shows the typical time trace of a neural *action potential*. At rest, the membrane potential is around -70 mV, dominated by the Nernst potential of potassium ions. An electrical stimulus can trigger an action potential when exceeding a threshold, typically -55 mV. Ion channels open, letting sodium ions entering the cell, and the membrane potential turns positive ("depolarization"). At around $+30$ mV, the sodium ion channels close, and the outflow of potassium ion reestablishes the baseline ("repolarization"), going through an overshoot ("hyperpolarization") in which the cell is refractory.

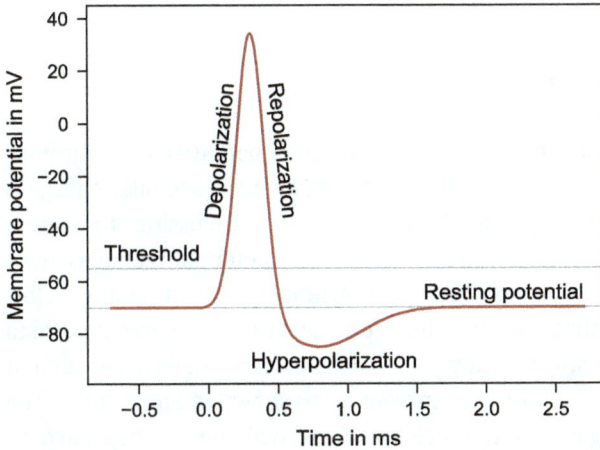

Figure 10.22: Action potential of an excitable biological cell (neuron, muscle cell) in response to a stimulus exceeding the threshold potential.

10.3.2 Neural recording by electrodes

Recording brain signals as sequences of action potentials are accessible on different levels. In electroencephalography (EEG), electrodes placed on the scalp record electrical signals from the surface layer of the brain. Typical amplitudes are between 10 and 100 μV. EEG supports the diagnosis of epilepsy, sleep disorders, some neurodegenerative diseases, but also tumors and stroke. In research, it is a low-impact method to learn about human brain activity. Frequency bands (Delta: < 4 Hz, Theta: 4 to 8 Hz, Alpha: 8 to 12 Hz, Beta: 12 to 30 Hz, and Gamma: > 30 Hz) associated with healthy and pathological characteristics help classify the brain waves. Electrodes made of Ag/AgCl connect through a conductive gel to the skin, recently also dry electrodes, e. g., using carbon fibers or silver, are available.

The next level of invasiveness is electrocorticography (ECoG), with an electrode array placed intracranially. ECoG allows for recording during surgery but may remain in place permanently, e. g., for some epilepsy patients. ECoG arrays can be part of a brain-computer interface (BCI).

Using electrodes in the brain tissue and breaking the blood-brain barrier is much more invasive but justified when deeper brain regions require access, mainly in combination with deep brain stimulation (DBS).

In research, invasive recording is routine in animal experiments or neuronal cell cultures by a microelectrode array (MEA) with up to 1000 microelectrodes. Electrode materials comprise ITO, titanium, gold, platinum, or carbon nanotubes. Generally, recording in the brain or with MEAs neglect the electrochemical potential (the DC potential with respect to a known reference potential).

10.3.3 Neural stimulation by electrodes

Electrical stimulation using an electrode can trigger an action potential at a neuron or muscle cell and initiate its propagation. An additional negative extracellular charge is needed to overcome the membrane potential threshold. When discussing stimulation by an electrode, pay attention not to confuse the membrane potential (voltage across the cell membrane) and the electrochemical scale concerning the SHE or another reference electrode. Practically, stimulation electrodes use a short cathodic current pulse to trigger the action potential. Investigating how to deliver the charge in terms of current and time leads to a diagram for the threshold current I_{th}, as shown in Figure 10.23. The threshold current for a very long pulse duration is associated with the rheobase current. The chronaxie t_c is the time at which I_{th} is twice I_{rh}. Most effective stimulation with low charge occurs with pulse duration around chronaxie [261]. For neural cells, t_c is around 100 µs; in muscle cells, the chronaxie is around two orders of magnitude higher.

Signal transmission in vivo uses multiple action potentials; the repetition rate encodes the intensity of the information. Electrical stimulation uses repetitive pulses with rates in the range of 10 to several 100 Hz. Using cathodic current pulses only could decrease the electrode potential out of the water window. Therefore, for more than half a century, biphasic, charge-balanced stimulation protocols have been in use [262]. In a biphasic protocol, an anodic current pulse follows the cathodic phase reversing charge-up of the double layer capacity and faradaic surface reactions. The disadvantage of a following anodic pulse is also partial cancellation of the stimulation effect. A short delay with several 10 µs in between the two pulses restore the efficiency to values like a monophasic, cathodic stimulation. Figure 10.24 shows a typical biphasic, charge-balanced stimulation protocol.

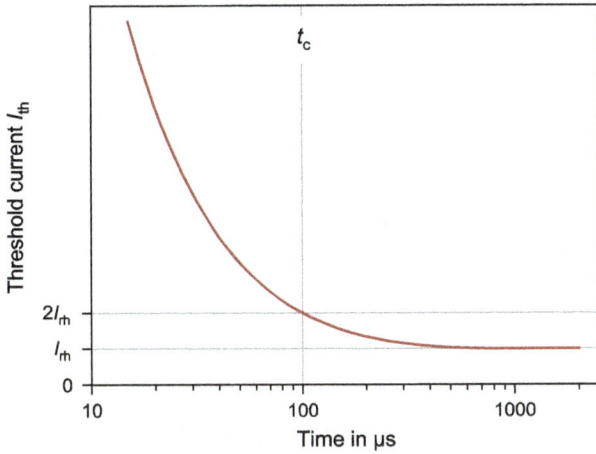

Figure 10.23: Neural stimulation by an electrode: chronaxie t_c is the pulse duration at which the threshold current to cause an action potential is twice the rheobase current I_{rh} (threshold current for very long pulse duration).

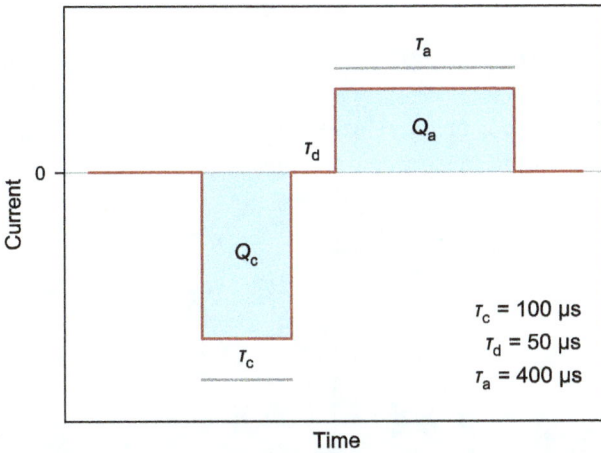

Figure 10.24: A typical biphasic stimulation protocol starting with the cathodic phase to trigger an action potential followed by a delay before the anodic pulse compensates the charge ($Q_a = -Q_c$). The pulse durations are typical values for stimulation of neuronal cells.

Neurotechnology is a fascinating subject bringing together biomedical application and electrochemistry. However, electrochemists need to understand that stimulation occurs at much lower timescales but higher current densities (around 1000 mA cm^{-2}!) than they usually deal with. Also, nonelectrochemists working in the field should be familiar with electrochemical methods and instrumentation. We recently addressed the latter in a tutorial [263]. A common source of confusion is the voltage between two electrodes versus the electrochemical potential scale – something which cannot be emphasized enough. From an electrochemistry point of view, any description of potentials occurring during stimulation should relate to a reference electrode (for neuroprosthetic applications, of course, there are good reasons not to have a reference electrode like Ag/AgCl in a patient). Cogan provided a sound overview on neural stimulation from an electrochemical perspective [264]. Doering et al. developed a setup measuring the potentials occurring during neural stimulation protocols with respect to a reference electrode and discussed the factors affecting the observed potentials [265].

Stimulation electrodes currently used in clinical applications rely on platinum, platinum alloy, or stainless steel electrodes. Besides charging the double layer capacity, platinum allows charge delivery from the surface reactions (e. g., formation of PtO). However, repetitive cycling in the region of PtO formation/removal risks platinum dissolution, as discussed in Section 8.4.3. Design of stimulation protocols should aim at staying in the potential window of the double layer region [266]. When oxygen reduction is part of the faradaic process, the fraction of irreversible charge in the cathodic phase depends on the oxygenation of the electrode environment [267].

Shannon developed a model to categorize the safety of different stimulation protocols [268]. He observed when comparing the charge density q and charge Q delivered per phase in a double-logarithmic plot (Figure 10.25), a line separates between safe and unsafe parameter space:

$$\lg q = k - \lg Q \tag{10.28}$$

Unsafe relates to either "overstimulation" or the release of toxic substances. From an electrochemical point of view, the second cause seems more likely.

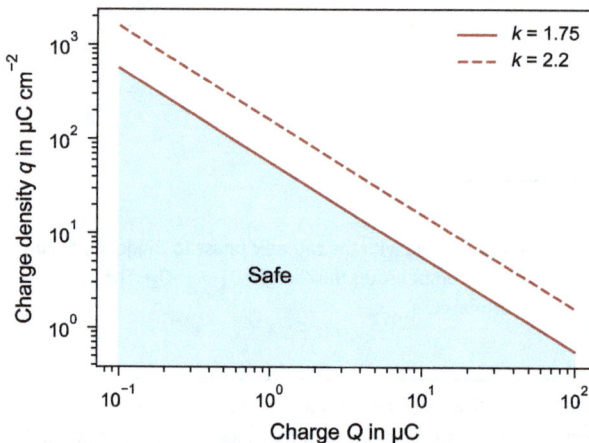

Figure 10.25: Shannon's plot to categorize different stimulation protocols according to their safety. The line represents $k = 1.75$ for platinum, and around 2.2 for stainless steel [266, 268].

Iridium oxide seems a promising alternative electrode material. The redox reaction Ir(III)/Ir(IV) provides a charge density without risking electrode dissolution or electrolysis (compare the cyclic voltammogram of an iridium oxide electrode, Figure 10.2). Initially, AIROF was in use; also, SIROF is possible [269]. For iridium oxide, a potential bias instead of an anodic counter pulse was proposed [270].

Academic works discuss alternative electrode materials such as carbon allotropes, electroactive polymers, or various metal oxides and nitrites. So far, no clear candidate was identified having the perspective to maintain its stimulation capability over many years in the human body without releasing toxic substances.

By now, clinical applications have to use the established electrodes like platinum. Here, an option is tuning the stimulation protocol. Since long charge-balancing between the cathodic and anodic pulse was standard, however, there is no clear electrochemical rationale behind it. Scheiner et al. found that anodic pulses require less charge to stay within the safe range [271]. Depending on geometry and environment, the optimum seems in between monophasic cathodic only and charge-balanced biphasic protocols. The key is considering irreversible faradaic processes like oxygen reduction. Kumsa et al. showed that in an oxygen-deficient environment, safe operation beyond Shannon's prediction is possible [266].

10.3.4 Neurotransmitter monitoring

While you read these lines and learn about electrochemistry, your brain uses both electrical and biochemical signals for information processing and hopefully remembering what you studied. The whole process occurs relatively inaccessible in everyone's mind, with all subconscious processes on top, including letting you continue to breathe while you read. Recording electrical signals of the brain enable one to observe intercellular communication through action potential. Of course, observing is far from understanding, and given the vast amount of information, we have to admit how little we know about the brain's functionality. Additionally, many neurotransmitters are active and have a specific and possibly different role depending on the brain's region. While electrical signals, at least integral values from the surface layer of the brain, are easily accessible by electroencephalography on the scalp, neurotransmitter concentrations are accessible by invasive methods only, generally restricted to research with animals.

Here, electrochemical methods come into play, selectively monitoring localized information about the different neurotransmitter and their temporal behavior. The timescale of interest is in the range of seconds or below. Accordingly, a fast but sensitive and selective detection principle is essential. In the last decades, fast-scan cyclic voltammetry (FSCV) with carbon-fiber microelectrodes evolved as the state-of-art method matching both the temporal and sensitivity demands, initially focusing on dopamine [272–274].

Measurements in the striatum of anesthetized or freely moving animals, typically rats, promise new insights into the biochemical background of behavior. Venton and Wightman coined the term "psychoanalytical electrochemistry" to emphasize the translational possibilities of the electrochemical measurements [275].

Dopamine detection

Dopamine is the primary neurotransmitter addressed by FSCV. Similarly, other catecholamines like norepinephrine (noradrenaline) or epinephrine (adrenaline) are accessible.

Carbon is the preferred electrode material, typically in cylindrical carbon-fiber microelectrodes with diameters between 5 and 30 μm and a length of 20 to 200 μm. Figure 10.26 shows a classical protocol providing ten readings per second.

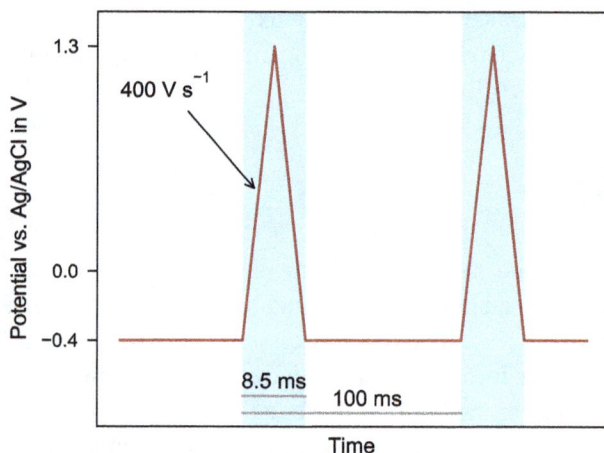

Figure 10.26: FSCV waveform for dopamine detection. After holding the electrode at the adsorption potential, the adsorbed dopamine gets oxidized to dopamine-o-quinone in the upward and reduced back in the downward scan.

The principle combines constant polarization at a hold potential at which dopamine adsorbs onto the electrode, followed by a cycle of quickly ramping up and down, oxidizing the adsorbed dopamine to dopamine-o-quinone and reducing it back:

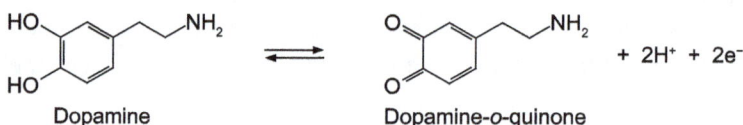

Dopamine   Dopamine-o-quinone $+ 2H^+ + 2e^-$

High scanrates not only provide sufficient temporal resolution but allow for higher currents. The upper limit for reasonable scanrates is, apart from instrumentation limitations, the increasing peak separation and, therefore, the shift of the dopamine oxidation peak towards the end of the water window. A reasonable value for dopamine is 300 or 400 V s^{-1}. The downside of the fast scan is a large background signal, often two orders of magnitude higher than the dopamine signal (Figure 10.27(A)). Therefore, signal evaluation requires subtraction of a blank curve (Figure 10.27(B)) revealing the oxidation and reduction peak.

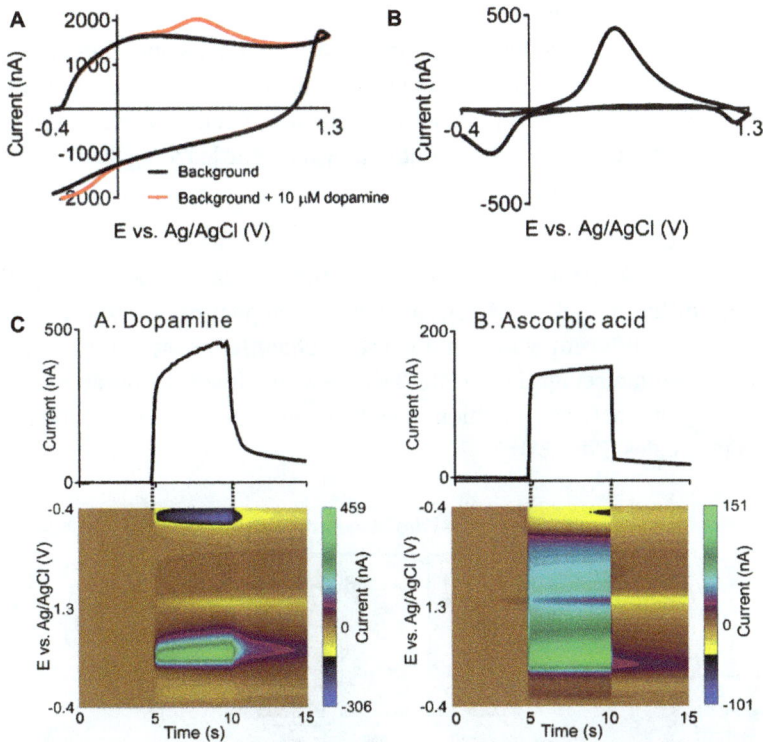

Figure 10.27: Fast-scan cyclic voltammogram with and without dopamine in PBS. Although 10 µM is a high concentration, the contribution from dopamine is minor (A). After background subtraction (B), the oxidation and reduction peaks are visible. Notable is the massive peak shift and the lower area under the reduction peak. Time trace of the current and color plot (C) for 10 µM dopamine and 20 µM ascorbic acid injected to PBS at 5 s, and PBS was injected at 10 s [276]. Republished with permission of the Royal Society of Chemistry, from Analyst, Fundamentals of fast-scan cyclic voltammetry for dopamine detection, Venton et al., vol. 145, 1158–1168, 2020; permission conveyed through Copyright Clearance Center, Inc.

A closer look at the substracted cyclic voltammogram shows an imbalance between oxidation and reduction. The area under the reduction peak is lower than for oxidation because dopamine oxidized to dopamine-*o*-quinone is attached less intensely to the electrode surface. Consequently, some molecules are desorbed before the reduction occurs.

FSCV with carbon-fiber microelectrodes provides a detection limit of few ten nM with a linear range of up to several µM matching well the typical in vivo concentrations between 0.1 to 1 µM during phasic activity depending on the brain region and the stimulus.

Michael et al. suggested color plots to visualize the temporal change of the cyclic voltammograms, today widely adopted by the community [277]. The y-axis represents the potential; the x-axis is the time. A color scale encodes the current. Figure 10.27 shows an example for dopamine and, in comparison, ascorbic acid, one of the ubiquitous interferents. Color plots provide a good overview of transient effects and allow easy identification of changes in the background current.

Venton and Cao provide an excellent introduction into FSCV for dopamine detection with carbon-fiber microelectrodes [276]. Rodeberg et al. discuss the experimental aspects of the measurements in animals [279]. Puthongkham and Venton review recent trends, including a brief discussion on data evaluation by machine learning [280].

FSCV for other substances

Varying the holding potential, the upper turning point, and the scanrate allow tuning the sensitivity toward different molecules, e. g., adenosine, octopamine, and serotonin. Advanced waveform with different slopes and constant-potential phases provide selectivity for various target molecules. Figure 10.28 shows a "modified-sawhorse waveform" (MSW) to distinguish between methionine-enkephalin, an endogenous opioid, and closely related molecules [278, 281].

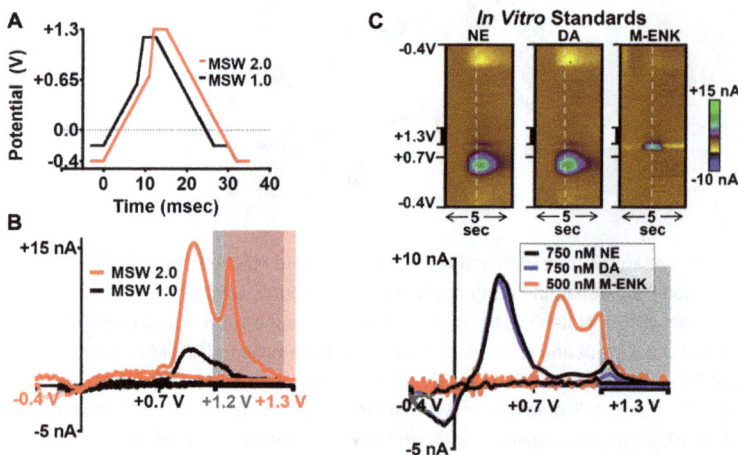

Figure 10.28: FSCV to selectively detect methionine-enkephalin (M-ENK). "Modified-sawhorse waveforms" (MSW) comprising different slopes and constant-potential phases (A). Current response (background subtracted) to 1 µM M-ENK (B). Comparison of 750 nM norepinephrine (NE), 750 nM dopamine (DA), and 500 nM M-ENK using the MSW 2.0 (C). Reprinted with permission from [278]. https://pubs.acs.org/doi/10.1021/acschemneuro.8b00351. Permission requests should be directed to the American Chemical Society.

Instrumentation

While conventional potentiostats with high-speed functionality allow for FSCV experiments in principle, researchers in the field prefer specific hardware and software. Most in vivo FSCV measurements use a two-electrode setup skipping the need for a potentiostat with separated reference/counter electrode. The WaveNeuro One FSCV Potentiostat System by Pine Research Instrumentation, USA, is a widely used commercial system. Apart from the electrical characteristics of the instrumentation, the setup has to handle and evaluate a massive amount of data, e. g., at least 10 CVs each second. Here, the High Definition Cyclic Voltammetry software by the University of North Carolina at Chapel Hill, USA, became the reference [282]. Several works address alternative platforms with a smaller form factor and lower price. One example is the TinyFSCV system [283].

10.3.5 Sensing by electrodes from neural implants

Sensors like the just-discussed carbon-fiber electrodes to monitor neurotransmitters are too invasive for clinical applications. Additionally, most chemical or biosensors are not long-term stable enough, allowing the observation of the microenvironment at neuroprosthesis at reasonable timescales. However, the situation at the electrode-tissue interface is of high importance for neurostimulators as electrode degradation, or the body response may cause the device's failure or even could affect the body.

A possible solution is to recruit the neuroprosthesis electrode itself as a sensor electrode. An extra benefit is that sensing of the electrode-tissue interface is ultimately in situ. Relying on the platinum electrochemistry of the typically used platinum or platinum-iridium electrode enables the detection of integral parameters rather than quantifying selectively one substance, which is often sufficient when dealing with long-term effects at electrodes. We demonstrated such a sensing approach in an animal model (rat) by measuring oxidizable and reducible species during an implantation period of 4 weeks [6].

The sensor protocol combines chronoamperometry with a potentiometric phase (compare active potentiometry, Section 5.1.5). Figure 10.29 shows a typical protocol with its response. Three chronoamperometric cycles condition the electrode to a PtO electrode (applied potential E_{PtO}), allowing detection of oxidizable substances at the potential E_{anod} followed by reduction of the PtO and measuring of the reducible species at E_{cath}. In the brain, the reducible species is mainly oxygen. Each step records an early and a late time point, representing a transient and a steady-state value. Accordingly, the protocol provides 20 data points for each 2-minute run, which may repeat continuously, intermittent to stimulation, or after a dormant time of days to weeks. We recently demonstrated that a similar approach allows sensing using the electrodes of a cochlear implant [284]. By now, the bottleneck of the approach is the need for a reference electrode apart from the regular implant limiting the immediate application to research studies with animals.

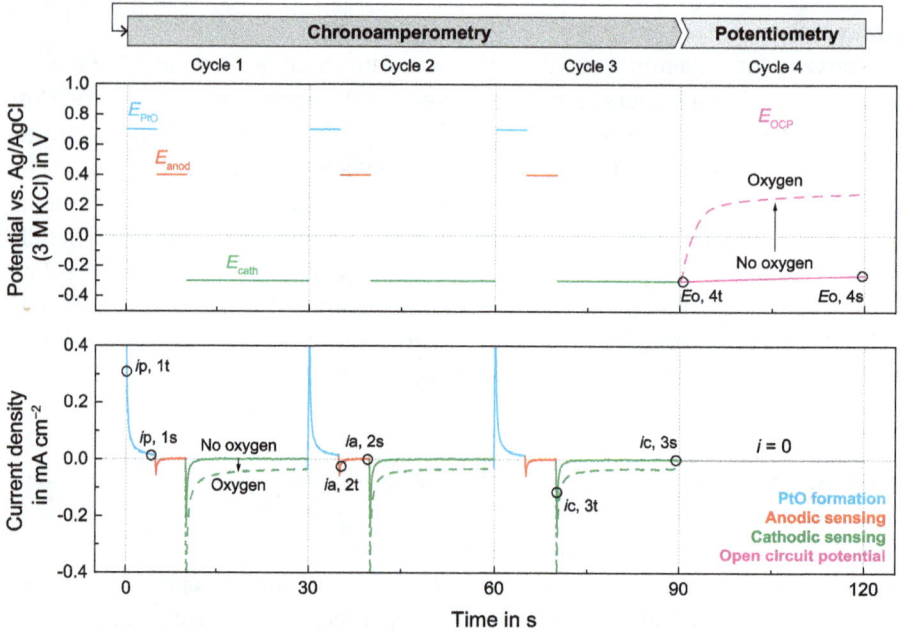

Figure 10.29: Sensing protocol using the existing electrode from a neural implant: the protocol combines three chronoamperometric and one potentiometric cycle measure oxidizable and reducible species exploiting the platinum surface reactions. The circles indicate the considered data points [6]. Reproduced from Weltin et al., New life for old wires: electrochemical sensor method for neural implants, 2020, DOI 10.1088/1741-2552/ab4c69, under CC BY 3.0 (https://creativecommons.org/licenses/by/3.0/).

11 Selected aspects: energy applications

Global warming has gained much more attention recently, although it is not new, leading to a worldwide technology shift toward fossil-free fuel and renewable energies. Hydrogen is probably the most attractive fuel, mainly because of its incredible specific energy ($33\,kW\,h\,kg^{-1}$) and its clean combustion to water. Storage is possible in tanks, such as compressed or liquified hydrogen, under high pressure or low temperature. Alternatively, chemical storage in the form of a chemical compound is possible and under ongoing discussion, which is the most economical way to build on the infrastructure. A prominent representative is ammonia (NH_3) because of the ease of transport, but also dangerous for the environment when accidentally released. Hydrogen storage is also possible as metal hydrates, most well-known in palladium.

The increasing share of renewable energy in the power mix brings new demand on electrical storage technology ranging from the subsecond range for frequency stabilization to load shifting in the hour range or beyond. Depending on their role, storage systems contribute on different levels in the energy grid. Stationary systems, mainly comprising lithium-ion or redox flow batteries, came up. The Moss Landing Vista Battery in California/USA is a massive lithium-ion energy storage system with an impressive 400 MW/1600 MW h specification. Vanadium flow batteries hold against, e. g., the Dalian Flow Battery Energy Storage in Liaoning/China, with a currently installed 100 MW/400 MW h capacity aiming at 200 MW/800 MW h. Since 2016, a sodium-sulfur (NAS) battery system rated for 50 MW/300 MW h at the Buzen Substation has been operating in Fukuoka, Japan. Such battery systems are directly attached to power plants.

However, ecological aspects are not the only drivers: portable devices require efficient energy storage with low weight. While a decade back, energy harvesting was a hot topic, the massive progress in batteries banned it to niches. Lithium-ion batteries play a role in many applications, from tiny smart devices to electromobility. Along with increasingly higher energy densities, safety issues became more prominent and played a crucial role in commercialization. Movie clips of burning smartphones and electric vehicles became viral, although the technology matured to high safety standards. Apart from safety, the availability of resources is crucial. Especially the growing electromobility needs enormous amounts of lithium and cobalt, given much responsibility to the researchers finding alternatives. Recent progress points to sodium as a possible replacement [285], by now better suited for stationary storage than application in cars.

While many systems have larger outer dimensions and seem far from micro- and nanoscale, the electrodes themselves have their critical structures typically in the micrometer range. In this chapter, the discussion is more general to introduce you to the overall topic. Once you work in the field, you can apply the fundamentals and methods you learned in the book's first two parts.

https://doi.org/10.1515/9783111488844-011

11.1 Energy conversion

11.1.1 Fuel cells

Fuel cells convert chemical energy from fuel into electrical energy. At the anode, hydrogen or methane gets oxidized. The electrolyte could be liquid or a solid membrane. In contrast to a battery, fuel cells benefit from the possibility of continuous fuel supply instead of a finite amount of stored energy. Different fuel cells evolved, classified by the electrolyte and the conducting ion with different operating temperatures (Table 11.1).

Table 11.1: Fuel cells, sorted by fuel and the operating temperature. The electrolyte with its conducting ion in brackets determines the electrochemical system.

Fuel cell	Temperature	Electrolyte	Fuel	Oxidant
Proton-exchange m. (PEMFC)	50–80 °C	Polymer (H^+)	H_2	O_2
Alkaline (AFC)	60–90 °C	KOH (OH^-)	H_2	O_2
Phosphoric acid (PAFC)	150–200 °C	H_3PO_4 (H^+)	H_2	O_2
Molten carbonate (MCFC)	600–700 °C	Li_2/K_2CO_3 (CO_3^{2-})	H_2/CH_4	O_2, CO_2
Solid oxide (SOFC)	800–1000 °C	ZrO_2/Yt_2O_3 (O^{2-})	H_2/CH_4	O_2
Direct methanol (DMFC)	30–80 °C	Polymer (H^+)	CH_3OH	O_2
Direct ammonia (DAFC)	600–800 °C	ZrO_2/Yt_2O_3 (O^{2-})	NH_3	O_2

Proton-exchange membrane fuel cell
In a *proton-exchange membrane fuel cell* (PEMFC), the electrolyte is a membrane between two gas permeable electrodes. The hydrogen gets oxidized a the anode

$$H_2 \longrightarrow 2H^+ + 2e^- \tag{11.1}$$

with reduction of oxygen from air as the counter process releasing water:

$$O_2 + 4H^+ + 4e^- \longrightarrow 2H_2O \tag{11.2}$$

The electrolyte is a solid polymer membrane, e. g., Nafion, a sulfonated tetrafluoroethylene-based fluoropolymer-copolymer, which acts as a proton conductor. The membrane should be thin and needs hydration for its operation. Figure 11.1(A) shows the configuration of a PEMFC schematically. Usually, the porous electrodes consist of a carrier and noble metal catalyst particles. A critical aspect of the PEMFC is water management.

While the open-circuit voltage can reach close to 1.2 V, the typical design aims for a cell voltage of around 0.7 V. Higher voltages are possible by stacking several cells with so-called bipolar plates bridging between the electrodes of opposite polarity and allowing gas supply (Figure 11.2).

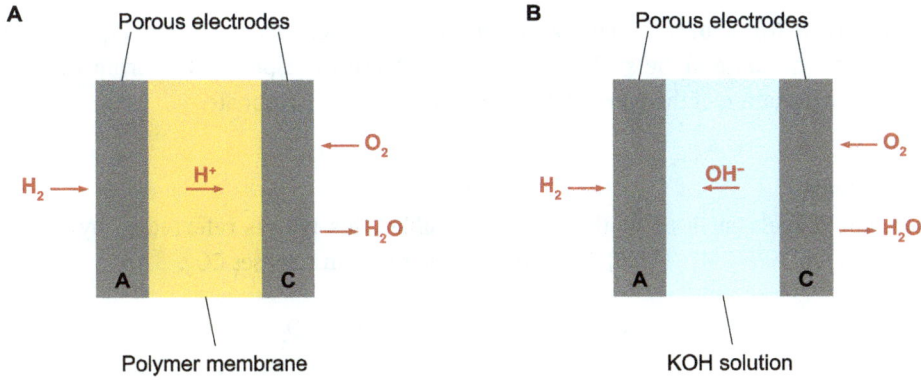

Figure 11.1: Hydrogen fuel cells for low-temperature operation: a proton-exchange membrane fuel cell (PEMFC) with a polymer membrane (A) and an alkaline fuel cell (AFC) with an aqueous KOH electrolyte (B).

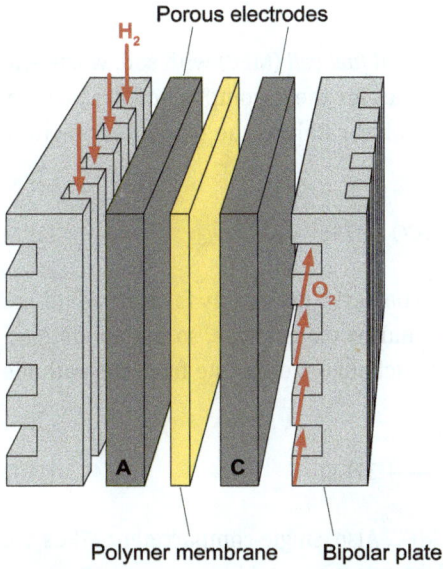

Figure 11.2: Proton-exchange membrane fuel cell: stacking of individual cells allows for higher voltages. The bipolar plates connect the neighboring electrodes and allow gas supply.

Alkaline fuel cell

Alkaline fuel cells (AFC) benefit from the faster kinetics of the oxygen reduction and, accordingly, more flexibility in choosing the catalyst at higher pH. The electrolyte is an aqueous KOH solution with a high concentration of up to 10 M. The electrode reactions are similar to the PEMFC but for an alkaline electrolyte:

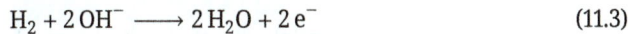

$$H_2 + 2\,OH^- \longrightarrow 2\,H_2O + 2\,e^- \tag{11.3}$$

$$O_2 + 4\,e^- + 2\,H_2O \longrightarrow 4\,OH^- \tag{11.4}$$

Figure 11.1(B) shows the configuration of an AFC schematically. The liquid electrolyte can circulate or be soaked in a matrix. A critical aspect is CO_2 from the air causing degradation of the electrolyte by the formation of carbonate.

Liquid fuels

Also, liquid fuels, such as methanol, are possible. This type is referred to by *direct methanol fuel cell* (DMFC) [286]. Here, the anodic reaction releases CO_2:

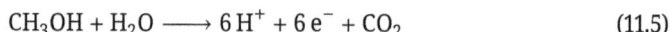

$$CH_3OH + H_2O \longrightarrow 6\,H^+ + 6\,e^- + CO_2 \tag{11.5}$$

In air-free applications, hydrogen peroxide, with its double-role as fuel and oxidizing agent, can be the sole energy carrier [287].

Microbial fuel cell

Another particular class of fuel cells is the *microbial fuel cell* (MFC) with soil, waste water, or agricultural waste as fuel. Possible substrates for the decomposition are acetate, glucose, or other sugars and molasses. The microbes catalysis the oxidation reaction, here with acetate as substrate:

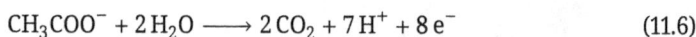

$$CH_3COO^- + 2\,H_2O \longrightarrow 2\,CO_2 + 7\,H^+ + 8\,e^- \tag{11.6}$$

The used microbes (microorganisms) are mainly bacteria, preferably forming a biofilm. Typically, the environment of the microbes should be oxygen-free, so the anode electrolyte needs to be separated, e. g., by a proton exchange membrane from the cathode at which oxygen reduction occurs:

$$O_2 + 4\,H^+ + 4\,e^- \longrightarrow 2\,H_2O \tag{11.7}$$

Figure 11.3 shows a schematic setup of an MFC. Also, single-compartment MFCs are possible, taking benefit from sedimentation of the microbes and low oxygen concentration at the bottom. Comparable to the enzymes in a biosensor (Section 10.1.3), different charge transfer principles are possible (Figure 11.4). The preferred principle is the direct electron transfer skipping the need for a redox mediator.

The review of Du et al. provides a comprehensive introduction into MFC [288], Munoz-Cupa et al. also covers more recent works [289].

11.1.2 Electrolysis

Energy conversion is an essential part of electrochemical energy systems. While in the previous section, the focus was on fuel cells, converting fuel into electrical energy, this

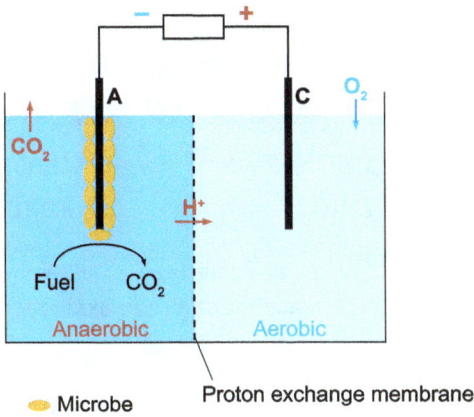

Figure 11.3: Schematic setup of a microbial fuel cell (MFC): the microbes locate at the anode in an anaerobic electrolyte, separated by a proton exchange membrane from the electrolyte at the cathode featuring oxygen reduction.

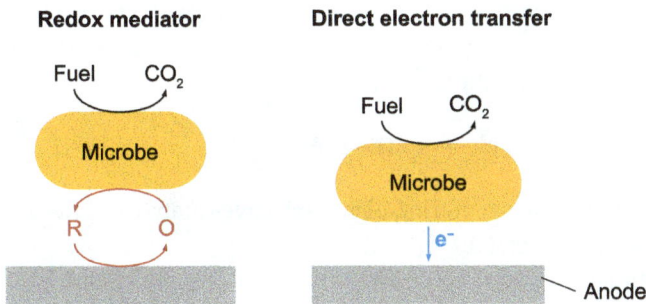

Figure 11.4: Charge transfer mechanism in microbial fuel cells: a redox mediator shuttles charge between the microbe and the anode, or direct electron transfer occurs depending on the microbe.

section deals with turning electrical power into fuel. Indeed, the described electrolysis systems reverse the principles used in fuel cells.

Water splitting

By principle, an electrolysis cell is the inverse of a fuel cell. Similarly, three major types of electrolysis cells exist. At low temperature (below $100\,°C$), liquid water reacts in the *proton-exchange membrane (PEM) electrolysis*:

$$2\,H^+ + 2\,e^- \longrightarrow H_2 \tag{11.8}$$

$$2\,H_2O \longrightarrow O_2 + 4\,H^+ + 4\,e^- \tag{11.9}$$

While PEM cells are very effective, the *alkaline electrolysis cell* (AEC) is another option with a cheaper electrode material, e. g., nickel electrodes instead of the platinum group metal catalyst in the PEM electrolyzer. Another difference is the more clean hydrogen as the diffusivity of gases is lower in alkaline electrolytes than the PEM reducing cross-diffusion between the two sides. Analog to the alkaline fuel cell, the reactions are:

$$2\,H_2O + 2\,e^- \longrightarrow H_2 + 2\,OH^- \tag{11.10}$$

$$4\,OH^- \longrightarrow O_2 + 4\,e^- + 2\,H_2O \tag{11.11}$$

At higher temperature (around 800 °C) the solid oxide electrolysis cell (SOEC) enables production of hydrogen from steam reversing the reactions of a solid oxide fuel cell:

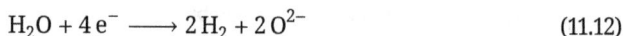

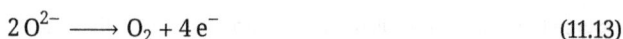

$$H_2O + 4\,e^- \longrightarrow 2\,H_2 + 2\,O^{2-} \tag{11.12}$$

$$2\,O^{2-} \longrightarrow O_2 + 4\,e^- \tag{11.13}$$

CO$_2$ electrolysis

Reduction of CO_2 is an attractive process, especially when combining with CO_2 capture. Possible products are CH_4, formic acid, or CO. By now, electrolysis systems for the latter are most matured [290]. The cathodic reaction

$$CO_2 + H_2O + 2\,e^- \longrightarrow CO + 2\,OH^- \tag{11.14}$$

is prone to inference with H_2 formation requiring highly selective catalysts. At the anode, oxidation of water occurs (equation (11.11)).

The energy storage technologies complete the electrochemical energy systems. Badwal et al. provided an excellent overview of the different technologies well suited as a more in-depth primer to the topic [291].

11.2 Energy storage

Long the only intention of electrochemical energy storage was to achieve portable operation of electrical devices. Lately, electrochemical storage competes and enhances classical storage principles like pumped hydroelectric energy storage in support of the power grid. Alongside electromobility, this is the area in which demand will increase the most in the coming decades.

11.2.1 Batteries

Primary cells

Primary cells, batteries that are not rechargeable, allowed portable power supply for more than a century used in torches, transistor radios, and later in the walkman. Still today, many devices fit the cylindrical cells in sizes like "AA" or "AAA." The setup goes back to the *Leclanché cell*, with a zink anode and a MnO_2 cathode. Originally, with a liquid electrolyte, the setup evolved to the classical "dry cell," with a paste of NH_4Cl as electrolyte and the MnO_2 mixed with carbon powder. The discharging reactions dissolve the electrodes:

$$Zn \longrightarrow Zn^{2+} + 2\,e^- \tag{11.15}$$

$$2\,MnO_2 + H_2O + 2\,e^- \longrightarrow Mn_2O_3 + 2\,OH^- \tag{11.16}$$

Instead of Mn_2O_3 also a hydroxide can form. The cell voltage is nominal 1.5 V but slightly drops upon discharge.

The *alkaline battery* is an improvement with an aqueous KOH solution as the electrolyte. In alkaline electrolytes, the anodic reaction differs, forming ZnO instead of dissolved zink ions:

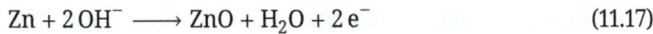

$$Zn + 2\,OH^- \longrightarrow ZnO + H_2O + 2\,e^- \tag{11.17}$$

The anode is zink powder soaked with the electrolyte. The cathode consists of MnO_2 and carbon powder. Figure 11.5 shows a typical arrangement. The performance due to the alkaline electrolyte supersedes the dry battery but with the risk of leakage. Therefore, housing materials are optimized, and the batteries should be removed from devices when empty or the expiry date is due. Overheating or burning may cause the formation of gas bubbles, why commercial cells contain safety vents close to the negative terminal.

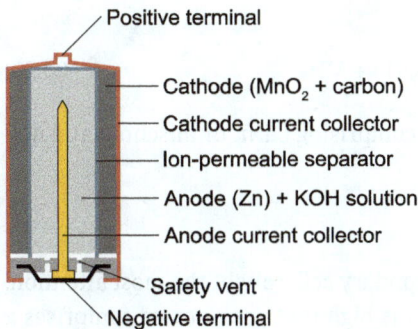

Positive terminal
Cathode (MnO_2 + carbon)
Cathode current collector
Ion-permeable separator
Anode (Zn) + KOH solution
Anode current collector
Safety vent
Negative terminal

Figure 11.5: Alkaline battery: the name refers to the KOH solution as the electrolyte. Both anode and cathode are powder-based.

The maximal charge (in the context of batteries named the capacity) of an alkaline battery is around 1.2 Ah for an "AAA" and 2.9 Ah for an "AA-" sized cell. However, the practically accessible amount is lower, depending on the discharging characteristic and the minimal accepted voltage in the application.

Secondary cells

Rechargeable batteries are so-called secondary cells. Apart from the lead-acid battery in cars (Section 1.3.1), the most common secondary cell was the Ni-Cd battery for long, followed by nickel-metal hydride (Ni-MH), avoiding the toxic cadmium. The Ni-MH battery offers slightly better performance, namely a less significant and reversible memory effect allowing recharging also before complete discharge. While Ni-Cd and Ni-MH batteries are still in use, lithium-ion batteries take over more and more applications of secondary cells.

> **!** While the positive terminal can be anode or cathode depending on the battery delivers current (galvanic cell) or gets recharged (electrolysis cell), it is common in the field of batteries to name the positive terminal of a secondary cell as cathode in general, referring to its operation as power supply. The negative electrode is the anode accordingly.

Ni-Cd and Ni-MH batteries have the same form factors as the dry cells and alkaline batteries discussed above. Both have a nickel cathode and rely on KOH as the electrolyte. At the positive terminal, the reactions are

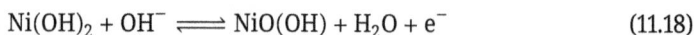

$$Ni(OH)_2 + OH^- \rightleftharpoons NiO(OH) + H_2O + e^- \tag{11.18}$$

In the case of Ni-Cd, the negative electrode is cadmium; the reactions are

$$Cd(OH)_2 + 2\,e^- \rightleftharpoons Cd + 2\,OH^- \tag{11.19}$$

In the Ni-MH cell, the formation and discharging of the metal hydride occurs at the negative electrode:

$$H_2O + M + e^- \rightleftharpoons OH^- + MH \tag{11.20}$$

The metal M forming the hydride is an alloy, e. g., comprising $LaNi_5$ or mischmetal, a naturally occurring mix of rare-earth metals.

Lithium-ion batteries

Recently, the lithium-ion battery (LIB), also a secondary cell, caught the most attention. If you read this book as an ebook, the likelihood is high that your device comprises a lithium-ion battery. Lithium is the lightest metal with the lowest standard reduction potential of $-3.04\,V$ for the Li^+/Li couple, enabling high specific energy and cell voltage. Critical is the metal's strong reaction with water, asking for nonaqueous systems. A common electrolyte is $LiPF_6$ in propylene carbonate as the solvent.

> **i** Energy (in the unit W h) and power (W) are the two essential quantities to compare electrochemical energy systems. The absolute numbers matter when relating the capabilities of a storage system to the demands in stationary applications next to wind farms or solar power plants. For mobile applications, weight is most

important. Here, specific energy ($W\,h\,kg^{-1}$) and specific power ($W\,kg^{-1}$) normalized to the weight are better characteristics. In detail, it matters whether the numbers relate to the electrochemical principle alone or cover the whole system. Other quantities are energy density ($W\,h\,l^{-1}$) and power density ($W\,l^{-1}$) relating to the volume. For the description of microsystems also normalization to the area is common: energy density ($mW\,h\,cm^{-2}$) and power density ($mW\,cm^{-2}$).

Figure 11.6 shows the setup of a typical lithium-ion battery with $LiCoO_2$ at the cathode. An ion-permeable separator prevents short circuits between anode and cathode.

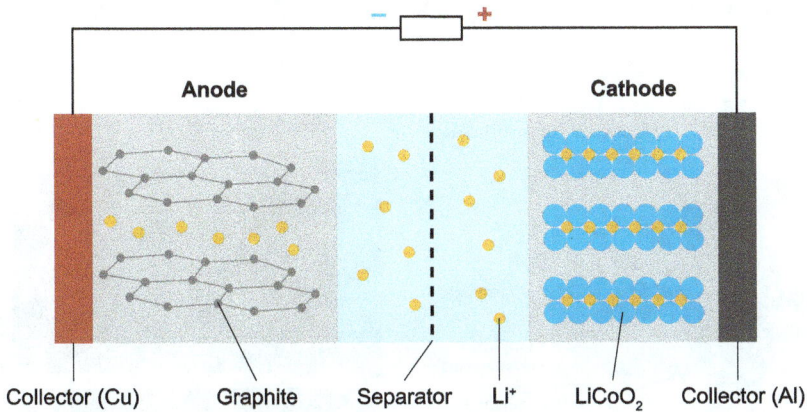

Figure 11.6: Lithium-ion battery (LIB): at the negative electrode, lithium intercalates into graphite; the cathode consists of $LiCoO_2$. A typical electrolyte would be $LiPF_6$ in propylene carbonate.

Using metallic lithium as the negative electrode bears the risk of dendrite formation, possibly causing short circuits. Therefore, most modern cells use graphite at the positive electrode into which lithium ions can intercalate:

$$Li_x\,C_6 \rightleftharpoons C_6 + x\,Li^+ + e^- \tag{11.21}$$

Several materials are possible as the cathode. Candidates are transition metal oxides with cobalt, manganese, or nickel. The most powerful material is $LiCoO_2$; $LiMn_2O_4$ is ecologically favorable. The charging and discharging reactions are:

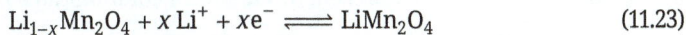

$$Li_{1-x}CoO_2 + x\,Li^+ + xe^- \rightleftharpoons LiCoO_2 \tag{11.22}$$

$$Li_{1-x}Mn_2O_4 + x\,Li^+ + xe^- \rightleftharpoons LiMn_2O_4 \tag{11.23}$$

Both materials lead to a nominal cell voltage of 3.7 V. Another possibility is $LiFePO_4$ offering higher safety and low ecological burden but with a nominal cell voltage of only 3.2 V.

A lithium-ion polymer battery (LiPo) contains a polymeric electrolyte, often classified as a semisolid electrolyte. Depending on the application, the disadvantage of lower conductivity than liquid electrolytes outweighs the lightweight setup.

Solid electrolytes based on inorganic lithium ion conducting materials are another interesting alternative, also to avoid the risk of fire or explosion. A commonly used solid electrolyte is lithium phosphorous oxynitride (LiPON). Substitution of some O atoms in Li_3PO_4 by N atoms causes disturbance of the structure enabling conductivity in the range of $1\,\mu S\,cm^{-1}$, still a low value compared to liquid electrolytes (Table 4.9).

Also, flexible microbatteries have been realized aiming at wearable applications. Figure 11.7 shows an all-solid-state battery with lithium boron oxynitride (LiBON) as the electrolyte.

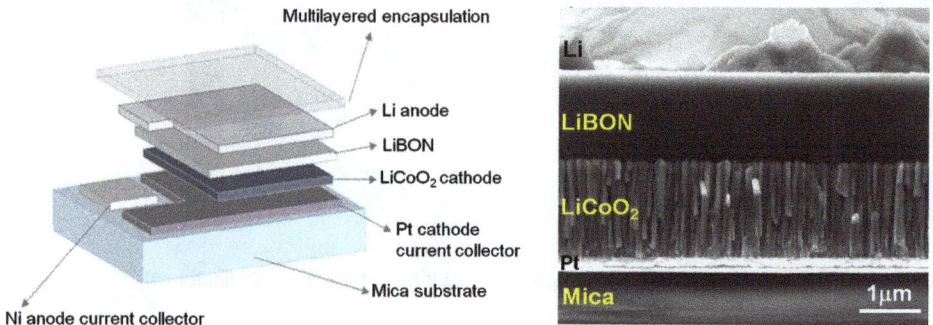

Figure 11.7: Flexible all-solid-state microbattery with LiBON electrolyte. Higher energy density is possible by stacking ten cells together, providing 5 mAh by a 3 cm^2 active electrode area [292]. Reprinted from Journal of Power Sources, 328, Song et al., High-performance flexible all-solid-state microbatteries based on solid electrolyte of lithium boron oxynitride, 311–317, Copyright 2016, with permission from Elsevier.

Sodium-ion batteries

A sodium-ion battery (SIB or sometimes NIB) usually shares the working principle of reversible intercalation with a lithium-ion battery but uses sodium instead of lithium ions. While the specific energy with sodium ions is lower than lithium ions, the abundance of sodium, in contrast to the limited lithium availability, makes it a promising alternative, especially for large-scale applications. Additionally, the principle does not depend on cobalt as it is often at the positive terminal of LIBs. Instead, iron can be used in the form of $NaFeO_2$. Also, variants with aqueous electrolytes are possible. Yabuuchi et al. provide a good introduction to the different SIB material combinations [293].

Anode-free sodium batteries

The anode-free sodium battery (AFSB) is another possibility, skipping the intercalation of the sodium ions at the negative terminal and essentially forming a metallic sodium

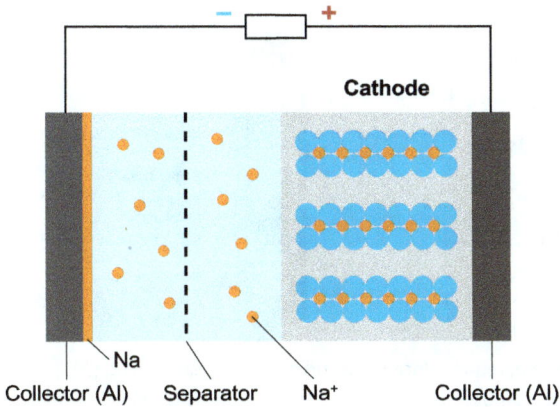

Figure 11.8: Simplified setup of an anode-free sodium battery (AFSB). In contrast to a LIB or SIB, sodium ions are stored as a metallic film instead of by intercalation at the negative terminal.

layer instead (Figure 11.8). This principle has certain drawbacks due to the high reactivity of the sodium and volume expansion during charging, but it attracts because of higher energy density and lower costs. Chen et al. compare all three LIBs, SIBs, and AFSBs from technical and economic perspectives [294].

Sodium-sulfur batteries

The sodium-sulfur (NAS) battery is a molten liquid battery operated around 300 °C that is not to be confused with sodium-ion batteries. The battery setup (Figure 11.9A) contains three compartments: the negative electrode with liquid sodium, a solid electrode (β-alumina ceramic), which allows the conduction of sodium ions, and the positive electrode with sulfur or sodium polysulfides respectively. The overall cell reaction

$$2\,Na + x\,S \rightleftharpoons Na_2S_x \tag{11.24}$$

yields a cell voltage between 1.74 and 2.08 V depending on the state of charge, resulting in different compositions of the sulfur electrode [295]. The individual discharging/charging reactions at the negative electrode are

$$2\,Na \rightleftharpoons 2\,Na^+ + 2\,e^- \tag{11.25}$$

and

$$x\,S + 2\,e^- \rightleftharpoons S_x^{2-} \tag{11.26}$$

at the positive electrode. Figure 11.9B shows a practical realization of a NAS battery cell. The required temperature and the corrosive behavior of sodium and sodium polysulfides limit the application of NAS batteries to stationary energy storage on a larger scale.

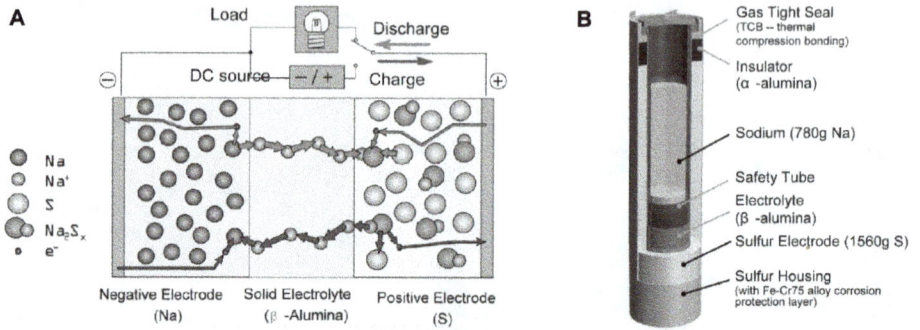

Figure 11.9: Sodium-sulfur (NAS) battery cell: operation principle (A) and practical realization (B) [295]. Reproduced with permission from Oshima et al., Int. J. Appl. Ceram. Technol., 1, 269–76 (2004), John Wiley and Sons.

11.2.2 Redox flow batteries

A fundamental advantage of fuel cells compared to batteries is the possibility of continuous fuel supply. The stored energy depends only on the fuel tank. Batteries allow much more straightforward and efficient recharge. A *redox flow battery* (RFB) unites both aspects. In contrast to the before discussed batteries, the electrolyte solution stores the charge. A tank hosts a large amount of electrolyte pumped into the cell for charging or discharging.

Figure 11.10 shows the setup of a *vanadium redox flow battery* (VRFB). The electrochemical cell connects to tanks storing the electrolyte. A pump system allows the exchange of the solution in the cell with graphite electrodes. The charging/discharging

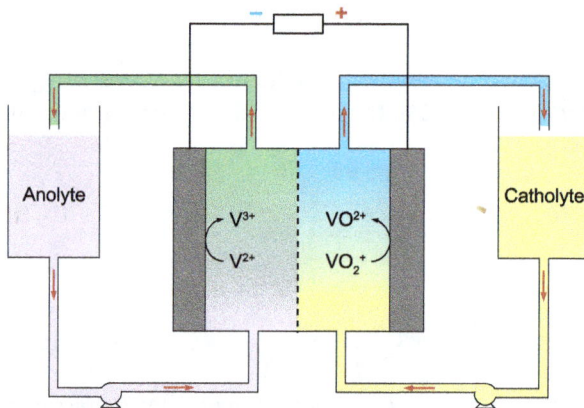

Figure 11.10: Vanadium redox flow battery (VRFB): the electrolyte stores the charge and flows through the electrochemical cell. The anode and cathode have different electrolytes: the anolyte and the catholyte.

reactions at the negative electrode relate to the V^{3+}/V^{2+} couple:

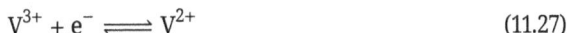

$$V^{3+} + e^- \rightleftharpoons V^{2+} \tag{11.27}$$

At the positive electrode, the vanadium changes its valency between (IV) and (V):

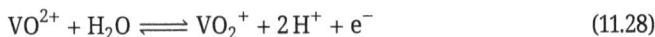

$$VO^{2+} + H_2O \rightleftharpoons VO_2^+ + 2H^+ + e^- \tag{11.28}$$

The electrolyte solutions are different at the anode and cathode. At the negative electrode, the anolyte contains V^{2+} ions, and at the positive electrode, the catholyte VO_2^+ ions, both in an aqueous H_2SO_4 solution. V_2O_5 is a good starting material for the electrolyte preparation [296].

The typcially low energy density demands large tanks limiting the usage of RFBs mainly to stationary applications. Apart from the VRFB, other systems like the polysulfide bromide flow battery and zinc-bromine flow battery exist. The latter offers slightly higher specific energy. An important advantage is that in an RFB, the power (defined by the electrochemical cell) and energy (determined by the tank volume) are unrelated.

Also, membrane-less flow batteries have been proposed [297, 298]. The principle benefits from laminar flow in microchannels, keeping the two electrolytes separated.

11.2.3 Supercapacitors

In contrast to batteries, supercapacitors (also ultracapacitors, or supercaps) do not need a chemical change of the electrode material (like in the nickel-metal hydride battery) or the electrolyte (like in a RFB). The primary charge storage mechanism is by charge separation in the double layer capacity; also, pseudo-capacities play a role, e. g., from adsorption/desorption processes. Sometimes intercalation processes or even faradaic processes account here, too. The term supercapacitor is not strictly defined; primarily, the charge or discharge characteristics distinguish between a battery or capacitor (Figure 11.11).

An electrical double layer capacitor (EDLC) stores charge in the electrolytic double-layer only. Planar electrodes would allow a capacity in the range of several 10 $\mu F\,cm^{-2}$. Therefore, typical EDLCs rely on microporous carbon electrodes with a huge roughness. Alternatively, carbon nanostructures help to increase the surface area.

Other systems also involve pseudo-capacitive effects from adsorption, increasing further the capacity. The lithium-ion capacitor (LIC) has a carbon cathode storing charge in the double layer. The anode is the same as in the lithium-ion battery allowing the intercalation of lithium ions in graphite.

Supercapacitors feature superior specific power but low specific energy. A disadvantage for longer storage durations is a substantial self-discharge. In contrast to classical capacitors, supercapacitors are unipolar with a strictly limited voltage range depending on the electrolyte's potential window.

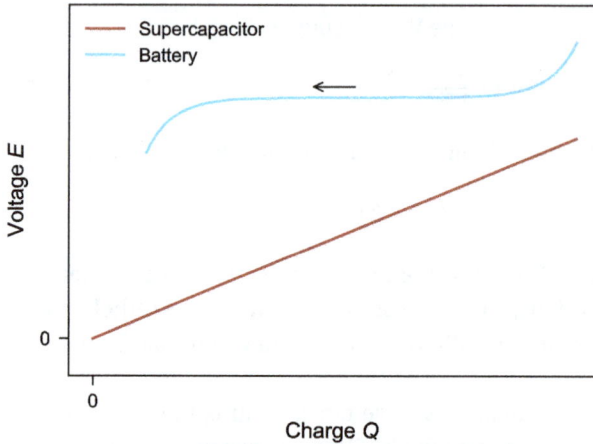

Figure 11.11: Comparison of the discharge curves: a supercapacitor shows a voltage proportional to the charge, while a battery provides an ideally constant voltage during discharge.

11.2.4 Ragone plot

Ragone presented in 1968 a diagram positioning different batteries for electrical mobility in a double logarithmic plot of specific power versus specific energy [299]. Figure 11.12 shows a modern interpretation comparing various electrochemical energy storage systems with the specific power on the x-axis. The *Ragone plot* is a powerful tool to compare different principles, but be aware that such diagrams are often tentative towards a specific system, especially when commercially available systems get mixed with superlatives reported in academic publications.

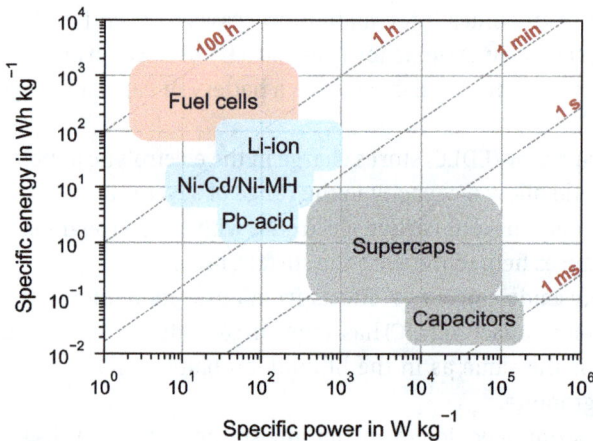

Figure 11.12: Ragone plot comparing different electrochemical energy storage systems and fuel cells. The diagonal lines indicate the time required for complete discharge. Of course, longer storage times are possible.

A Reference data

We live in a golden age of easy access to all different reference data available in free or commercial databases (thermodynamic data, electrolyte parameters, diffusion constants, etc.). The book's webpage (www.electrochemical-methods.org) provides some links to both publicly available and closed-access resources.

Previous chapters contain some tables to emphasize specific aspects knowing that without a particular need, no one would browse through parameter resources:
- Table 2.1: exchange current densities
- Table 4.1: reference electrode potentials
- Table 4.2: temperature dependency of Ag/AgCl reference electrodes
- Table 4.3: ion mobilities for infinite dilution
- Table 4.4: metal electrodes
- Table 4.5: carbon electrodes
- Table 4.6: aqueous electrolytes
- Table 4.8: nonaqueous solvents
- Table 4.9: nonaqueous electrolytes
- Table 4.10: redox couples for aqueous electrolytes
- Table 9.5: material combinations for underpotential deposition of metals

Additionally, some popular recipes are listed:
- Table 4.7: phosphate-buffered saline (PBS) solutions
- Table 9.1: Watts bath for nickel plating
- Table 9.2: sulfamate bath for nickel plating
- Table 9.3: acid copper bath
- Table 9.4: AEIROF electrodeposition solution according to Yamanaka

Here, reactions relevant to applications discussed in this book are alphabetically, or much more helpful, sorted according to their standard reduction potential.

Solubility of dissolved gases is a domain with less comprehensive resources, especially when it comes to salt dependency beyond seawater and at temperatures apart from room temperature, e. g., the 37 °C relevant in life science applications. Section A.2 contains the most needed data.

A.1 Standard reduction potentials

The tables contain standard reduction potentials E^0 ("standard electrode potentials" or the "electrochemical series") compiled from different sources. The data is for unit activities ($1\,\mathrm{mol\,l^{-1}}$) at room temperature (25 °C). The notation for metals "Me(Hg)" refers to measurements at mercury electrodes with the dissolution of the metal ions into the electrode.

https://doi.org/10.1515/9783111488844-012

A.1.1 Alphabetically sorted

Reaction	Potential vs. SHE
$Ag^+ + e^- \rightleftharpoons Ag$	0.799 V
$AgBr + e^- \rightleftharpoons Ag + Br^-$	0.071 V
$AgCN + e^- \rightleftharpoons Ag + CN^-$	−0.017 V
$AgCl + e^- \rightleftharpoons Ag + Cl^-$	0.222 V
$AgI + e^- \rightleftharpoons Ag + I^-$	−0.152 V
$Al^{3+} + 3e^- \rightleftharpoons Al$	−1.662 V
$Au^+ + e^- \rightleftharpoons Au$	1.692 V
$Au^{3+} + 2e^- \rightleftharpoons Au^+$	1.401 V
$CO_2 + 2H^+ + 2e^- \rightleftharpoons HCOOH$	−0.199 V
$Cd^{2+} + 2e^- \rightleftharpoons Cd$	−0.403 V
$Cd^{2+} + 2e^- \rightleftharpoons Cd(Hg)$	−0.352 V
$Cl_2 + 2e^- \rightleftharpoons 2Cl^-$	1.358 V
$Co^{2+} + 2e^- \rightleftharpoons Co$	−0.280 V
$Cr^{2+} + 2e^- \rightleftharpoons Cr$	−0.913 V
$Cr^{3+} + e^- \rightleftharpoons Cr^{2+}$	−0.407 V
$Cu^{2+} + 2e^- \rightleftharpoons Cu$	0.342 V
$Cu^{2+} + 2e^- \rightleftharpoons Cu(Hg)$	0.345 V
$F_2 + 2H^+ + 2e^- \rightleftharpoons 2HF$	3.053 V
$[Fc]^+ + e^- \rightleftharpoons Fc$	0.400 V
$Fe^{2+} + 2e^- \rightleftharpoons Fe$	−0.447 V
$Fe^{3+} + e^- \rightleftharpoons Fe^{2+}$	0.771 V
$Fe(CN)_6^{3-} e^- \rightleftharpoons Fe(CN)_6^{4-}$	0.361 V
$2H^+ + 2e^- \rightleftharpoons H_2$	0.000 V
$H_2O_2 + 2H^+ + 2e^- \rightleftharpoons 2H_2O$	1.776 V
$2H_2O + 2e^- \rightleftharpoons H_2 + 2OH^-$	−0.828 V
$Hg_2^{(+)} + 2e^- \rightleftharpoons Hg$	0.851 V
$Hg_2Br_2 + 2e^- \rightleftharpoons 2Hg + 2Br^-$	0.139 V
$Hg_2Cl_2 + 2e^- \rightleftharpoons 2Hg + 2Cl^-$	0.268 V
$HgO + H_2O + 2e^- \rightleftharpoons Hg + 2OH^-$	0.098 V
$Hg_2SO_4 + 2e^- \rightleftharpoons 2Hg + SO_4^{2-}$	0.613 V
$I_2 + 2e^- \rightleftharpoons 2I^-$	0.536 V
$Ir^{3+} + 3e^- \rightleftharpoons Ir$	1.156 V
$[IrCl_6]^{2-} + e^- \rightleftharpoons [IrCl_6]^{3-}$	0.867 V
$[IrCl_6]^{3-} + 3e^- \rightleftharpoons Ir + 6Cl^-$	0.77 V
$Ir_2O_3 + 3H_2O + 6e^- \rightleftharpoons 2Ir + 6OH^-$	0.098 V
$K^+ + e^- \rightleftharpoons K$	−2.931 V
$Li^+ + e^- \rightleftharpoons Li$	−3.040 V
$Mg^{2+} + 2e^- \rightleftharpoons Mg$	−2.372 V
$Mn^{2+} + 2e^- \rightleftharpoons Mn$	−1.185 V
$Mn^{3+} + e^- \rightleftharpoons Mn^{2+}$	1.542 V
$MnO_2 + 4H^+ + 2e^- \rightleftharpoons Mn^{2+} + 2H_2O$	1.224 V

Reaction	Potential vs. SHE
$Mo_3^+ + 3\,e^- \rightleftharpoons Mo$	−0.200 V
$N_2O + 2\,H^+ + 2\,e \rightleftharpoons N_2 + H_2O$	1.766 V
$2\,NO + 2\,H^+ + 2\,e^- \rightleftharpoons N_2O + H_2O$	1.591 V
$2\,NO + H_2O + 2\,e^- \rightleftharpoons N_2O + 2\,OH^-$	0.76 V
$Na^+ + e^- \rightleftharpoons Na$	−2.714 V
$Nb^{3+} + 3\,e^- \rightleftharpoons Nb$	−1.099 V
$Ni^{2+} + 2\,e^- \rightleftharpoons Ni$	−0.257 V
$O_2 + 4\,H^+ + 4\,e^- \rightleftharpoons 2\,H_2O$	1.229 V
$O_2 + 2\,H^+ + 2\,e^- \rightleftharpoons H_2O_2$	0.695 V
$O_2 + 2\,H_2O + 4\,e^- \rightleftharpoons 4\,OH^-$	0.401 V
$O_2 + 2\,H_2O + 2\,e^- \rightleftharpoons H_2O_2 + 2\,OH^-$	−0.146 V
$O_3 + 2\,H^+ + 2\,e^- \rightleftharpoons O_2 + H_2O$	2.076 V
$Pb^{2+} + 2\,e^- \rightleftharpoons Pb$	−0.126 V
$Pb^{2+} + 2\,e^- \rightleftharpoons Pb(Hg)$	−0.121 V
$PbO_4 + SO_4{}^{2-} + 4\,H^+ + 2\,e^- \rightleftharpoons PbSO_4 + 2\,H_2O$	1.691 V
$PbSO_4 + 2\,e^- \rightleftharpoons Pb + SO_4{}^{2-}$	−0.359 V
$Pd^{2+} + 2\,e^- \rightleftharpoons Pd$	0.951 V
$Pt^{2+} + 2\,e^- \rightleftharpoons Pt$	1.188 V
$[PtCl_4]^{2-} + 2\,e^- \rightleftharpoons Pt + 4\,Cl^-$	0.755 V
$[PtCl_6]^{2-} + 2\,e^- \rightleftharpoons [PtCl_4]^{2-} + 2\,Cl^-$	0.726 V
$PtO_2 + 2\,H^+ + 2\,e^- \rightleftharpoons PtO + H_2O$	1.01 V
$PtO_3 + 2\,H^+ + 2\,e^- \rightleftharpoons PtO_2 + H_2O$	1.7 V
$Ru^{2+} + 2\,e^- \rightleftharpoons Ru$	0.455 V
$[Ru(NH_3)_6]^{3+} + e^- \rightleftharpoons [Ru(NH_3)_6]^2$	0.10 V
$RuO_2 + 4\,H^+ + 2\,e^- \rightleftharpoons Ru^{2+} + 2\,H_2O$	1.120 V
$Sn^{2+} + 2\,e^- \rightleftharpoons Sn$	−0.138 V
$Ti^{2+} + 2\,e^- \rightleftharpoons Ti$	−1.630 V
$V^{3+} + e^- \rightleftharpoons V^{2+}$	−0.255 V
$VO^{2+} + 2\,H^+ + e^- \rightleftharpoons VO^{2+} + H_2O$	0.991 V
$Zn^{2+} + 2\,e^- \rightleftharpoons Zn$	−0.762 V
$Zn^{2+} + 2\,e^- \rightleftharpoons Zn(Hg)$	−0.763 V

A.1.2 Sorted by potential

Reaction	Potential vs. SHE
$Li^+ + e^- \rightleftharpoons Li$	−3.040 V
$K^+ + e^- \rightleftharpoons K$	−2.931 V
$Na^+ + e^- \rightleftharpoons Na$	−2.714 V
$Mg^{2+} + 2\,e^- \rightleftharpoons Mg$	−2.372 V
$Al^{3+} + 3\,e^- \rightleftharpoons Al$	−1.662 V
$Ti^{2+} + 2\,e^- \rightleftharpoons Ti$	−1.630 V
$Mn^{2+} + 2\,e^- \rightleftharpoons Mn$	−1.185 V
$Nb^{3+} + 3\,e^- \rightleftharpoons Nb$	−1.099 V

Reaction	Potential vs. SHE
$Cr^{2+} + 2e^- \rightleftharpoons Cr$	−0.913 V
$2 H_2O + 2e^- \rightleftharpoons H_2 + 2 OH^-$	−0.828 V
$Zn^{2+} + 2e^- \rightleftharpoons Zn(Hg)$	−0.763 V
$Zn^{2+} + 2e^- \rightleftharpoons Zn$	−0.762 V
$Fe^{2+} + 2e^- \rightleftharpoons Fe$	−0.447 V
$Cr^{3+} + e^- \rightleftharpoons Cr^{2+}$	−0.407 V
$Cd^{2+} + 2e^- \rightleftharpoons Cd$	−0.403 V
$PbSO_4 + 2e^- \rightleftharpoons Pb + SO_4^{2-}$	−0.359 V
$Cd^{2+} + 2e^- \rightleftharpoons Cd(Hg)$	−0.352 V
$Co^{2+} + 2e^- \rightleftharpoons Co$	−0.280 V
$Ni^{2+} + 2e^- \rightleftharpoons Ni$	−0.257 V
$V^{3+} + e^- \rightleftharpoons V^{2+}$	−0.255 V
$Mo_3^+ + 3e^- \rightleftharpoons Mo$	−0.200 V
$CO_2 + 2 H^+ + 2e^- \rightleftharpoons HCOOH$	−0.199 V
$AgI + e^- \rightleftharpoons Ag + I^-$	−0.152 V
$O_2 + 2 H_2O + 2e^- \rightleftharpoons H_2O_2 + 2 OH^-$	−0.146 V
$Sn^{2+} + 2e^- \rightleftharpoons Sn$	−0.138 V
$Pb^{2+} + 2e^- \rightleftharpoons Pb$	−0.126 V
$Pb^{2+} + 2e^- \rightleftharpoons Pb(Hg)$	−0.121 V
$AgCN + e^- \rightleftharpoons Ag + CN^-$	−0.017 V
$2 H^+ + 2e^- \rightleftharpoons H_2$	0.000 V
$AgBr + e^- \rightleftharpoons Ag + Br^-$	0.071 V
$HgO + H_2O + 2e^- \rightleftharpoons Hg + 2 OH^-$	0.098 V
$Ir_2O_3 + 3 H_2O + 6e^- \rightleftharpoons 2 Ir + 6 OH^-$	0.098 V
$[Ru(NH_3)_6]^{3+} + e^- \rightleftharpoons [Ru(NH_3)_6]^2$	0.10 V
$Hg_2Br_2 + 2e^- \rightleftharpoons 2 Hg + 2 Br^-$	0.139 V
$AgCl + e^- \rightleftharpoons Ag + Cl^-$	0.222 V
$Hg_2Cl_2 + 2e^- \rightleftharpoons 2 Hg + 2 Cl^-$	0.268 V
$Cu^{2+} + 2e^- \rightleftharpoons Cu$	0.342 V
$Cu^{2+} + 2e^- \rightleftharpoons Cu(Hg)$	0.345 V
$Fe(CN)_6^{3-} e^- \rightleftharpoons Fe(CN)_6^{4-}$	0.361 V
$[Fc]^+ + e^- \rightleftharpoons Fc$	0.400 V
$O_2 + 2 H_2O + 4e^- \rightleftharpoons 4 OH^-$	0.401 V
$Ru^{2+} + 2e^- \rightleftharpoons Ru$	0.455 V
$I_2 + 2e^- \rightleftharpoons 2 I^-$	0.536 V
$Hg_2SO_4 + 2e^- \rightleftharpoons 2 Hg + SO_4^{2-}$	0.613 V
$O_2 + 2 H^+ + 2e^- \rightleftharpoons H_2O_2$	0.695 V
$[PtCl_6]^{2-} + 2e^- \rightleftharpoons [PtCl_4]^{2-} + 2 Cl^-$	0.726 V
$[PtCl_4]^{2-} + 2e^- \rightleftharpoons Pt + 4 Cl^-$	0.755 V
$2 NO + H_2O + 2e^- \rightleftharpoons N_2O + 2 OH^-$	0.76 V
$[IrCl_6]^{3-} + 3e^- \rightleftharpoons Ir + 6 Cl^-$	0.77 V
$Fe^{3+} + e^- \rightleftharpoons Fe^{2+}$	0.771 V
$Ag^+ + e^- \rightleftharpoons Ag$	0.799 V
$Hg_2^{(+)} + 2e^- \rightleftharpoons Hg$	0.851 V
$[IrCl_6]^{2-} + e^- \rightleftharpoons [IrCl_6]^{3-}$	0.867 V

Reaction	Potential vs. SHE
$Pd^{2+} + 2e^- \rightleftharpoons Pd$	0.951 V
$VO^{2+} + 2H^+ + e^- \rightleftharpoons VO^{2+} + H_2O$	0.991 V
$PtO_2 + 2H^+ + 2e^- \rightleftharpoons PtO + H_2O$	1.01 V
$RuO_2 + 4H^+ + 2e^- \rightleftharpoons Ru^{2+} + 2H_2O$	1.120 V
$Ir^{3+} + 3e^- \rightleftharpoons Ir$	1.156 V
$Pt^{2+} + 2e^- \rightleftharpoons Pt$	1.188 V
$MnO_2 + 4H^+ + 2e^- \rightleftharpoons Mn^{2+} + 2H_2O$	1.224 V
$O_2 + 4H^+ + 4e^- \rightleftharpoons 2H_2O$	1.229 V
$Cl_2 + 2e^- \rightleftharpoons 2Cl^-$	1.358 V
$Au^{3+} + 2e^- \rightleftharpoons Au^+$	1.401 V
$Mn^{3+} + e^- \rightleftharpoons Mn^{2+}$	1.542 V
$2NO + 2H^+ + 2e^- \rightleftharpoons N_2O + H_2O$	1.591 V
$PbO_4 + SO_4^{2-} + 4H^+ + 2e^- \rightleftharpoons PbSO_4 + 2H_2O$	1.691 V
$Au^+ + e^- \rightleftharpoons Au$	1.692 V
$PtO_3 + 2H^+ + 2e^- \rightleftharpoons PtO_2 + H_2O$	1.7 V
$N_2O + 2H^+ + 2e \rightleftharpoons N_2 + H_2O$	1.766 V
$H_2O_2 + 2H^+ + 2e^- \rightleftharpoons 2H_2O$	1.776 V
$O_3 + 2H^+ + 2e^- \rightleftharpoons O_2 + H_2O$	2.076 V
$F_2 + 2H^+ + 2e^- \rightleftharpoons 2HF$	3.053 V

A.2 Dissolved gases

The concentration of gas dissolved in a liquid phase depends, apart from the gas/liquid combination, on the partial pressure of the gas. Here, we consider aqueous liquid phases only; however, the same concept would apply to nonaqueous systems.

A.2.1 Solubility

The partial pressure of a gas depends on the mole fraction x of the particular gas in a gas mixture. Assuming an ideal gas (compression factor is 1), the mole fraction and the volume fraction are the same. For example, $x = 0.2095$ is the oxygen fraction in the atmosphere.

The composition of the earth's atmosphere is 20.95 % oxygen, 78.08 % nitrogen, and 0.93 % argon. The exact numbers depend on the humidity and, thereby, on the temperature. The next most available gas is carbon dioxide, which alarmingly increased from below 320 ppm in 1960, approaching 420 ppm on this book's release date measured by the generally accepted data source, the Mauna Loa Observatory, Hawaii.[1]

i

---

1 https://gml.noaa.gov/ccgg/trends/, see also Figure 1.10.

Directly at the phase boundary between the liquid and the gas mixture, the gas phase is assumed to be saturated with water vapor. Therefore, the vapor pressure p_{vap} should be subtracted from the total pressure p_{tot} for calculation of the partial pressure p_{gas} of the particular gas:

$$p_{gas} = x \cdot (p_{tot} - p_{vap}) \tag{A.1}$$

The vapor pressure of water depends strongly on the temperature (Table A.1).

Table A.1: Vapor pressure p_{vap} of water at different temperatures.

Temperature	Vapor pressure
21 °C	2.49 kPa
25 °C	3.17 kPa
37 °C	6.28 kPa

Henry's law tells the concentration c of the dissolved gas depending on the partial pressure:

$$c = \frac{p_{gas}}{k_H} \tag{A.2}$$

k_H is the Henry's law constant, often reported as the inverse k_H^{-1}, also referred to by *Henry solubility*. Table A.2 contains some typical values. Please note the unique role of CO_2 and its equilibrium with H_2CO_3 (equation (10.8)), while the other gases are inert in an aqueous solution.

Table A.2: Henry solubility k_H^{-1} in µM kPa^{-1} calculated for various gases according to [300].

Temperature	Oxygen	Hydrogen	Nitrogen	Argon	CO$_2$
21 °C	13.36	7.85	6.76	14.64	371.36
25 °C	12.46	7.66	6.36	13.65	332.20
37 °C	10.48	7.26	5.49	11.41	246.96

Older literature sometimes uses the Bunsen (absorption) coefficient α instead of the Henry solubility. It relates the volume of the dissolved gas to the solvent volume at fixed conditions (originally standard pressure and 0 °C).

A.2.2 Salting-out effect

A common source of inaccuracy is neglecting the electrolytes salt content when discussing dissolved gas. Depending on composition and concentration, a "salting-out effect" occurs, lowering the dissolved gas concentration tremendously. A convenient way to describe the effect is the c/c_0 factor with c_0 the gas concentration in pure water and c in the solution. In some cases, e. g., in HCl solutions, increased solubility compared to pure water is possible. Table A.3 list values for typically used electrolytes based on the model of Weisenberger and Schumpe [301]. The book's webpage provides further resources.

Table A.3: Salting-out of dissolved gases, the values are the dimensionless c/c_0 parameter with c_0 the gas concentration in pure water and c the concentration in the solution. Values calculated based on [301]. PBS refers to a phosphate-buffered saline solution according to recipe II in Table 4.7.

Electrolyte	Temperature	Oxygen	Hydrogen	Nitrogen	Argon	CO_2
	21 °C	0.875	0.944	0.875	0.856	0.929
0.5 M H_2SO_4	25 °C	0.879	0.948	0.882	0.862	0.933
	37 °C	0.892	0.960	0.905	0.880	0.946
	21 °C	0.924	1.022	0.923	0.897	1.000
1 M HCl	25 °C	0.929	1.028	0.934	0.905	1.006
	37 °C	0.947	1.045	0.965	0.930	1.025
	21 °C	0.890	0.912	0.89	0.884	0.907
PBS	25 °C	0.892	0.914	0.893	0.886	0.909
	37 °C	0.896	0.917	0.900	0.892	0.913
	21 °C	0.663	0.733	0.662	0.644	0.717
1 M KOH	25 °C	0.667	0.737	0.67	0.649	0.722
	37 °C	0.679	0.749	0.693	0.667	0.735
	21 °C	0.708	0.764	0.708	0.693	0.751
0.5 M K_2SO_4	25 °C	0.711	0.767	0.714	0.697	0.755
	37 °C	0.721	0.776	0.732	0.711	0.765
	21 °C	0.710	0.785	0.710	0.690	0.768
1 M NaCl	25 °C	0.714	0.790	0.718	0.696	0.773
	37 °C	0.728	0.803	0.742	0.715	0.788
	21 °C	0.747	0.826	0.747	0.726	0.809
1 M KCl	25 °C	0.752	0.831	0.755	0.732	0.814
	37 °C	0.766	0.845	0.781	0.752	0.829

B Instrumentation

Chapter 3 discusses the instrumentation, mainly the potentiostatic circuit, to get a better understanding of what happens when applying the different methods. A strong focus is on the particularities of digital signal generation and data acquisition. The text assumes basic knowledge of electronics, namely operational amplifier circuits. This chapter provides a primer to the operational amplifier and discusses some fundamental circuits for readers not familiar with electronics or electrical engineering. You will develop a level to understand simple potentiostat circuits and follow the discussion in Chapter 3. The knowledge may be a good starting point to develop further, allowing you to build simple potentiostatic circuits on your own; however, the practical aspects of setting up circuits on a breadboard are beyond this text. The chapter also clarifies ground and virtual ground to decipher the meaning of auxiliary connections at your device and fundamentally understand what often appears as a side note in device manuals.

B.1 Operational amplifier primer

The operational amplifier (opamp) is an integrated circuit designed initially for analog mathematical operations (adding, integrating). Essentially, it is a differential amplifier with a huge gain. A typical opamp contains 20 or more transistors monolithically integrated. Often, the integrated circuit includes multiple opamps. Figure B.1 shows the full circuit diagram symbol with the five connections of an opamp.

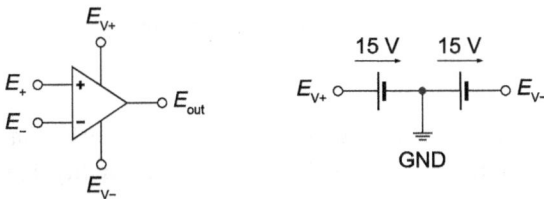

Figure B.1: The operational amplifier has five terminals: positive/negative input (E_+, E_-), the output E_{out}, and positive/negative supply voltage (E_{V+}, E_{V-}). The bipolar supply voltage is symmetrical to GND (0 V), here with ±15 V. In a circuit diagram, the supply voltage terminals are often not drawn.

The amplifier has two inputs, the noninverting input (E_+) and the inverting input (E_-). Without external components, the opamp amplifies the difference voltage ($E_+ - E_-$) by its open-loop gain (V_0). The amplified voltage is between the output pin and the circuit's ground (GND):

$$E_{out} = V_0 \cdot (E_+ - E_-) \tag{B.1}$$

https://doi.org/10.1515/9783111488844-013

For its operation, the opamp requires symmetrical, bipolar supply voltages[1] (E_{V+}, E_{V-}). The integrated circuit has no ground pin but assumes the mid between the two supply voltage pins as 0 V. Most opamps can handle higher supply voltages such as ±15 V than the typical 5 V or less available in most digital circuits. The supply voltage terminals are often not drawn for simplification, especially when discussing general circuits rather than a specific device.

Open-loop amplifier and voltage follower

The open-loop amplifier (Figure B.2) has the inverting input connected to GND, amplifying the input voltage at the noninverting input:

$$E_{out} = V_0 \cdot E_{in} \tag{B.2}$$

Typical values for the open-loop gain are 100,000 or even larger. Less than a millivolt difference at the input causes saturation of the output at either the positive or negative limit[2] depending on the input difference's sign. Practically the open-loop amplifier acts as a comparator with the output flipping between the positive or negative limit depending on the input's sign.

Open-loop amplifier (comparator)　　　　**Voltage follower (electrometer)**

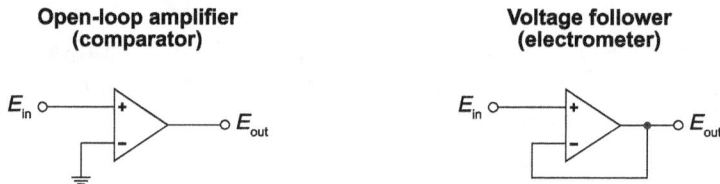

Figure B.2: Voltage amplification: the open-loop amplifier with very high (infinite) amplification practically works as a comparator; the opposite is the voltage follower with amplification 1.

Therefore, opamps are typically operated with a feedback loop. The feedback from the output connects to the inverting input (*negative feedback*) to ensure stability. The most simple negative feedback configuration is the voltage follower or electrometer amplifier (Figure B.2). The output directly connects to the inverting input.

In the general transfer function, the inverting input is the same as the output:

$$E_{out} = V_0 \cdot (E_+ - E_-) = V_0 \cdot (E_{in} - E_{out}) \tag{B.3}$$

---

1 There are specific single-supply opamps designed for unipolar operation. Although in principle, any opamp can work with a single supply voltage.

2 For most opamps, the highest/lowest output voltage is slightly lower than the range given by the supply voltages. Especially with lower supply voltage levels, e. g., ±3.3 V, that might affect the overall performance. For such applications, specific rail-to-rail opamps allow outputs down to/up to the supply voltages.

Rearranging and assuming the open-loop gain very high ($V_0 \to \infty$) leads to the transfer function of the voltage follower:

$$E_{out} = \frac{1}{1 + \frac{1}{V_0}} \cdot E_{in} \tag{B.4}$$

$$E_{out} = E_{in} \tag{B.5}$$

Using the limit transition for the open-loop gain toward infinity allows the derivation of all transfer functions. However, for circuits with negative feedback, two simple rules are enough to understand the circuit's functionality and derive the transfer function directly:

1. There is no current flow into the inputs of the opamp.
2. The opamp keeps both inputs at the same voltage (as long as the operation is within the device specification).

These rules facilitate the derivation of the transfer function of the voltage follower: the noninverting input connects to E_{in}. As both inputs have the same voltage (rule 2), the inverting input is on the level E_{in}, too. Because of the direct feedback connection to the output pin, also the E_{out} is on the level of E_{in}.

The role of a voltage follower is the buffering of the input voltage. Its performance mainly depends on the leakage current in real devices into the noninverting input. Selecting the parts for a circuit should primarily address this demand. A different opamp type is often the best choice for the voltage follower than for opamps driving substantial output currents like the control amplifier in a potentiostat. Opamps designed for applications in electrometers can show input bias currents below 100 fA.

Inverting amplifier and inverting adder

The same rules lead to the transfer function of the inverting amplifier (Figure B.3). The noninverting input connects to GND (0 V). As there is no voltage difference between the inputs, the inverting input is at 0 V, too. There is no direct connection, and the 0 V level is only valid while the opamp operates within the specifications. This level in the circuit is referred to as virtual ground (VGND), indicated in the figure by the red dot. With rule 1 (no current flow into the opamps' inputs), it implies that the sum of the currents at this point is

$$E_{in} \cdot R_1 + E_{out} \cdot R_f = 0 \tag{B.6}$$

Rearrangement leads to the transfer function of the inverting amplifier:

$$E_{out} = -\frac{R_f}{R_1} \cdot E_{in} \tag{B.7}$$

Inverting amplifier

Inverting adder

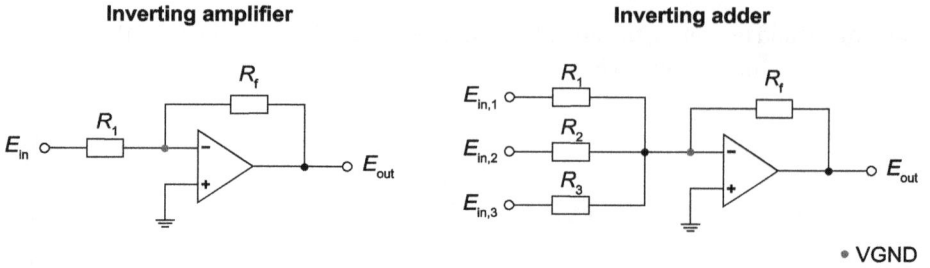

• VGND

Figure B.3: Inverting circuits: the positive input connects to the ground and forces the negative input to the virtual ground (VGND). The output signal is inverted and scaled by the ratio of the resistor in the feedback loop and the input line(s).

Similar argumentation results in the transfer function of the inverting adder:

$$E_{out} = -\left(\frac{R_f}{R_1} \cdot E_{in,1} + \frac{R_f}{R_2} \cdot E_{in,2} + \frac{R_f}{R_3} \cdot E_{in,3} \right) \tag{B.8}$$

Current follower

The introduction of a resistor between the circuit and the working electrode allows measurement of the cell current by reading the voltage drop at this resistor proportional to the current passing through. Such a resistor, often called a shunt resistor, should be small not to influence the working electrode potential, but this also lowers the voltage drop making its measurement critical. A better solution is introducing a current follower (Figure B.4) that allows current measurement with zero voltage drop.

Current follower

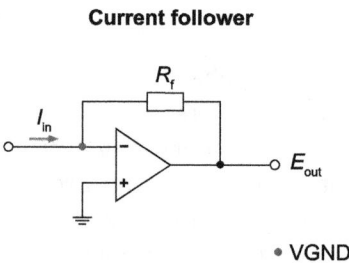

• VGND

Figure B.4: The current follower is called a transimpedance amplifier (TIA) or zero-resistance amperemeter (ZRA), converting a current sink into a voltage source. It allows current measurement without voltage drop.

The noninverting input of the current follower connects GND leading to VGND at the inverting input. Any input current through the circuit (I_{in}) must pass through the resistor in the feedback loop. With the inverting input on VGND, the voltage drop at the resistor is proportional to the input current, leading to the transfer function:

$$E_{out} = R_f \cdot I_{in} \tag{B.9}$$

The role of a current buffer is the transfer of current from a part of the circuit with low output impedance to another circuitry with high input impedance. Accordingly, the current buffer is a transimpedance amplifier (TIA), the common name used in the block diagrams of integrated potentiostats.

In the context of electrochemical noise analysis (Section 6.3.3), the current follower is named zero-resistance amperemeter (ZRA).

Integrator

The integrator plays an essential role in coulometry and cyclic voltammetry with a staircase excitation combined with current integration. The circuit resembles the inverting amplifier with a capacitor in the feedback loop instead of the resistor (Figure B.5). The transfer function depends on the time constant RC, which describes how the output voltage relates to the input:

$$E_{out} = -\int_0^t E_{in} \cdot \frac{1}{RC} dt \tag{B.10}$$

Practical considerations are the saturation charge given by C and the maximal output voltage the opamp can handle. In a potentiostat, the selected current range, the time constant of the integrator, and the saturation charge need careful consideration. Especially if there is a drift correction, the actual saturation charge could be less than the nominal saturation charge depending on the drift.

Integrator

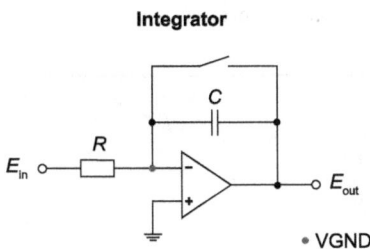

Figure B.5: Integrator circuit: the characteristic parameter is the time constant RC with the value of C defining the saturation charge. The switch allows discharging.

An integrator typically has a discharge possibility indicated as a switch in Figure B.5. Appropriate discharge is vital for coulometry experiments. Potentiostats offering linear scans or cyclic voltammetry with current integration usually discharge the integrator automatically after each potential setup.

B.2 Ground and virtual ground

Electronic circuit diagrams use the ground (GND) symbol to indicate the zero level in the circuit's voltage scale. In general, the level of GND does not relate to any reference system with an absolute potential scale. In contrast, the electrochemical scale always refers to the NHE, which is related to other absolute scales (Figure 2.2). The difference between the circuit's zero level and the potentials in the electrochemical cell is a common source of misunderstanding when discussing electrical circuits together with electrochemistry.

Virtual ground (VGND) is on the same level as GND but not galvanically connected. With operational amplifier circuits, the zero-difference between the two input terminals of the amplifier generates VGND level at one input when the other terminal is on GND. This condition is valid as long as the operational amplifier works within its specifications. The circuit's functionality needs to avoid any connection between VGND and GND.

In mixed-signal circuits with analog and digital components, the analog front-end is preferably spatially separated from the digital part because the digital signals may influence low-level analog signals. Analog-digital converters and digital-analog converters translate in between the two worlds. Board designers place the components separated and may use separate ground planes for the analog and digital sections. The ground levels in the two compartments are analog ground (AGND) and digital ground (DGND). Depending on the circuit concept, AGND and DGND can be hardwired by a ground plane bridge in the board layout or through a joint of the connections at the power supply. If direct wires are not acceptable, optoisolators or transformer can link the analog and digital parts.

! Some potentiostats provide both AGND and DGND as an external connection option. In this case, you should use AGND when, e. g., measuring the analog outputs with an oscilloscope and DGND for digital input/output. Although AGND and DGND may be internally connected, distinguishing them carefully when connecting external devices help avoid ground loops (Section 3.5).

Bibliography

[1]   A. J. Bard, L. R. Faulkner, *Electrochemical methods: fundamentals and applications*, Wiley, New York, **2001**.

[2]   D. A. Wells, *The Science of Common Things: A Familiar Explanation of the First Principles of Physical Science*, Ivison, Blakeman, Taylor, New York, **1857**.

[3]   W. R. Grove, "LXXII. On a gaseous voltaic battery," *The London, Edinburgh, and Dublin Philosophical Magazine and Journal of Science* **1842**, *21*, 417–420, DOI https://doi.org/10.1080/14786444208621600.

[4]   A. K. Shukla, T. P. Kumar, "ECS classics: pillars of modern electrochemistry," *The Electrochemical Society Interface* **2008**, *17*, 31–39, DOI https://doi.org/10.1149/2.f01083if.

[5]   J. Kieninger, A. Weltin, H. Flamm, G. A. Urban, "Microsensor systems for cell metabolism – from 2D culture to organ-on-chip," *Lab on a Chip* **2018**, *18*, 1274–1291, DOI https://doi.org/10.1039/c7lc00942a.

[6]   A. Weltin, D. Ganatra, K. Knig, K. Joseph, U. G. Hofmann, G. A. Urban, J. Kieninger, "New life for old wires: electrochemical sensor method for neural implants," *Journal of Neural Engineering* **2020**, *17*, 016007, DOI https://doi.org/10.1088/1741-2552/ab4c69.

[7]   R. F. Keeling, C. D. Keeling, Atmospheric Monthly In Situ CO2 Data – Mauna Loa Observatory, Hawaii (Archive 2021-09-07), In Scripps CO2 Program Data. UC San Diego Library Digital Collections, **2017**, DOI 10.6075/J08W3BHW.

[8]   D. Lüthi, M. L. Floch, B. Bereiter, T. Blunier, J.-M. Barnola, U. Siegenthaler, D. Raynaud, J. Jouzel, H. Fischer, K. Kawamura, T. F. Stocker, "High-resolution carbon dioxide concentration record 650,000–800,000 years before present," *Nature* **2008**, *453*, 379–382, DOI https://doi.org/10.1038/nature06949.

[9]   J. Inczédy, T. Lengyel, A. Ure, A. Gelencsér, *Compendium of Analytical Nomenclature: Definitive Rules 1997*, IUPAC Chemical Data Series, Blackwell Science, **1998**.

[10]   C. M. A. Brett, A. M. O. Brett, *Electrochemistry: principles, methods, and applications*, Oxford University Press, Oxford New York, **1993**.

[11]   V. S. Bagotskii, *Fundamentals of electrochemistry*, Wiley-Interscience, Hoboken, N.J., **2006**.

[12]   C. H. Hamann, A. Hamnett, W. Vielstich, *Electrochemistry*, Wiley-VCH, Weinheim, **2007**.

[13]   S. Trasatti, "The absolute electrode potential: an explanatory note (recommendations 1986)," *Pure and Applied Chemistry* **1986**, *58*, 955–966, DOI https://doi.org/10.1351/pac198658070955.

[14]   A. Kahn, "Fermi level, work function and vacuum level," *Materials Horizons* **2016**, *3*, 7–10, DOI https://doi.org/10.1039/c5mh00160a.

[15]   G. Burstein, "A hundred years of Tafel's equation: 1905–2005," *Corrosion Science* **2005**, *47*, 2858–2870, DOI https://doi.org/10.1016/j.corsci.2005.07.002.

[16]   D. Shoup, A. Szabo, "Chronoamperometric current at finite disk electrodes," *Journal of Electroanalytical Chemistry and Interfacial Electrochemistry* **1982**, *140*, 237–245, DOI https://doi.org/10.1016/0022-0728(82)85171-1.

[17]   P. J. Mahon, K. B. Oldham, "Diffusion-controlled chronoamperometry at a disk electrode," *Analytical Chemistry* **2005**, *77*, 6100–6101, DOI https://doi.org/10.1021/ac051034y.

[18]   P. Daubinger, J. Kieninger, T. Unmüssig, G. A. Urban, "Electrochemical characteristics of nanostructured platinum electrodes – a cyclic voltammetry study," *Physical Chemistry Chemical Physics* **2014**, *16*, 8392–8399, DOI https://doi.org/10.1039/c4cp00342j.

[19]   J. C. Myland, K. B. Oldham, "Uncompensated resistance. 1. The effect of cell geometry," *Analytical Chemistry* **2000**, *72*, 3972–3980, DOI https://doi.org/10.1021/ac0001535.

[20]   O. S. Hoilett, J. F. Walker, B. M. Balash, N. J. Jaras, S. Boppana, J. C. Linnes, "KickStat: a coin-sized potentiostat for high-resolution electrochemical analysis," *Sensors* **2020**, *20*, 2407, DOI https://doi.org/10.3390/s20082407.

[21]   D. Bill, M. Jasper, A. Weltin, G. A. Urban, S. J. Rupitsch, J. Kieninger, "Electrochemical methods in the cloud: FreiStat, an IoT-enabled embedded potentiostat," *Analytical Chemistry* **2023**, *95*, 13003–13009, DOI https://doi.org/10.1021/acs.analchem.3c02114.

https://doi.org/10.1515/9783111488844-014

[22] F. Liebisch, A. Weltin, J. Marzioch, G. A. Urban, J. Kieninger, "Zero-consumption Clark-type microsensor for oxygen monitoring in cell culture and organ-on-chip systems," *Sensors and Actuators. B, Chemical* **2020**, *322*, 128652, DOI https://doi.org/10.1016/j.snb.2020.128652.

[23] J. D. Benck, B. A. Pinaud, Y. Gorlin, T. F. Jaramillo, "Substrate selection for fundamental studies of electrocatalysts and photoelectrodes: inert potential windows in acidic, neutral, and basic electrolyte," *PLoS ONE* **2014**, *9*, e107942, DOI https://doi.org/10.1371/journal.pone.0107942.

[24] C. Köhler, L. Bleck, M. Frei, R. Zengerle, S. Kerzenmacher, "Poisoning of highly porous platinum electrodes by amino acids and tissue fluid constituents," *ChemElectroChem* **2015**, *2*, 1785–1793, DOI https://doi.org/10.1002/celc.201500215.

[25] W. J. Hamer, Y.-C. Wu, "Osmotic coefficients and mean activity coefficients of uni-univalent electrolytes in water at 25 °C," *Journal of Physical and Chemical Reference Data* **1972**, *1*, 1047–1100, DOI https://doi.org/10.1063/1.3253108.

[26] R. G. Bates, V. E. Bower, "Standard potential of the silver-silver-chloride electrode from 0 degrees to 95 degrees C and the thermodynamic properties of dilute hydrochloric acid solutions," *Journal of Research of the National Bureau of Standards* **1954**, *53*, 283, DOI https://doi.org/10.6028/jres.053.037.

[27] A. Economou, P. R. Fielden, "Mercury film electrodes: developments, trends and potentialities for electroanalysis," *The Analyst* **2003**, *128*, 205–213, DOI https://doi.org/10.1039/b201130c.

[28] J. Mauzeroll, E. A. Hueske, A. J. Bard, "Scanning electrochemical microscopy. 48. Hg/Pt hemispherical ultramicroelectrodes: fabrication and characterization," *Analytical Chemistry* **2003**, *75*, 3880–3889, DOI https://doi.org/10.1021/ac034088l.

[29] S. Sharma, *Carbon for Micro and Nano Devices*, De Gruyter, Berlin, Boston, **2024**, DOI https://doi.org/10.1515/9783110620634.

[30] Annu, S. Sharma, R. Jain, A. N. Raja, "Review – pencil graphite electrode: an emerging sensing material," *Journal of the Electrochemical Society* **2019**, *167*, 037501, DOI https://doi.org/10.1149/2.0012003jes.

[31] J. V. Macpherson, "A practical guide to using boron doped diamond in electrochemical research," *Physical Chemistry Chemical Physics* **2014**, *17*, 2935–2949, DOI https://doi.org/10.1039/c4cp04022h.

[32] B. Wessling, A. Besmehn, W. Mokwa, U. Schnakenberg, "Reactively sputtered iridium oxide," *Journal of the Electrochemical Society* **2007**, *154*, F83, DOI https://doi.org/10.1149/1.2713691.

[33] K. Yamanaka, "Anodically electrodeposited iridium oxide films (AEIROF) from alkaline solutions for electrochromic display devices," *Japanese Journal of Applied Physics* **1989**, *28*, 632–637, DOI https://doi.org/10.1143/jjap.28.632.

[34] M. R. Akanda, A. M. Osman, M. K. Nazal, M. A. Aziz, "Review – recent advancements in the utilization of indium tin oxide (Ito) in electroanalysis without surface modification," *Journal of the Electrochemical Society* **2020**, *167*, 037534, DOI https://doi.org/10.1149/1945-7111/ab64bd.

[35] J. Ufheil, K. Borgwarth, J. Heinze, "Introduction to the principles of ultramicroheptodes in ring-disk interactions," *Analytical Chemistry* **2002**, *74*, 1316–1321, DOI https://doi.org/10.1021/ac010912z.

[36] O. Sklyar, J. Ufheil, J. Heinze, G. Wittstock, "Application of the boundary element method numerical simulations for characterization of heptode ultramicroelectrodes in SECM experiments," *Electrochimica Acta* **2003**, *49*, 117–128, DOI https://doi.org/10.1016/j.electacta.2003.04.007.

[37] L. Danis, D. Polcari, A. Kwan, S. M. Gateman, J. Mauzeroll, "Fabrication of carbon, gold, platinum, silver, and mercury ultramicroelectrodes with controlled geometry," *Analytical Chemistry* **2015**, *87*, 2565–2569, DOI https://doi.org/10.1021/ac503767n.

[38] J. Heinze, "Ultramicroelectrodes in electrochemistry," *Angewandte Chemie. International Edition in English* **1993**, *32*, 1268–1288, DOI https://doi.org/10.1002/anie.199312681.

[39] M. Ue, K. Ida, S. Mori, "Electrochemical properties of organic liquid electrolytes based on quaternary onium salts for electrical double-layer capacitors," *Journal of the Electrochemical Society* **1994**, *141*, 2989–2996-2989-2996, DOI https://doi.org/10.1149/1.2059270.

[40] V. Bocharova, A. P. Sokolov, "Perspectives for polymer electrolytes: a view from fundamentals of ionic conductivity," *Macromolecules* **2020**, *53*, 4141–4157, DOI https://doi.org/10.1021/acs.macromol.9b02742.

[41]   S. Tanimoto, A. Ichimura, "Discrimination of inner- and outer-sphere electrode reactions by cyclic voltammetry experiments," *Journal of Chemical Education* **2013**, *90*, 778–781, DOI https://doi.org/10.1021/ed200604m.

[42]   D. Himmel, S. K. Goll, I. Leito, I. Krossing, "A unified pH scale for all phases," *Angewandte Chemie. International Edition in English* **2010**, *49*, 6885–6888, DOI https://doi.org/10.1002/anie.201000252.

[43]   V. Radtke, D. Stoica, I. Leito, F. Camões, I. Krossing, B. Anes, M. Roziková, L. Deleebeeck, S. Veltzé, T. Näykki, F. Bastkowski, A. Heering, N. Dániel, R. Quendera, L. Liv, E. Uysal, N. Lawrence, "A unified pH scale for all solvents: part I – intention and reasoning (IUPAC technical report)," *Pure and Applied Chemistry* **2021**, *93*, 1049–1060, DOI https://doi.org/10.1515/pac-2019-0504.

[44]   P. Zimmermann, A. Weltin, G. A. Urban, J. Kieninger, "Active potentiometry for dissolved oxygen monitoring with platinum electrodes," *Sensors* **2018**, *18*, 2024, DOI https://doi.org/10.3390/s18082404.

[45]   B. Uka, J. Kieninger, G. A. Urban, A. Weltin, "Electrochemical microsensor for microfluidic glyphosate monitoring in water using MIP-based concentrators," *ACS Sensors* **2021**, *6*, 2738–2746, DOI https://doi.org/10.1021/acssensors.1c00884.

[46]   J. Fedorowski, W. R. LaCourse, "A review of pulsed electrochemical detection following liquid chromatography and capillary electrophoresis," *Analytica Chimica Acta* **2015**, *861*, 1–11, DOI https://doi.org/10.1016/j.aca.2014.08.035.

[47]   R. S. Nicholson, I. Shain, "Theory of stationary electrode polarography. Single scan and cyclic methods applied to reversible, irreversible, and kinetic systems," *Analytical Chemistry* **1964**, *36*, 706–723, DOI https://doi.org/10.1021/ac60210a007.

[48]   M. Rudolph, "A fast implicit finite difference algorithm for the digital simulation of electrochemical processes," *Journal of Electroanalytical Chemistry and Interfacial Electrochemistry* **1991**, *314*, 13–22, DOI https://doi.org/10.1016/0022-0728(91)85425-o.

[49]   J. H. Brown, "Development and use of a cyclic voltammetry simulator to introduce undergraduate students to electrochemical simulations," *Journal of Chemical Education* **2015**, *92*, 1490–1496, DOI https://doi.org/10.1021/acs.jchemed.5b00225.

[50]   S. Wang, J. Wang, Y. Gao, "Development and use of an open-source, user-friendly package to simulate voltammetry experiments," *Journal of Chemical Education* **2017**, *94*, 1567–1570, DOI https://doi.org/10.1021/acs.jchemed.6b00986.

[51]   R. S. Nicholson, "Theory and application of cyclic voltammetry for measurement of electrode reaction kinetics," *Analytical Chemistry* **1965**, *37*, 1351–1355, DOI https://doi.org/10.1021/ac60230a016.

[52]   J. Heinze, "Cyclic voltammetry – "Electrochemical spectroscopy". New analytical methods (25)," *Angewandte Chemie. International Edition in English* **1984**, *23*, 831–847, DOI https://doi.org/10.1002/anie.198408313.

[53]   J. Osteryoung, "Pulse voltammetry," *Journal of Chemical Education* **1983**, *60*, 296, DOI https://doi.org/10.1021/ed060p296.

[54]   W. Xu, T. Lu, C. Liu, W. Xing, "Supplement to the theory of normal pulse voltammetry and its application to the kinetic study of methanol oxidation on a polycrystalline platinum electrode," *The Journal of Physical Chemistry B* **2005**, *109*, 7872–7877, DOI https://doi.org/10.1021/jp0444058.

[55]   J. G. Osteryoung, R. A. Osteryoung, "Square wave voltammetry," *Analytical Chemistry* **1985**, *57*, 101A–110A, DOI https://doi.org/10.1021/ac00279a789.

[56]   J. A. Rodrigues, C. M. Rodrigues, P. J. Almeida, I. M. Valente, L. M. Gonçalves, R. G. Compton, A. A. Barros, "Increased sensitivity of anodic stripping voltammetry at the hanging mercury drop electrode by ultracathodic deposition," *Analytica Chimica Acta* **2011**, *701*, 152–156, DOI https://doi.org/10.1016/j.aca.2011.05.031.

[57]   D. Britz, "Setting the record straight on reciprocal derivative chronopotentiometry," *International Journal of Electrochemical Science* **2006**, *1*, 379–382, DOI https://doi.org/10.1016/s1452-3981(23)17166-5.

[58] S. Bi, J. Yu, "Investigations on cyclic reciprocal derivative chronopotentiometry. Part 1. Theory for a reversible reaction," *Journal of Electroanalytical Chemistry* **1996**, *405*, 51–58, DOI https://doi.org/10.1016/0022-0728(95)04377-2.

[59] L. Chen, Z. Wang, J. Chen, S. Bi, "Theoretical investigation on cyclic reciprocal derivative chronopotentiometry Kinetic study of totally irreversible electrode processes," *Electrochimica Acta* **2007**, *52*, 8020–8030, DOI https://doi.org/10.1016/j.electacta.2007.06.071.

[60] J. Wang, B. Tian, "Chronopotentiometric analysis of highly resistive media," *Analytical Chemistry* **2000**, *72*, 3241–3244, DOI https://doi.org/10.1021/ac000230w.

[61] A. Molina, J. González, F. Saavedra, L. M. Abrantes, "Cyclic reciprocal derivative chronopotentiometry. Applications to the detection and characterisation of adsorption processes," *Electrochimica Acta* **1999**, *45*, 761–773, DOI https://doi.org/10.1016/s0013-4686(99)00255-8.

[62] S. Musa, D. R. Rand, C. Bartic, W. Eberle, B. Nuttin, G. Borghs, "Coulometric detection of irreversible electrochemical reactions occurring at Pt microelectrodes used for neural stimulation," *Analytical Chemistry* **2011**, *83*, 4012–4022, DOI https://doi.org/10.1021/ac103037u.

[63] W. J. Albery, S. Bruckenstein, "Ring-disc electrodes. Part 2. – Theoretical and experimental collection effciencies," *Transactions of the Faraday Society* **1966**, *62*, 1920–1931, DOI https://doi.org/10.1039/tf9666201920.

[64] T. Schmidt, U. Paulus, H. Gasteiger, R. Behm, "The oxygen reduction reaction on a Pt/carbon fuel cell catalyst in the presence of chloride anions," *Journal of Electroanalytical Chemistry* **2001**, *508*, 41–47, DOI https://doi.org/10.1016/s0022-0728(01)00499-5.

[65] J. Herranz, A. Garsuch, H. A. Gasteiger, "Using rotating ring disc electrode voltammetry to quantify the superoxide radical stability of aprotic Li–air battery electrolytes," *The Journal of Physical Chemistry C* **2012**, *116*, 19084–19094, DOI https://doi.org/10.1021/jp304277z.

[66] A. Yaroshchuk, T. Luxbacher, "Interpretation of electrokinetic measurements with porous films: role of electric conductance and streaming current within porous structure," *Langmuir* **2010**, *26*, 10882–10889, DOI https://doi.org/10.1021/la100777z.

[67] A. J. Bard, F. R. F. Fan, J. Kwak, O. Lev, "Scanning electrochemical microscopy. Introduction and principles," *Analytical Chemistry* **1989**, *61*, 132–138, DOI https://doi.org/10.1021/ac00177a011.

[68] M. A. Edwards, S. Martin, A. L. Whitworth, J. V. Macpherson, P. R. Unwin, "Scanning electrochemical microscopy: principles and applications to biophysical systems," *Physiological Measurement* **2006**, *27*, R63–R108, DOI https://doi.org/10.1088/0967-3334/27/12/r01.

[69] B. Liu, S. A. Rotenberg, M. V. Mirkin, "Scanning electrochemical microscopy of living cells: different redox activities of nonmetastatic and metastatic human breast cells," *Proceedings of the National Academy of Sciences* **2000**, *97*, 9855–9860, DOI https://doi.org/10.1073/pnas.97.18.9855.

[70] A. Schulte, W. Schuhmann, "Single-cell microelectrochemistry," *Angewandte Chemie. International Edition in English* **2007**, *46*, 8760–8777, DOI https://doi.org/10.1002/anie.200604851.

[71] J. Clausmeyer, W. Schuhmann, "Nanoelectrodes: applications in electrocatalysis, single-cell analysis and high-resolution electrochemical imaging," *TrAC Trends in Analytical Chemistry* **2016**, *79*, 46–59, DOI https://doi.org/10.1016/j.trac.2016.01.018.

[72] K. Itaya, E. Tomita, "Scanning tunneling microscope for electrochemistry – a new concept for the in situ scanning tunneling microscope in electrolyte solutions," *Surface Science* **1988**, *201*, L507–L512, DOI https://doi.org/10.1016/0039-6028(88)90489-x.

[73] C.-S. Lu, O. Lewis, "Investigation of film-thickness determination by oscillating quartz resonators with large mass load," *Journal of Applied Physics* **1972**, *43*, 4385–4390, DOI https://doi.org/10.1063/1.1660931.

[74] K. K. Kanazawa, J. G. Gordon, "Frequency of a quartz microbalance in contact with liquid," *Analytical Chemistry* **1985**, *57*, 1770–1771, DOI https://doi.org/10.1021/ac00285a062.

[75] K. Ogle, V. Baudu, L. Garrigues, X. Philippe, "Localized electrochemical methods applied to cut edge corrosion," *Journal of The Electrochemical Society* **2000**, *147*, 3654–3660, DOI https://doi.org/10.1149/1.1393954.

[76] S. V. Lamaka, M. G. Taryba, M. L. Zheludkevich, M. G. S. Ferreira, "Novel solid-contact ion-selective microelectrodes for localized potentiometric measurements," *Electroanalysis* **2009**, *21*, 2447–2453, DOI https://doi.org/10.1002/elan.200900258.

[77] V. A. Nazarov, M. G. Taryba, E. A. Zdrachek, K. A. Andronchyk, V. V. Egorov, S. V. Lamaka, "Sodium- and chloride-selective microelectrodes optimized for corrosion studies," *Journal of Electroanalytical Chemistry* **2013**, *706*, 13–24, DOI https://doi.org/10.1016/j.jelechem.2013.07.034.

[78] M. Strebl, M. Bruns, S. Virtanen, "Editors' choice – respirometric in situ methods for real-time monitoring of corrosion rates: part I. Atmospheric corrosion," *Journal of the Electrochemical Society* **2020**, *167*, 021510, DOI https://doi.org/10.1149/1945-7111/ab6c61.

[79] M. G. Strebl, M. P. Bruns, G. Schulze, S. Virtanen, "Respirometric in situ methods for real-time monitoring of corrosion rates: part II. Immersion," *Journal of the Electrochemical Society* **2021**, *168*, 011502, DOI https://doi.org/10.1149/1945-7111/abdb4a.

[80] M. G. Strebl, M. P. Bruns, S. Virtanen, "Respirometric in situ methods for real-time monitoring of corrosion rates: Part III. Deconvolution of electrochemical polarization curves," *Journal of The Electrochemical Society* **2023**, *170*, 061503, DOI https://doi.org/10.1149/1945-7111/acd665.

[81] M. G. Strebl, M. P. Bruns, S. Virtanen, "Coupling respirometric HER and ORR monitoring with electrochemical measurements," *Electrochimica Acta* **2022**, *412*, 140152, DOI https://doi.org/10.1016/j.electacta.2022.140152.

[82] U. Bertocci, F. Huet, "Noise resistance applied to corrosion measurements: III. Influence of the instrumental noise on the measurements," *Journal of The Electrochemical Society* **1997**, *144*, 2786–2793, DOI https://doi.org/10.1149/1.1837896.

[83] S. Ritter, F. Huet, R. A. Cottis, "Guideline for an assessment of electrochemical noise measurement devices," *Materials and Corrosion* **2012**, *63*, 297–302, DOI https://doi.org/10.1002/maco.201005839.

[84] D.-H. Xia, S. Song, Y. Behnamian, W. Hu, Y. F. Cheng, J.-L. Luo, F. Huet, "Review – electrochemical noise applied in corrosion science: theoretical and mathematical models towards quantitative analysis," *Journal of the Electrochemical Society* **2020**, *167*, 081507, DOI https://doi.org/10.1149/1945-7111/ab8de3.

[85] H. A. A Al-Mazeedi, R. A. Cottis, "A practical evaluation of electrochemical noise parameters as indicators of corrosion type," *Electrochimica Acta* **2004**, *49*, 2787–2793, DOI https://doi.org/10.1016/j.electacta.2004.01.040.

[86] A. Aballe, M. Bethencourt, F. Botana, M. Marcos, "Wavelet transform-based analysis for electrochemical noise," *Electrochemistry Communications* **1999**, *1*, 266–270, DOI https://doi.org/10.1016/s1388-2481(99)00053-3.

[87] A. M. Homborg, T. Tinga, X. Zhang, E. P. M. van Westing, P. J. Oonincx, J. H. W. de Wit, J. M. C. Mol, "Time–frequency methods for trend removal in electrochemical noise data," *Electrochimica Acta* **2012**, *70*, 199–209, DOI https://doi.org/10.1016/j.electacta.2012.03.062.

[88] R. A. Cottis, "Interpretation of electrochemical noise data," *Corrosion* **2001**, *57*, 265–285, DOI https://doi.org/10.5006/1.3290350.

[89] R. Moshrefi, M. G. Mahjani, M. Jafarian, "Application of wavelet entropy in analysis of electrochemical noise for corrosion type identification," *Electrochemistry Communications* **2014**, *48*, 49–51, DOI https://doi.org/10.1016/j.elecom.2014.08.005.

[90] J. J. A. Lozeman, P. Führer, W. Olthuis, M. Odijk, "Spectroelectrochemistry, the future of visualizing electrode processes by hyphenating electrochemistry with spectroscopic techniques," *The Analyst* **2020**, *145*, 2482–2509, DOI https://doi.org/10.1039/c9an02105a.

[91] C. M. A. Brett, L. Dias, B. Trindade, R. Fischer, S. Mies, "Characterisation by EIS of ternary Mg alloys synthesised by mechanical alloying," *Electrochimica Acta* **2006**, *51*, 1752–1760, DOI https://doi.org/10.1016/j.electacta.2005.02.124.

[92] Y. Song, D. Shan, R. Chen, E.-H. Han, "Corrosion characterization of Mg–8Li alloy in NaCl solution," *Corrosion Science* **2009**, *51*, 1087–1094, DOI https://doi.org/10.1016/j.corsci.2009.03.011.

[93]   I. Pivac, F. Barbir, "Inductive phenomena at low frequencies in impedance spectra of proton exchange membrane fuel cells – A review," *Journal of Power Sources* **2016**, *326*, 112–119, DOI https://doi.org/10.1016/j.jpowsour.2016.06.119.

[94]   N. Hensle, D. Brinker, S. Metz, T. Smolinka, A. Weber, "On the role of inductive loops at low frequencies in PEM electrolysis," *Electrochemistry Communications* **2023**, *155*, 107585, DOI https://doi.org/10.1016/j.elecom.2023.107585.

[95]   D. Brinker, N. Hensle, J. H. de la Viña, I. Franzetti, L. V. Bühre, U. A. Andaluri, C. Menke, T. Smolinka, A. Weber, "Inductive loops in impedance spectra of PEM water electrolyzers," *Journal of Power Sources* **2024**, *622*, 235375, DOI https://doi.org/10.1016/j.jpowsour.2024.235375.

[96]   J. E. B. Randles, "Kinetics of rapid electrode reactions," *Discussions of the Faraday Society* **1947**, *1*, 11, DOI https://doi.org/10.1039/df9470100011.

[97]   B. A. Boukamp, "A linear Kronig–Kramers transform test for immittance data validation," *Journal of the Electrochemical Society* **1995**, *142*, 1885–1894-1885-1894, DOI https://doi.org/10.1149/1.2044210.

[98]   M. Schönleber, D. Klotz, E. Ivers-Tiffée, "A method for improving the robustness of linear Kramers–Kronig validity tests," *Electrochimica Acta* **2014**, *131*, 20–27, DOI https://doi.org/10.1016/j.electacta.2014.01.034.

[99]   International Organization for Standardization, Corrosion of metals and alloys – Vocabulary, Standard ISO 8044:2020(en), ISO/TC 156, **2020**.

[100]   G. S. Frankel, T. Li, J. R. Scully, "Perspective – localized corrosion: passive film breakdown vs pit growth stability," *Journal of the Electrochemical Society* **2017**, *164*, C180–C181, DOI https://doi.org/10.1149/2.1381704jes.

[101]   T. Li, J. R. Scully, G. S. Frankel, "Localized corrosion: passive film breakdown vs pit growth stability: part II. A model for critical pitting temperature," *Journal of the Electrochemical Society* **2018**, *165*, C484–C491, DOI https://doi.org/10.1149/2.0591809jes.

[102]   T. Li, J. R. Scully, G. S. Frankel, "Localized corrosion: passive film breakdown vs. pit growth stability: part III. A unifying set of principal parameters and criteria for pit stabilization and salt film formation," *Journal of the Electrochemical Society* **2018**, *165*, C762–C770, DOI https://doi.org/10.1149/2.0251811jes.

[103]   T. Li, J. R. Scully, G. S. Frankel, "Localized corrosion: passive film breakdown vs. pit growth stability: part IV. The role of salt film in pit growth: a mathematical framework," *Journal of the Electrochemical Society* **2019**, *166*, C115–C124, DOI https://doi.org/10.1149/2.0211906jes.

[104]   T. Li, J. R. Scully, G. S. Frankel, "Localized corrosion: passive film breakdown vs pit growth stability: part V. Validation of a new framework for pit growth stability using one-dimensional artificial pit electrodes," *Journal of the Electrochemical Society* **2019**, *166*, C3341–C3354, DOI https://doi.org/10.1149/2.0431911jes.

[105]   D. Enning, J. Garrelfs, "Corrosion of iron by sulfate-reducing bacteria: new views of an old problem," *Applied and Environmental Microbiology* **2014**, *80*, 1226–1236, DOI https://doi.org/10.1128/aem.02848-13.

[106]   H. H. Uhlig, P. F. King, "The flade potential of iron passivated by various inorganic corrosion inhibitors," *Journal of the Electrochemical Society* **1959**, *106*, 1, DOI https://doi.org/10.1149/1.2427255.

[107]   K. Ogle, "Atomic emission spectroelectrochemistry: real-time rate measurements of dissolution, corrosion, and passivation," *Corrosion* **2019**, *75*, 1398–1419, DOI https://doi.org/10.5006/3336.

[108]   S. V. Lambeets, E. J. Kautz, M. G. Wirth, G. J. Orren, A. Devaraj, D. E. Perea, "Nanoscale perspectives of metal degradation via in situ atom probe tomography," *Topics in Catalysis* **2020**, *63*, 1606–1622, DOI https://doi.org/10.1007/s11244-020-01367-z.

[109]   C. Gabrielli, S. Joiret, M. Keddam, H. Perrot, N. Portail, P. Rousseau, V. Vivier, "Development of a coupled SECM-EQCM technique for the study of pitting corrosion on iron," *Journal of the Electrochemical Society* **2006**, *153*, B68, DOI https://doi.org/10.1149/1.2161574.

[110] G. Jerkiewicz, G. Vatankhah, J. Lessard, M. P. Soriaga, Y.-S. Park, "Surface-oxide growth at platinum electrodes in aqueous H2SO4 reexamination of its mechanism through combined cyclic-voltammetry, electrochemical quartz-crystal nanobalance, and auger electron spectroscopy measurements," *Electrochimica Acta* **2004**, *49*, 1451–1459, DOI https://doi.org/10.1016/j.electacta.2003.11.008.

[111] H. Angerstein-Kozlowska, B. Conway, W. Sharp, "The real condition of electrochemically oxidized platinum surfaces, part I. Resolution of component processes," *Journal of Electroanalytical Chemistry and Interfacial Electrochemistry* **1973**, *43*, 9–36, DOI https://doi.org/10.1016/s0022-0728(73)80307-9.

[112] M. Alsabet, M. Grden, G. Jerkiewicz, "Comprehensive study of the growth of thin oxide layers on Pt electrodes under well-defined temperature, potential, and time conditions," *Journal of Electroanalytical Chemistry* **2006**, *589*, 120–127, DOI https://doi.org/10.1016/j.jelechem.2006.01.022.

[113] A. A. McMath, J. v. Drunen, J. Kim, G. Jerkiewicz, "Identification and analysis of electrochemical instrumentation limitations through the study of platinum surface oxide formation and reduction," *Analytical Chemistry* **2016**, *88*, 3136–3143, DOI https://doi.org/10.1021/acs.analchem.5b04239.

[114] T. Biegler, D. Rand, R. Woods, "Limiting oxygen coverage on platinized platinum; relevance to determination of real platinum area by hydrogen adsorption," *Journal of Electroanalytical Chemistry and Interfacial Electrochemistry* **1971**, *29*, 269–277, DOI https://doi.org/10.1016/s0022-0728(71)80089-x.

[115] B. E. Conway, H. Angerstein-Kozlowska, W. B. A. Sharp, E. E. Criddle, "Ultrapurification of water for electrochemical and surface chemical work by catalytic pyrodistillation," *Analytical Chemistry* **1973**, *45*, 1331–1336, DOI https://doi.org/10.1021/ac60330a025.

[116] J. Hoare, "On the interaction of oxygen with platinum," *Electrochimica Acta* **1982**, *27*, 1751–1761, DOI https://doi.org/10.1016/0013-4686(82)80174-6.

[117] D. Rand, R. Woods, "A study of the dissolution of platinum, palladium, rhodium and gold electrodes in 1 m sulphuric acid by cyclic voltammetry," *Journal of Electroanalytical Chemistry and Interfacial Electrochemistry* **1972**, *35*, 209–218, DOI https://doi.org/10.1016/s0022-0728(72)80308-5.

[118] A. A. Topalov, I. Katsounaros, M. Auinger, S. Cherevko, J. C. Meier, S. O. Klemm, K. J. J. Mayrhofer, "Dissolution of platinum: limits for the deployment of electrochemical energy conversion?," *Angewandte Chemie. International Edition in English* **2012**, *51*, 12613–12615, DOI https://doi.org/10.1002/anie.201207256.

[119] A. A. Topalov, S. Cherevko, A. R. Zeradjanin, J. C. Meier, I. Katsounaros, K. J. J. Mayrhofer, "Towards a comprehensive understanding of platinum dissolution in acidic media," *Chemical Science* **2014**, *5*, 631–638, DOI https://doi.org/10.1039/c3sc52411f.

[120] S. Cherevko, N. Kulyk, K. J. Mayrhofer, "Durability of platinum-based fuel cell electrocatalysts: dissolution of bulk and nanoscale platinum," *Nano Energy* **2016**, *29*, 275–298, DOI https://doi.org/10.1016/j.nanoen.2016.03.005.

[121] S. Geiger, S. Cherevko, K. J. Mayrhofer, "Dissolution of platinum in presence of chloride traces," *Electrochimica Acta* **2015**, *179*, 24–31, DOI https://doi.org/10.1016/j.electacta.2015.03.059.

[122] R. K. Shepherd, P. M. Carter, Y. L. Enke, A. K. Wise, J. B. Fallon, "Chronic intracochlear electrical stimulation at high charge densities results in platinum dissolution but not neural loss or functional changes in vivo," *Journal of Neural Engineering* **2019**, *16*, 026009, DOI https://doi.org/10.1088/1741-2552/aaf66b.

[123] M. Doering, J. Kieninger, G. A. Urban, A. Weltin, "Electrochemical microelectrode degradation monitoring: in situ investigation of platinum corrosion at neutral pH," *Journal of Neural Engineering* **2022**, DOI https://doi.org/10.1088/1741-2552/ac47da.

[124] S. Trasatti, "Work function, electronegativity, and electrochemical behaviour of metals III. Electrolytic hydrogen evolution in acid solutions," *Journal of Electroanalytical Chemistry and Interfacial Electrochemistry* **1972**, *39*, 163–184, DOI https://doi.org/10.1016/s0022-0728(72)80485-6.

[125] P. Quaino, F. Juarez, E. Santos, W. Schmickler, "Volcano plots in hydrogen electrocatalysis – uses and abuses," *Beilstein Journal of Nanotechnology* **2014**, *5*, 846–854, DOI https://doi.org/10.3762/bjnano.5.96.

[126] A. R. Zeradjanin, J.-P. Grote, G. Polymeros, K. J. J. Mayrhofer, "A critical review on hydrogen evolution electrocatalysis: re-exploring the volcano-relationship," *Electroanalysis* **2016**, *28*, 2256–2269, DOI https://doi.org/10.1002/elan.201600270.

[127] J. K. Nørskov, J. Rossmeisl, A. Logadottir, L. Lindqvist, J. R. Kitchin, T. Bligaard, H. Jónsson, "Origin of the overpotential for oxygen reduction at a fuel-cell cathode," *The Journal of Physical Chemistry B* **2004**, *108*, 17886–17892, DOI https://doi.org/10.1021/jp047349j.

[128] M. Madou, *Fundamentals of microfabrication and nanotechnology: 1. Solid-state physics, fluidics, and analytical techniques in micro- and nanotechnology*, CRC Press, Boca Raton, FL, **2012**.

[129] M. Madou, *Fundamentals of microfabrication and nanotechnology: 2. Manufacturing techniques for microfabrication and nanotechnology*, CRC Press, Boca Raton, FL, **2012**.

[130] W. Menz, J. Mohr, O. Paul, *Microsystem technology*, Wiley-VCH, Weinheim New York, **2001**.

[131] K. Parker, "The formulation of electroless nickel-phosphorus plating baths," *Plating and Surface Finishing* **1987**, *74*, 60–65,.

[132] F. Muench, "Electroless plating of metal nanomaterials," *ChemElectroChem* **2021**, *8*, 2993–3012, DOI https://doi.org/10.1002/celc.202100285.

[133] E. Kukharenka, M. M. Farooqui, L. Grigore, M. Kraft, N. Hollinshead, "Electroplating moulds using dry film thick negative photoresist," *Journal of Micromechanics and Microengineering* **2003**, *13*, S67, DOI https://doi.org/10.1088/0960-1317/13/4/311.

[134] I. Rose, C. Whittington, *Nickel Plating Handbook*, Nickel Institute, Bruessels, Belgium, **2014**.

[135] D. Edelstein, J. Heidenreich, R. Goldblatt, W. Cote, C. Uzoh, N. Lustig, P. Roper, T. McDevitt, W. Motsiff, A. Simon, J. Dukovic, R. Wachnik, H. Rathore, R. Schulz, L. Su, S. Luce, J. Slattery, "Full copper wiring in a sub-0.25 μm CMOS ULSI technology," *International Electron Devices Meeting. IEDM Technical Digest* **1997**, 773–776, DOI https://doi.org/10.1109/iedm.1997.650496.

[136] P. C. Andricacos, C. Uzoh, J. O. Dukovic, J. Horkans, H. Deligianni, "Damascene copper electroplating for chip interconnections," *IBM Journal of Research and Development* **1998**, *42*, 567–574, DOI https://doi.org/10.1147/rd.425.0567.

[137] P. M. Vereecken, R. A. Binstead, H. Deligianni, P. C. Andricacos, "The chemistry of additives in damascene copper plating," *IBM Journal of Research and Development* **2005**, *49*, 3–18, DOI https://doi.org/10.1147/rd.491.0003.

[138] T. P. Moffat, D. Josell, "Extreme bottom-up superfilling of through-silicon-vias by damascene processing: suppressor disruption, positive feedback and Turing patterns," *Journal of the Electrochemical Society* **2012**, *159*, D208–D216, DOI https://doi.org/10.1149/2.040204jes.

[139] H. Baltruschat, W. Vielstich, "On the mechanism of silver dissolution and deposition in aqueous KCN-KCl electrolytes," *Journal of Electroanalytical Chemistry and Interfacial Electrochemistry* **1983**, *154*, 141–153, DOI https://doi.org/10.1016/s0022-0728(83)80537-3.

[140] J. Kieninger, Y. Tamari, B. Enderle, G. Jobst, J. Sandvik, E. Pettersen, G. Urban, "Sensor access to the cellular microenvironment using the sensing cell culture flask," *Biosensors* **2018**, *8*, 44, DOI https://doi.org/10.3390/bios8020044.

[141] A. M. Feltham, M. Spiro, "Platinized platinum electrodes," *Chemical Reviews* **1971**, *71*, DOI https://doi.org/10.1021/cr60270a002.

[142] S. H. Sun, D. Q. Yang, D. Villers, G. X. Zhang, E. Sacher, J. P. Dodelet, "Template- and surfactant-free room temperature synthesis of self-assembled 3D pt nanoflowers from single-crystal nanowires," *Advanced Materials* **2008**, *20*, 571–574, DOI https://doi.org/10.1002/adma.200701408.

[143] A. Chen, P. Holt-Hindle, "Platinum-based nanostructured materials: synthesis, properties, and applications," *Chemical Reviews* **2010**, *110*, 3767–3804, DOI https://doi.org/10.1021/cr9003902.

[144] T. Unmüssig, A. Weltin, S. Urban, P. Daubinger, G. A. Urban, J. Kieninger, "Non-enzymatic glucose sensing based on hierarchical platinum micro-/nanostructures," *Journal of Electroanalytical Chemistry* **2018**, *816*, 215–222, DOI https://doi.org/10.1016/j.jelechem.2018.03.061.

[145] M. Al-Halhouli, J. Kieninger, O. Yurchenko, G. Urban, "Mass transport and catalytic activity in hierarchical/non-hierarchical and internal/external nanostructures: a novel comparison using 3D simulation," *Applied Catalysis. A, General* **2016**, *517*, 12–20, DOI https://doi.org/10.1016/j.apcata.2016.02.030.

[146] M. Halhouli, J. Kieninger, P. Daubinger, O. Yurchenko, G. Urban, "Sensitivity and selectivity of porous electrodes in heterogeneous liquid-based catalytic reactions: 3D simulation study," *Journal of the Electrochemical Society* **2016**, *163*, E273–E281, DOI https://doi.org/10.1149/2.0151610jes.

[147] S. A. M. Marzouk, S. Ufer, R. P. Buck, T. A. Johnson, L. A. Dunlap, W. E. Cascio, "Electrodeposited iridium oxide pH electrode for measurement of extracellular myocardial acidosis during acute ischemia," *Analytical Chemistry* **1998**, *70*, 5054–5061, DOI https://doi.org/10.1021/ac980608e.

[148] S. A. M. Marzouk, "Improved electrodeposited iridium oxide pH sensor fabricated on etched titanium substrates," *Analytical Chemistry* **2003**, *75*, 1258–1266, DOI https://doi.org/10.1021/ac0261404.

[149] I. A. Ges, B. L. Ivanov, D. K. Schaffer, E. A. Lima, A. A. Werdich, F. J. Baudenbacher, "Thin-film IrOx pH microelectrode for microfluidic-based microsystems," *Biosensors & Bioelectronics* **2005**, *21*, 248–256, DOI https://doi.org/10.1016/j.bios.2004.09.021.

[150] A. N. Bezbaruah, T. C. Zhang, "Fabrication of anodically electrodeposited iridium oxide film pH microelectrodes for microenvironmental studies," *Analytical Chemistry* **2002**, *74*, 5726–5733, DOI https://doi.org/10.1021/ac020326l.

[151] E. Hull, R. Piech, W. W. Kubiak, "Iridium oxide film electrodes for anodic stripping voltammetry," *Electroanalysis* **2008**, *20*, 2070–2075, DOI https://doi.org/10.1002/elan.200804295.

[152] H. A. Elsen, C. F. Monson, M. Majda, "Effects of electrodeposition conditions and protocol on the properties of iridium oxide pH sensor electrodes," *Journal of the Electrochemical Society* **2009**, *156*, F1, DOI https://doi.org/10.1149/1.3001924.

[153] R. D. Meyer, S. F. Cogan, T. H. Nguyen, R. D. Rauh, "Electrodeposited iridium oxide for neural stimulation and recording electrodes," *IEEE Transactions on Neural Systems and Rehabilitation Engineering* **2001**, *9*, 2, DOI https://doi.org/10.1109/7333.918271.

[154] H. Yang, S. K. Kang, C. A. Choi, H. Kim, D.-H. Shin, Y. S. Kim, Y. T. Kim, "An iridium oxide reference electrode for use in microfabricated biosensors and biochips," *Lab on a Chip* **2003**, *4*, 42–46, DOI https://doi.org/10.1039/b309899k.

[155] Y. Jung, J. Lee, Y. Tak, "Electrochromic mechanism of IrO2 prepared by pulsed anodic electrodeposition," *Electrochemical and Solid-State Letters* **2004**, *7*, H5, DOI https://doi.org/10.1149/1.1634083.

[156] E. Herrero, L. J. Buller, H. D. Abruña, "Underpotential deposition at single crystal surfaces of au, pt, ag and other materials," *Chemical Reviews* **2001**, *101*, 1897–1930, DOI https://doi.org/10.1021/cr9600363.

[157] V. Sudha, M. V. Sangaranarayanan, "Underpotential deposition of metals: structural and thermodynamic considerations," *The Journal of Physical Chemistry B* **2002**, *106*, 2699–2707, DOI https://doi.org/10.1021/jp013544b.

[158] N. Mayet, K. Servat, K. B. Kokoh, T. W. Napporn, "Probing the surface of noble metals electrochemically by underpotential deposition of transition metals," *Surfaces* **2019**, *2*, 257–276, DOI https://doi.org/10.3390/surfaces2020020.

[159] R. W. Johnson, A. Hultqvist, S. F. Bent, "A brief review of atomic layer deposition: from fundamentals to applications," *Materials Today* **2014**, *17*, 236–246, DOI https://doi.org/10.1016/j.mattod.2014.04.026.

[160] J. L. Stickney, "The chalkboard: electrochemical atomic layer deposition," *Interface Magazine* **2011**, *20*, 28–30, DOI https://doi.org/10.1149/2.f02112if.

[161] B. W. Gregory, M. L. Norton, J. L. Stickney, "Thin-layer electrochemical studies of the underpotential deposition of cadmium and tellurium on polycrystalline au, pt and cu electrodes," *Journal of Electroanalytical Chemistry and Interfacial Electrochemistry* **1990**, *293*, 85–101, DOI https://doi.org/10.1016/0022-0728(90)80054-a.

[162] B. W. Gregory, J. L. Stickney, "Electrochemical atomic layer epitaxy (ECALE)," *Journal of Electroanalytical Chemistry and Interfacial Electrochemistry* **1991**, *300*, 543–561, DOI https://doi.org/10.1016/0022-0728(91)85415-l.

[163] B. W. Gregory, D. W. Suggs, J. L. Stickney, "Conditions for the deposition of CdTe by electrochemical atomic layer epitaxy," *Journal of the Electrochemical Society* **1991**, *138*, 1279–1284-1279-1284, DOI https://doi.org/10.1149/1.2085773.

[164] I. Villegas, P. Napolitano, "Development of a continuous-flow system for the growth of compound semiconductor thin films via electrochemical atomic layer epitaxy," *Journal of the Electrochemical Society* **1999**, *146*, 117–124-117-124, DOI https://doi.org/10.1149/1.1391573.

[165] L. T. Viyannalage, R. Vasilic, N. Dimitrov, "Epitaxial growth of Cu on Au(111) and Ag(111) by surface limited redox replacement – an electrochemical and STM study," *The Journal of Physical Chemistry C* **2007**, *111*, 4036–4041, DOI https://doi.org/10.1021/jp067168c.

[166] J. Y. Kim, Y.-G. Kim, J. L. Stickney, "Cu nanofilm formation by electrochemical atomic layer deposition (ALD) in the presence of chloride ions," *Journal of Electroanalytical Chemistry* **2008**, *621*, 205–213, DOI https://doi.org/10.1016/j.jelechem.2007.10.005.

[167] C. Thambidurai, Y.-G. Kim, N. Jayaraju, V. Venkatasamy, J. L. Stickney, "Copper nanofilm formation by electrochemical ALD," *Journal of the Electrochemical Society* **2009**, *156*, D261, DOI https://doi.org/10.1149/1.3134555.

[168] J. Dornhof, G. A. Urban, J. Kieninger, "Deposition of copper nanofilms by surface-limited redox replacement of underpotentially deposited lead on polycrystalline gold," *Journal of the Electrochemical Society* **2018**, *166*, D3001–D3005, DOI https://doi.org/10.1149/2.0011901jes.

[169] Y.-G. Kim, J. Y. Kim, D. Vairavapandian, J. L. Stickney, "Platinum nanofilm formation by EC-ALE via redox replacement of UPD copper: studies using in-situ scanning tunneling microscopy," *The Journal of Physical Chemistry B* **2006**, *110*, 17998–18006, DOI https://doi.org/10.1021/jp063766f.

[170] M. Fayette, Y. Liu, D. Bertrand, J. Nutariya, N. Vasiljevic, N. Dimitrov, "From Au to Pt via surface limited redox replacement of Pb UPD in one-cell configuration," *Langmuir* **2011**, *27*, 5650–5658, DOI https://doi.org/10.1021/la200348s.

[171] N. Jayaraju, D. Vairavapandian, Y. G. Kim, D. Banga, J. L. Stickney, "Electrochemical atomic layer deposition (E-ALD) of Pt nanofilms using SLRR cycles," *Journal of the Electrochemical Society* **2012**, *159*, D616–D622, DOI https://doi.org/10.1149/2.053210jes.

[172] M. A. Hossain, K. D. Cummins, Y.-S. Park, M. P. Soriaga, J. L. Stickney, "Layer-by-layer deposition of Pd on Pt(111) electrode: an electron spectroscopy–electrochemistry study," *Electrocatalysis* **2012**, *3*, 183–191, DOI https://doi.org/10.1007/s12678-012-0102-5.

[173] L. B. Sheridan, J. Czerwiniski, N. Jayaraju, D. K. Gebregziabiher, J. L. Stickney, D. B. Robinson, M. P. Soriaga, "Electrochemical atomic layer deposition (E-ALD) of palladium nanofilms by surface limited redox replacement (SLRR), with EDTA complexation," *Electrocatalysis* **2012**, *3*, 96–107, DOI https://doi.org/10.1007/s12678-012-0080-7.

[174] L. B. Sheridan, D. K. Gebregziabiher, J. L. Stickney, D. B. Robinson, "Formation of palladium nanofilms using electrochemical atomic layer deposition (E-ALD) with chloride complexation," *Langmuir* **2013**, *29*, 1592–1600, DOI https://doi.org/10.1021/la303816z.

[175] C. Thambidurai, Y.-G. Kim, J. L. Stickney, "Electrodeposition of Ru by atomic layer deposition (ALD)," *Electrochimica Acta* **2008**, *53*, 6157–6164, DOI https://doi.org/10.1016/j.electacta.2008.01.003.

[176] R. Vasilic, L. T. Viyannalage, N. Dimitrov, "Epitaxial growth of Ag on Au(111) by galvanic displacement of Pb and Tl monolayers," *Journal of the Electrochemical Society* **2006**, *153*, C648, DOI https://doi.org/10.1149/1.2218769.

[177] K. Venkatraman, R. Gusley, L. Yu, Y. Dordi, R. Akolkar, "Electrochemical atomic layer deposition of copper: a lead-free process mediated by surface-limited redox replacement of underpotentially deposited zinc," *Journal of the Electrochemical Society* **2016**, *163*, D3008–D3013, DOI https://doi.org/10.1149/2.0021612jes.

[178] K. Venkatraman, R. Gusley, A. Lesak, R. Akolkar, "Electrochemistry-enabled atomic layer deposition of copper: investigation of the deposit growth rate and roughness," *Journal of Vacuum Science & Technology. A. Vacuum, Surfaces, and Films* **2019**, *37*, 020901, DOI https://doi.org/10.1116/1.5079560.

[179] J. Nutariya, M. Fayette, N. Dimitrov, N. Vasiljevic, "Growth of Pt by surface limited redox replacement of underpotentially deposited hydrogen," *Electrochimica Acta* **2013**, *112*, 813–823, DOI https://doi.org/10.1016/j.electacta.2013.01.052.

[180] S. Ambrozik, N. Dimitrov, "The deposition of Pt via electroless surface limited redox replacement," *Electrochimica Acta* **2015**, *169*, 248–255, DOI https://doi.org/10.1016/j.electacta.2015.04.043.

[181] G. V. Lowry, R. J. Hill, S. Harper, A. F. Rawle, C. O. Hendren, F. Klaessig, U. Nobbmann, P. Sayre, J. Rumble, "Guidance to improve the scientific value of zeta-potential measurements in nanoEHS," *Environmental Science: Nano* **2016**, *3*, 953–965, DOI https://doi.org/10.1039/c6en00136j.

[182] L. Besra, M. Liu, "A review on fundamentals and applications of electrophoretic deposition (EPD)," *Progress in Materials Science* **2007**, *52*, 1–61, DOI https://doi.org/10.1016/j.pmatsci.2006.07.001.

[183] R. J. Geise, J. M. Adams, N. J. Barone, A. M. Yacynych, "Electropolymerized films to prevent interferences and electrode fouling in biosensors," *Biosensors & Bioelectronics* **1991**, *6*, 151–160, DOI https://doi.org/10.1016/0956-5663(91)87039-e.

[184] R. D. O'Neill, G. Rocchitta, C. P. McMahon, P. A. Serra, J. P. Lowry, "Designing sensitive and selective polymer/enzyme composite biosensors for brain monitoring in vivo," *TrAC Trends in Analytical Chemistry* **2008**, *27*, 78–88, DOI https://doi.org/10.1016/j.trac.2007.11.008.

[185] J. P. Lowry, R. D. O'Neill, "Partial characterization in vitro of glucose oxidase-modified poly(phenylenediamine)-coated electrodes for neurochemical analysis in vivo," *Electroanalysis* **1994**, *6*, 369–379, DOI https://doi.org/10.1002/elan.1140060504.

[186] M. J. Donahue, A. Sanchez-Sanchez, S. Inal, J. Qu, R. M. Owens, D. Mecerreyes, G. G. Malliaras, D. C. Martin, "Tailoring PEDOT properties for applications in bioelectronics," *Materials Science & Engineering. R, Reports* **2020**, *140*, 100546, DOI https://doi.org/10.1016/j.mser.2020.100546.

[187] B. Bhattacharyya, J. Munda, M. Malapati, "Advancement in electrochemical micro-machining," *International Journal of Machine Tools and Manufacture* **2004**, *44*, 1577–1589, DOI https://doi.org/10.1016/j.ijmachtools.2004.06.006.

[188] J. Gou, Z. Liang, B. Wang, "Experimental design and optimization of dispersion process for single-walled carbon nanotube bucky paper," *International Journal of Nanoscience* **2004**, *03*, 293–307, DOI https://doi.org/10.1142/s0219581x04002085.

[189] D. A. Dikin, S. Stankovich, E. J. Zimney, R. D. Piner, G. H. B. Dommett, G. Evmenenko, S. T. Nguyen, R. S. Ruoff, "Preparation and characterization of graphene oxide paper," *Nature* **2007**, *448*, 457–460, DOI https://doi.org/10.1038/nature06016.

[190] I. Dumitrescu, P. R. Unwin, J. V. Macpherson, "Electrochemistry at carbon nanotubes: perspective and issues," *Chemical Communications* **2009**, *0*, 6886–6901, DOI https://doi.org/10.1039/b909734a.

[191] M. Pumera, "The electrochemistry of carbon nanotubes: fundamentals and applications," *Chemistry – A European Journal* **2009**, *15*, 4970–4978, DOI https://doi.org/10.1002/chem.200900421.

[192] P. R. Unwin, A. G. Güell, G. Zhang, "Nanoscale electrochemistry of sp2 carbon materials: from graphite and graphene to carbon nanotubes," *Accounts of Chemical Research* **2016**, *49*, 2041–2048, DOI https://doi.org/10.1021/acs.accounts.6b00301.

[193] K. Gong, S. Chakrabarti, L. Dai, "Electrochemistry at carbon nanotube electrodes: is the nanotube tip more active than the sidewall?," *Angewandte Chemie* **2008**, *120*, 5526–5530, DOI https://doi.org/10.1002/ange.200801744.

[194] L. Wang, Z. Sofer, M. Pumera, "Will any crap we put into graphene increase its electrocatalytic effect?," *ACS Nano* **2020**, *14*, 21–25, DOI https://doi.org/10.1021/acsnano.9b00184.

[195] R. P. Feynman, "There's plenty of room at the bottom [data storage]," *Journal of Microelectromechanical Systems* **1992**, *1*, 60–66, DOI https://doi.org/10.1109/84.128057.

[196] C. Bieg, K. Fuchsberger, M. Stelzle, "Introduction to polymer-based solid-contact ion-selective electrodes – basic concepts, practical considerations, and current research topics," *Analytical and Bioanalytical Chemistry* **2017**, *409*, 45–61, DOI https://doi.org/10.1007/s00216-016-9945-6.

[197] L. Burke, D. Whelan, "A voltammetric investigation of the charge storage reactions of hydrous iridium oxide layers," *Journal of Electroanalytical Chemistry and Interfacial Electrochemistry* **1984**, *162*, 121–141, DOI https://doi.org/10.1016/s0022-0728(84)80159-x.

[198] S. Głab, A. Hulanicki, G. Edwall, F. Ingman, "Metal-metal oxide and metal oxide electrodes as pH sensors," *Critical Reviews in Analytical Chemistry* **1989**, *21*, 29–47, DOI https://doi.org/10.1080/10408348908048815.

[199] W. Olthuis, M. Robben, P. Bergveld, M. Bos, W. v. d. Linden, "pH sensor properties of electrochemically grown iridium oxide," *Sensors and Actuators. B, Chemical* **1990**, *2*, 247–256, DOI https://doi.org/10.1016/0925-4005(90)80150-x.

[200] L. C. Clark Jr., "Monitor and control of blood and tissue oxygen tensions," *Transactions – American Society for Artificial Internal Organs* **1956**, *2*, 41–48,.

[201] J. J. W. Ross, Method and apparatus for electrolytically determining a species in a fluid, US Patent 3,260,656, **1966**.

[202] G. Jobst, G. Urban, A. Jachimowicz, F. Kohl, O. Tilado, I. Lettenbichler, G. Nauer, "Thin-film Clark-type oxygen sensor based on novel polymer membrane systems for in vivo and biosensor applications," *Biosensors & Bioelectronics* **1993**, *8*, 123–128, DOI https://doi.org/10.1016/0956-5663(93)85024-i.

[203] J. W. Severinghaus, A. F. Bradley, "Electrodes for blood pO2 and pCO2 determination," *Journal of Applied Physiology* **1958**, *13*, 515–520, DOI https://doi.org/10.1152/jappl.1958.13.3.515.

[204] D. R. Thevenot, K. Tóth, R. A. Durst, G. S. Wilson, "Electrochemical biosensors: recommended definitions and classification," *Pure and Applied Chemistry* **1999**, *71*, 2333–2348, DOI https://doi.org/10.1351/pac199971122333.

[205] L. C. Clark, C. Lyons, "Electrode systems for continuous monitoring in cardiovascular surgery," *Annals of the New York Academy of Sciences* **1962**, *102*, 29–45, DOI https://doi.org/10.1111/j.1749-6632.1962.tb13623.x.

[206] F. Ricci, G. Palleschi, "Sensor and biosensor preparation, optimisation and applications of Prussian Blue modified electrodes," *Biosensors & Bioelectronics* **2005**, *21*, 389–407, DOI https://doi.org/10.1016/j.bios.2004.12.001.

[207] N. A. Sitnikova, A. V. Borisova, M. A. Komkova, A. A. Karyakin, "Superstable advanced hydrogen peroxide transducer based on transition metal hexacyanoferrates," *Analytical Chemistry* **2011**, *83*, 2359–2363, DOI https://doi.org/10.1021/ac1033352.

[208] N. A. Sitnikova, M. A. Komkova, I. V. Khomyakova, E. E. Karyakina, A. A. Karyakin, "Transition metal hexacyanoferrates in electrocatalysis of H2O2 reduction: an exclusive property of prussian blue," *Analytical Chemistry* **2014**, *86*, 4131–4134, DOI https://doi.org/10.1021/ac500595v.

[209] G. Jobst, I. Moser, M. Varahram, P. Svasek, E. Aschauer, Z. Trajanoski, P. Wach, P. Kotanko, F. Skrabal, G. Urban, "Thin-film microbiosensors for glucose-lactate monitoring," *Analytical Chemistry* **1996**, *68*, 3173–3179, DOI https://doi.org/10.1021/ac950630x.

[210] I. Moser, G. Jobst, G. A. Urban, "Biosensor arrays for simultaneous measurement of glucose, lactate, glutamate, and glutamine," *Biosensors & Bioelectronics* **2002**, *17*, 297–302, DOI https://doi.org/10.1016/s0956-5663(01)00298-6.

[211] I. Moser, G. Jobst, "Pre-calibrated biosensors for single-use applications," *Chemie Ingenieur Technik* **2013**, *85*, 172–178, DOI https://doi.org/10.1002/cite.201200129.

[212] J. H. Luong, J. D. Glennon, A. Gedanken, S. K. Vashist, "Achievement and assessment of direct electron transfer of glucose oxidase in electrochemical biosensing using carbon nanotubes, graphene, and their nanocomposites," *Mikrochimica Acta* **2017**, *184*, 369–388, DOI https://doi.org/10.1007/s00604-016-2049-3.

[213] F. Schachinger, H. Chang, S. Scheiblbrandner, R. Ludwig, "Amperometric biosensors based on direct electron transfer enzymes," *Molecules* **2021**, *26*, 4525, DOI https://doi.org/10.3390/molecules26154525.

[214] R. L. Brazg, L. J. Klaff, A. M. Sussman, "New generation blood glucose monitoring system exceeds international accuracy standards," *Journal of Diabetes Science and Technology* **2016**, *10*, 1414–1415, DOI https://doi.org/10.1177/1932296816652902.

[215] J. Horak, C. Dincer, H. Bakirci, G. Urban, "Sensitive, rapid and quantitative detection of substance P in serum samples using an integrated microfluidic immunochip," *Biosensors & Bioelectronics* **2014**, *58*, 186–192, DOI https://doi.org/10.1016/j.bios.2014.02.058.

[216] J. Horak, C. Dincer, H. Bakirci, G. Urban, "A disposable dry film photoresist-based microcapillary immunosensor chip for rapid detection of Epstein–Barr virus infection," *Sensors and Actuators. B, Chemical* **2014**, *191*, 813–820, DOI https://doi.org/10.1016/j.snb.2013.10.019.

[217] J. Horak, C. Dincer, E. Qelibari, H. Bakirci, G. Urban, "Polymer-modified microfluidic immunochip for enhanced electrochemical detection of troponin I," *Sensors and Actuators. B, Chemical* **2015**, *209*, 478–485, DOI https://doi.org/10.1016/j.snb.2014.12.006.

[218] C. Kokkinos, A. Economou, M. I. Prodromidis, "Electrochemical immunosensors: critical survey of different architectures and transduction strategies," *TrAC Trends in Analytical Chemistry* **2016**, *79*, 88–105, DOI https://doi.org/10.1016/j.trac.2015.11.020.

[219] S. Partel, C. Dincer, S. Kasemann, J. Kieninger, J. Edlinger, G. Urban, "Lift-off free fabrication approach for periodic structures with tunable nano gaps for interdigitated electrode arrays," *ACS Nano* **2016**, *10*, 1086–1092, DOI https://doi.org/10.1021/acsnano.5b06405.

[220] D. G. Sanderson, L. B. Anderson, "Filar electrodes: steady-state currents and spectroelectrochemistry at twin interdigitated electrodes," *Analytical Chemistry* **1985**, *57*, 2388–2393, DOI https://doi.org/10.1021/ac00289a050.

[221] A. J. Bard, J. A. Crayston, G. P. Kittlesen, T. V. Shea, M. S. Wrighton, "Digital simulation of the measured electrochemical response of reversible redox couples at microelectrode arrays: consequences arising from closely spaced ultramicroelectrodes," *Analytical Chemistry* **1986**, *58*, 2321–2331, DOI https://doi.org/10.1021/ac00124a045.

[222] K. Aoki, "Theory of the steady-state current of a redox couple at interdigitated array electrodes of which pairs are insulated electrically by steps," *Journal of Electroanalytical Chemistry and Interfacial Electrochemistry* **1989**, *270*, 35–41, DOI https://doi.org/10.1016/0022-0728(89)85026-0.

[223] O. Niwa, M. Morita, H. Tabei, "Electrochemical behavior of reversible redox species at interdigitated array electrodes with different geometries: consideration of redox cycling and collection efficiency," *Analytical Chemistry* **1990**, *62*, 447–452, DOI https://doi.org/10.1021/ac00204a006.

[224] O. Niwa, M. Morita, H. Tabei, "Highly sensitive and selective voltammetric detection of dopamine with vertically separated interdigitated array electrodes," *Electroanalysis* **1991**, *3*, 163–168, DOI https://doi.org/10.1002/elan.1140030305.

[225] M. Takahashi, M. Morita, O. Niwa, H. Tabei, "Highly sensitive high-performance liquid chromatography detection of catecholamine with interdigitated array microelectrodes," *Journal of Electroanalytical Chemistry* **1992**, *335*, 253–263, DOI https://doi.org/10.1016/0022-0728(92)80246-z.

[226] U. Wollenberger, M. Paeschke, R. Hintsche, "Interdigitated array microelectrodes for the determination of enzyme activities," *Analyst* **1994**, *119*, 1245–1249, DOI https://doi.org/10.1039/an9941901245.

[227] O. Niwa, "Electroanalysis with interdigitated array microelectrodes," *Electroanalysis* **1995**, *7*, 606–613, DOI https://doi.org/10.1002/elan.1140070702.

[228] B. Wolfrum, M. Zevenbergen, S. Lemay, "Nanofluidic redox cycling amplification for the selective detection of catechol," *Analytical Chemistry* **2008**, *80*, 972–977, DOI https://doi.org/10.1021/ac7016647.

[229] J. I. Heo, D. S. Shim, G. T. Teixidor, S. Oh, M. J. Madou, H. Shin, "Carbon interdigitated array nanoelectrodes for electrochemical applications," *Journal of the Electrochemical Society* **2011**, *158*, J76, DOI https://doi.org/10.1149/1.3531952.

[230] S. Partel, S. Kasemann, P. Choleva, C. Dincer, J. Kieninger, G. Urban, "Novel fabrication process for sub-micron interdigitated electrode arrays for highly sensitive electrochemical detection," *Sensors and Actuators. B, Chemical* **2014**, *205*, 193–198, DOI https://doi.org/10.1016/j.snb.2014.08.065.

[231] B. Wolfrum, E. Kätelhön, A. Yakushenko, K. J. Krause, N. Adly, M. Hüske, P. Rinklin, "Nanoscale electrochemical sensor arrays: redox cycling amplification in dual-electrode systems," *Accounts of Chemical Research* **2016**, *49*, 2031–2040, DOI https://doi.org/10.1021/acs.accounts.6b00333.

[232] R. R. Kamath, M. J. Madou, "Three-dimensional carbon interdigitated electrode arrays for redox-amplification," *Analytical Chemistry* **2014**, *86*, 2963–2971, DOI https://doi.org/10.1021/ac4033356.

[233] P. Ebbesen, E. O. Pettersen, T. A. Gorr, G. Jobst, K. Williams, J. Kieninger, R. H. Wenger, S. Pastorekova, L. Dubois, P. Lambin, B. G. Wouters, T. V. D. Beucken, C. T. Supuran, L. Poellinger, P. Ratcliffe, A. Kanopka, A. Görlach, M. Gasmann, A. L. Harris, P. Maxwell, A. Scozzafava, "Taking advantage of tumor cell adaptations to hypoxia for developing new tumor markers and treatment strategies," *Journal of Enzyme Inhibition and Medicinal Chemistry* **2009**, *24*, 1–39, DOI https://doi.org/10.1080/14756360902784425.

[234] E. O. Pettersen, P. Ebbesen, R. G. Gieling, K. J. Williams, L. Dubois, P. Lambin, C. Ward, J. Meehan, I. H. Kunkler, S. P. Langdon, A. H. Ree, K. Flatmark, H. Lyng, M. J. Calzada, L. d. Peso, M. O. Landazuri, A. Görlach, H. Flamm, J. Kieninger, G. Urban, A. Weltin, D. C. Singleton, S. Haider, F. M. Buffa, A. L. Harris, A. Scozzafava, C. T. Supuran, I. Moser, G. Jobst, M. Busk, K. Toustrup, J. Overgaard, J. Alsner, J. Pouyssegur, J. Chiche, N. Mazure, I. Marchiq, S. Parks, A. Ahmed, M. Ashcroft, S. Pastorekova, Y. Cao, K. M. Rouschop, B. G. Wouters, M. Koritzinsky, H. Mujcic, D. Cojocari, "Targeting tumour hypoxia to prevent cancer metastasis. From biology, biosensing and technology to drug development: the METOXIA consortium," *Journal of Enzyme Inhibition and Medicinal Chemistry* **2014**, *30*, 689–721, DOI https://doi.org/10.3109/14756366.2014.966704.

[235] J. Kieninger, K. Aravindalochanan, J. A. Sandvik, E. O. Pettersen, G. A. Urban, "Pericellular oxygen monitoring with integrated sensor chips for reproducible cell culture experiments," *Cell Proliferation* **2014**, *47*, 180–188, DOI https://doi.org/10.1111/j.1365-2184.2013.12089.x.

[236] H. Flamm, J. Kieninger, A. Weltin, G. Urban, "Superoxide microsensor integrated into a sensing cell culture flask microsystem using direct oxidation for cell culture application," *Biosensors & Bioelectronics* **2015**, *65*, 354–359, DOI https://doi.org/10.1016/j.bios.2014.10.062.

[237] J. Marzioch, J. Kieninger, A. Weltin, H. Flamm, K. Aravindalochanan, J. A. Sandvik, E. O. Pettersen, Q. Peng, G. A. Urban, "On-chip photodynamic therapy – monitoring cell metabolism using electrochemical microsensors," *Lab on a Chip* **2018**, *18*, 3353–3360, DOI https://doi.org/10.1039/c8lc00799c.

[238] R. Skloot, *The Immortal Life of Henrietta Lacks*, Crown Publishers, New York, **2010**.

[239] J. Dornhof, J. Kieninger, H. Muralidharan, J. Maurer, G. A. Urban, A. Weltin, "Microfluidic organ-on-chip system for multi-analyte monitoring of metabolites in 3D cell cultures," *Lab on a Chip* **2022**, DOI https://doi.org/10.1039/D1LC00689D.

[240] A. Weltin, K. Slotwinski, J. Kieninger, I. Moser, G. Jobst, M. Wego, R. Ehret, G. A. Urban, "Cell culture monitoring for drug screening and cancer research: a transparent, microfluidic, multi-sensor microsystem," *Lab on a Chip* **2014**, *14*, 138–146, DOI https://doi.org/10.1039/c3lc50759a.

[241] J. Dornhof, J. Kieninger, H. Muralidharan, J. Maurer, G. A. Urban, A. Weltin, "Microfluidic organ-on-chip system for multi-analyte monitoring of metabolites in 3D cell cultures," *Lab on a Chip* **2022**, *22*, 225–239, DOI https://doi.org/10.1039/d1lc00689d.

[242] J. Dornhof, V. Zieger, J. Kieninger, D. Frejek, R. Zengerle, G. A. Urban, S. Kartmann, A. Weltin, "Bioprinting-based automated deposition of single cancer cell spheroids into oxygen sensor microelectrode wells," *Lab on a Chip* **2022**, *22*, 4369–4381, DOI https://doi.org/10.1039/d2lc00705c.

[243] A. Weltin, J. Kieninger, B. Enderle, A.-K. Gellner, B. Fritsch, G. A. Urban, "Polymer-based, flexible glutamate and lactate microsensors for in vivo applications," *Biosensors & Bioelectronics* **2014**, *61*, 192–199, DOI https://doi.org/10.1016/j.bios.2014.05.014.

[244] A. Weltin, B. Enderle, J. Kieninger, G. A. Urban, "Multiparametric, flexible microsensor platform for metabolic monitoring in vivo," *IEEE Sensors Journal* **2014**, *14*, 3345–3351, DOI https://doi.org/10.1109/jsen.2014.2323220.

[245] A. Weltin, S. Hammer, F. Noor, Y. Kaminski, J. Kieninger, G. A. Urban, "Accessing 3D microtissue metabolism: lactate and oxygen monitoring in hepatocyte spheroids," *Biosensors & Bioelectronics* **2017**, *87*, 941–948, DOI https://doi.org/10.1016/j.bios.2016.07.094.

[246] A. Ring, H. Sorg, A. Weltin, D. J. Tilkorn, J. Kieninger, G. Urban, J. Hauser, "In-vivo monitoring of infection via implantable microsensors: a pilot study," *Biomedical Engineering/Biomedizinische Technik* **2018**, *63*, 421–426, DOI https://doi.org/10.1515/bmt-2016-0250.

[247] A. Weltin, J. Kieninger, G. A. Urban, "Microfabricated, amperometric, enzyme-based biosensors for in vivo applications," *Analytical and Bioanalytical Chemistry* **2016**, *408*, 4503–4521, DOI https://doi.org/10.1007/s00216-016-9420-4.

[248] S. Urban, A. Weltin, H. Flamm, J. Kieninger, B. J. Deschner, M. Kraut, R. Dittmeyer, G. A. Urban, "Electrochemical multisensor system for monitoring hydrogen peroxide, hydrogen and oxygen in direct synthesis microreactors," *Sensors and Actuators. B, Chemical* **2018**, *273*, 973–982, DOI https://doi.org/10.1016/j.snb.2018.07.014.

[249] S. Urban, B. J. Deschner, L. L. Trinkies, J. Kieninger, M. Kraut, R. Dittmeyer, G. A. Urban, A. Weltin, "In situ mapping of H 2, O 2, and H 2 O 2 in microreactors: a parallel, selective multianalyte detection method," *ACS Sensors* **2021**, DOI https://doi.org/10.1021/acssensors.0c02509.

[250] L. L. Trinkies, A. Düll, J. Zhang, S. Urban, B. J. Deschner, M. Kraut, B. P. Ladewig, A. Weltin, J. Kieninger, R. Dittmeyer, "Investigation of mass transport processes in a microstructured membrane reactor for the direct synthesis of hydrogen peroxide," *Chemical Engineering Science* **2022**, *248*, 117145, DOI https://doi.org/10.1016/j.ces.2021.117145.

[251] J. T. Santini, M. J. Cima, R. Langer, "A controlled-release microchip," *Nature* **1999**, *397*, 335–338, DOI https://doi.org/10.1038/16898.

[252] J.-M. Pernaut, J. R. Reynolds, "Use of conducting electroactive polymers for drug delivery and sensing of bioactive molecules. A redox chemistry approach," *The Journal of Physical Chemistry B* **2000**, *104*, 4080–4090, DOI https://doi.org/10.1021/jp994274o.

[253] V. Pillay, T.-S. Tsai, Y. E. Choonara, L. C. d. Toit, P. Kumar, G. Modi, D. Naidoo, L. K. Tomar, C. Tyagi, V. M. K. Ndesendo, "A review of integrating electroactive polymers as responsive systems for specialized drug delivery applications," *Journal of Biomedical Materials Research. Part A* **2014**, *102*, 2039–2054, DOI https://doi.org/10.1002/jbm.a.34869.

[254] J. G. Hardy, D. J. Mouser, N. Arroyo-Currás, S. Geissler, J. K. Chow, L. Nguy, J. M. Kim, C. E. Schmidt, "Biodegradable electroactive polymers for electrochemically-triggered drug delivery," *Journal of Physical Chemistry. B* **2014**, *2*, 6809–6822, DOI https://doi.org/10.1039/c4tb00355a.

[255] P.-Y. Li, R. Sheybani, C. A. Gutierrez, J. T. W. Kuo, E. Meng, "A parylene bellows electrochemical actuator," *Journal of Microelectromechanical Systems* **2010**, *19*, 215–228, DOI https://doi.org/10.1109/jmems.2009.2032670.

[256] P.-Y. Li, J. Shih, R. Lo, S. Saati, R. Agrawal, M. S. Humayun, Y.-C. Tai, E. Meng, "An electrochemical intraocular drug delivery device," *Sensors and Actuators. A, Physical* **2008**, *143*, 41–48, DOI https://doi.org/10.1016/j.sna.2007.06.034.

[257] C. Neagu, J. Gardeniers, M. Elwenspoek, J. Kelly, "An electrochemical microactuator: principle and first results," *Journal of Microelectromechanical Systems* **1996**, *5*, 2–9, DOI https://doi.org/10.1109/84.485209.

[258] C. Lu, Y. Yang, J. Wang, R. Fu, X. Zhao, L. Zhao, Y. Ming, Y. Hu, H. Lin, X. Tao, Y. Li, W. Chen, "High-performance graphdiyne-based electrochemical actuators," *Nature Communications* **2018**, *9*, 752, DOI https://doi.org/10.1038/s41467-018-03095-1.

[259] Q. Pei, O. Inganaes, "Electrochemical applications of the bending beam method. 2. Electroshrinking and slow relaxation in polypyrrole," *The Journal of Physical Chemistry* **1993**, *97*, 6034–6041, DOI https://doi.org/10.1021/j100124a041.

[260] H. Xu, C. Wang, C. Wang, J. Zoval, M. Madou, "Polymer actuator valves toward controlled drug delivery application," *Biosensors & Bioelectronics* **2006**, *21*, 2094–2099, DOI https://doi.org/10.1016/j.bios.2005.10.020.

[261] W. M. Grill, R. F. Kirsch, "Neuroprosthetic applications of electrical stimulation," *Assistive Technology* **2010**, *12*, 6–20, DOI https://doi.org/10.1080/10400435.2000.10132006.

[262] J. C. Lilly, J. R. Hughes, E. C. Alvord Jr., T. W. Galkin, "Brief, noninjurious electric waveform for stimulation of the brain," *Science* **1955**, *121*, 468–469, DOI https://doi.org/10.1126/science.121.3144.468.

[263] A. Weltin, J. Kieninger, "Electrochemical methods for neural interface electrodes," *Journal of Neural Engineering* **2021**, *18*, 052001, DOI https://doi.org/10.1088/1741-2552/ac28d5.

[264] S. F. Cogan, "Neural stimulation and recording electrodes," *Annual Review of Biomedical Engineering* **2008**, *10*, 275–309, DOI https://doi.org/10.1146/annurev.bioeng.10.061807.160518.

[265] M. Doering, J. Kieninger, J. Kübler, U. G. Hofmann, S. J. Rupitsch, G. A. Urban, A. Weltin, "Advanced electrochemical potential monitoring for improved understanding of electrical neurostimulation protocols," *Journal of Neural Engineering* **2023**, *20*, 036036, DOI https://doi.org/10.1088/1741-2552/acdd9d.

[266] D. W. Kumsa, F. W. Montague, E. M. Hudak, J. T. Mortimer, "Electron transfer processes occurring on platinum neural stimulating electrodes: pulsing experiments for cathodic-first/charge-balanced/biphasic pulses for $0.566 < k > 2.3$ in oxygenated and deoxygenated sulfuric acid," *Journal of Neural Engineering* **2016**, *13*, 056001, DOI https://doi.org/10.1088/1741-2560/13/5/056001.

[267] E. M. Hudak, D. W. Kumsa, H. B. Martin, J. T. Mortimer, "Electron transfer processes occurring on platinum neural stimulating electrodes: calculated charge-storage capacities are inaccessible during applied stimulation," *Journal of Neural Engineering* **2017**, *14*, 046012, DOI https://doi.org/10.1088/1741-2552/aa6945.

[268] R. Shannon, "A model of safe levels for electrical stimulation," *IEEE Transactions on Biomedical Engineering* **1992**, *39*, 424–426, DOI https://doi.org/10.1109/10.126616.

[269] S. F. Cogan, J. Ehrlich, T. D. Plante, A. Smirnov, D. B. Shire, M. Gingerich, J. F. Rizzo, "Sputtered iridium oxide films for neural stimulation electrodes," *Journal of Biomedical Materials Research. Part B, Applied Biomaterials* **2009**, *89B*, 353–361, DOI https://doi.org/10.1002/jbm.b.31223.

[270] P. R. Troyk, D. Detlefsen, S. F. Cogan, J. Ehrlich, M. Bak, D. B. McCreery, L. Bullara, E. Schmidt, ""Safe" charge-injection waveforms for iridium oxide (AIROF) microelectrodes," *The 26th Annual International Conference of the IEEE Engineering in Medicine and Biology Society* **2004**, *2*, 4141–4144, DOI https://doi.org/10.1109/iembs.2004.1404155.

[271] A. Scheiner, J. T. Mortimer, U. Roessmann, "Imbalanced biphasic electrical stimulation: muscle tissue damage," *Annals of Biomedical Engineering* **1990**, *18*, 407–425, DOI https://doi.org/10.1007/bf02364157.

[272] M. Armstrong-James, K. Fox, Z. Kruk, J. Millar, "Quantitative ionophoresis of catecholamines using multibarrel carbon fibre microelectrodes," *Journal of Neuroscience Methods* **1981**, *4*, 385–406, DOI https://doi.org/10.1016/0165-0270(81)90008-x.

[273] J. E. Baur, E. W. Kristensen, L. J. May, D. J. Wiedemann, R. M. Wightman, "Fast-scan voltammetry of biogenic amines," *Analytical Chemistry* **1988**, *60*, 1268–1272, DOI https://doi.org/10.1021/ac00164a006.

[274] D. L. Robinson, B. J. Venton, M. L. Heien, R. M. Wightman, "Detecting subsecond dopamine release with fast-scan cyclic voltammetry in vivo," *Clinical Chemistry* **2003**, *49*, 1763–1773, DOI https://doi.org/10.1373/49.10.1763.

[275] B. J. Venton, R. M. Wightman, "Psychoanalytical electrochemistry: dopamine and behavior," *Analytical Chemistry* **2003**, *75*, 414 A–421 A, DOI https://doi.org/10.1021/ac031421c.

[276] B. J. Venton, Q. Cao, "Fundamentals of fast-scan cyclic voltammetry for dopamine detection," *Analyst* **2020**, *145*, 1158–1168, DOI https://doi.org/10.1039/c9an01586h.

[277] D. Michael, E. R. Travis, R. M. Wightman, "Color images for fast-scan CV measurements in biological systems," *Analytical Chemistry* **1998**, *70*, 586A–592A, DOI https://doi.org/10.1021/ac9819640.

[278] S. E. Calhoun, C. J. Meunier, C. A. Lee, G. S. McCarty, L. A. Sombers, "Characterization of a multiple-scan-rate voltammetric waveform for real-time detection of met-enkephalin," *ACS Chemical Neuroscience* **2019**, *10*, 2022–2032, DOI https://doi.org/10.1021/acschemneuro.8b00351.

[279] N. T. Rodeberg, S. G. Sandberg, J. A. Johnson, P. E. M. Phillips, R. M. Wightman, "Hitchhiker's guide to voltammetry: acute and chronic electrodes for in vivo fast-scan cyclic voltammetry," *ACS Chemical Neuroscience* **2017**, *8*, 221–234, DOI https://doi.org/10.1021/acschemneuro.6b00393.

[280] P. Puthongkham, B. J. Venton, "Recent advances in fast-scan cyclic voltammetry," *Analyst* **2020**, *145*, 1087–1102, DOI https://doi.org/10.1039/c9an01925a.

[281] A. C. Schmidt, L. E. Dunaway, J. G. Roberts, G. S. McCarty, L. A. Sombers, "Multiple scan rate voltammetry for selective quantification of real-time enkephalin dynamics," *Analytical Chemistry* **2014**, *86*, 7806–7812, DOI https://doi.org/10.1021/ac501725u.

[282] E. S. Bucher, K. Brooks, M. D. Verber, R. B. Keithley, C. Owesson-White, S. Carroll, P. Takmakov, C. J. McKinney, R. M. Wightman, "Flexible software platform for fast-scan cyclic voltammetry data acquisition and analysis," *Analytical Chemistry* **2013**, *85*, 10344–10353, DOI https://doi.org/10.1021/ac402263x.

[283] S. D. Adams, E. H. Doeven, S. J. Tye, K. E. Bennet, M. Berk, A. Z. Kouzani, "TinyFSCV: FSCV for the masses," *IEEE Transactions on Neural Systems and Rehabilitation Engineering* **2020**, *28*, 133–142, DOI https://doi.org/10.1109/tnsre.2019.2956479.

[284] A. Weltin, J. Kieninger, G. A. Urban, S. Buchholz, S. Arndt, N. Rosskothen-Kuhl, "Standard cochlear implants as electrochemical sensors: Intracochlear oxygen measurements in vivo," *Biosensors and Bioelectronics* **2022**, *199*, 113859, DOI https://doi.org/10.1016/j.bios.2021.113859.

[285] K. M. Abraham, "How comparable are sodium-ion batteries to lithium-ion counterparts?," *ACS Energy Letters* **2020**, *5*, 3544–3547, DOI https://doi.org/10.1021/acsenergylett.0c02181.

[286] A. Hamnett, "Mechanism and electrocatalysis in the direct methanol fuel cell," *Catalysis Today* **1997**, *38*, 445–457, DOI https://doi.org/10.1016/s0920-5861(97)00054-0.

[287] L. An, T. Zhao, X. Yan, X. Zhou, P. Tan, "The dual role of hydrogen peroxide in fuel cells," *Science Bulletin* **2015**, *60*, 55–64, DOI https://doi.org/10.1007/s11434-014-0694-7.

[288] Z. Du, H. Li, T. Gu, "A state of the art review on microbial fuel cells: a promising technology for wastewater treatment and bioenergy," *Biotechnology Advances* **2007**, *25*, 464–482, DOI https://doi.org/10.1016/j.biotechadv.2007.05.004.

[289] C. Munoz-Cupa, Y. Hu, C. Xu, A. Bassi, "An overview of microbial fuel cell usage in wastewater treatment, resource recovery and energy production," *Science of the Total Environment* **2021**, *754*, 142429, DOI https://doi.org/10.1016/j.scitotenv.2020.142429.

[290] R. I. Masel, Z. Liu, H. Yang, J. J. Kaczur, D. Carrillo, S. Ren, D. Salvatore, C. P. Berlinguette, "An industrial perspective on catalysts for low-temperature CO2 electrolysis," *Nature Nanotechnology* **2021**, *16*, 118–128, DOI https://doi.org/10.1038/s41565-020-00823-x.

[291] S. P. S. Badwal, S. S. Giddey, C. Munnings, A. I. Bhatt, A. F. Hollenkamp, "Emerging electrochemical energy conversion and storage technologies," *Frontiers in Chemistry* **2014**, *2*, 79, DOI https://doi.org/10.3389/fchem.2014.00079.

[292] S.-W. Song, K.-C. Lee, H.-Y. Park, "High-performance flexible all-solid-state microbatteries based on solid electrolyte of lithium boron oxynitride," *Journal of Power Sources* **2016**, *328*, 311–317, DOI https://doi.org/10.1016/j.jpowsour.2016.07.114.

[293] N. Yabuuchi, K. Kubota, M. Dahbi, S. Komaba, "Research development on sodium-ion batteries," *Chemical Reviews* **2014**, *114*, 11636–11682, DOI https://doi.org/10.1021/cr500192f.

[294] Y. Chen, C. Ye, N. Zhang, J. Liu, H. Li, K. Davey, S.-Z. Qiao, "Prospects for practical anode-free sodium batteries," *Materials Today* **2024**, *73*, 260–274, DOI https://doi.org/10.1016/j.mattod.2024.01.002.

[295] T. Oshima, M. Kajita, A. Okuno, "Development of sodium-sulfur batteries," *International Journal of Applied Ceramic Technology* **2004**, *1*, 269–276, DOI https://doi.org/10.1111/j.1744-7402.2004.tb00179.x.

[296] J. Martin, K. Schafner, T. Turek, "Preparation of electrolyte for vanadium redox-flow batteries based on vanadium pentoxide," *Energy Technology* **2020**, *8*, 2000522, DOI https://doi.org/10.1002/ente.202000522.

[297] W. A. Braff, M. Z. Bazant, C. R. Buie, "Membrane-less hydrogen bromine flow battery," *Nature Communications* **2013**, *4*, 2346, DOI https://doi.org/10.1038/ncomms3346.

[298] M. O. Bamgbopa, S. Almheiri, H. Sun, "Prospects of recently developed membraneless cell designs for redox flow batteries," *Renewable & Sustainable Energy Reviews* **2017**, *70*, 506–518, DOI https://doi.org/10.1016/j.rser.2016.11.234.

[299] D. V. Ragone, "Review of battery systems for electrically powered vehicles," *SAE Technical Paper Series* **1968**, DOI https://doi.org/10.4271/680453.

[300] E. Wilhelm, R. Battino, R. J. Wilcock, "Low-pressure solubility of gases in liquid water," *Chemical Reviews* **1977**, *77*, 219–262, DOI https://doi.org/10.1021/cr60306a003.

[301] S. Weisenberger, A. Schumpe, "Estimation of gas solubilities in salt solutions at temperatures from 273 K to 363 K," *AIChE Journal* **1996**, *42*, 298–300, DOI https://doi.org/10.1002/aic.690420130.

Index

https://doi.org/10.1515/9783111488844-015

Nomenclature

Acronyms

AC	Alternating current
AdSV	Adsorptive stripping voltammetry
AEC	Alkaline electrolysis cell
AEIROF	Anodically electrodeposited iridium oxide film
AESEC	Atomic emission spectroelectrochemistry
AFC	Alkaline fuel cell
AFM	Atomic force microscopy
AFSB	Anode-free sodium battery
AGND	Analog ground (electrical circuit)
AIROF	Activated iridium oxide film
ALD	Atomic layer deposition
ALE	Atomic layer epitaxy
AN	Acetonitrile
APAD	Activated pulsed amperometric detection
AP	Anodic protection
APT	Atom probe tomography
ASTM	Global standardization organization, formerly American Society for Testing and Materials
ASV	Anodic stripping voltammetry
BCI	Brain-computer interface
BDD	Boron-doped diamond
BSA	Bovine serum albumin
CA	Chronoamperometry
CC	Chronocoulometry
CE	Capillary electrophoresis
CE	Counter electrode
CEIROF	Cathodically electrodeposited iridium oxide film
CGM	Continuous glucose monitoring
CI	Cochlear implant
CI	Corrosion inhibitor
CMOS	Complementary metal-oxide-semiconductor
CMP	Chemical mechanical polishing
CNF	Carbon nanofibre
CNT	Carbon nanotubes
CP	Cathodic protection
CPE	Constant phase element
CPT	Critical pitting temperature
CRDC	Cyclic reciprocal derivative chronopotentiometry
CSV	Cathodic stripping voltammetry
CV	Cyclic voltammetry
CV	Cyclic voltammogram
CVD	Chemical vapor deposition
DAB	Diaminobenzene
DA	Dopamine
DBS	Deep brain stimulation
DBS	Dodecyl benzenesulfonate

https://doi.org/10.1515/9783111488844-016

DC	Direct current
dec	Decade
DET	Direct electron transfer
DFT	Density-functional theory
DGND	Digital ground (electrical circuit)
DME	Dropping mercury electrode
DPV	Differential pulse voltammetry
E-ALD	Electrochemical atomic layer deposition
EC-AFM	Electrochemical atomic force microscopy
EC	Enzyme Commission (in EC number)
ECM	Electrochemical machining
ECMM	Electrochemical micromachining
ECN	Eelectrochemical current noise
ECoG	Electrocorticography
EDLC	Electrical double layer capacitor
EDL	Electrical double layer
EDTA	Ethylenediaminetetraacetic acid
EEG	Electroencephalography
EIS	Electrochemical impedance spectroscopy
EMM	Electrochemical micromachining
ENA	Electrochemical noise analysis
ENEPIG	Electroless Nickel Electroless Palladium Immersion Gold (plating)
ENIG	Electroless Nickel Immersion Gold (plating)
EPD	Electrophoretic deposition
EPN	Electrochemical potential noise
EQCM	Electrochemical quartz crystal microbalance
ESTM	Electrochemical scanning tunneling microscope
FAD	Flavin adenine dinucleotide
FcCA	Ferrocene carboxylic acid
Fc	Ferrocene
FcMeOH	Ferrocene methanol
FEM	Finite element method (simulation)
FRA	Frequency response analyzer
FSCV	Fast-scan cyclic voltammetry
GC	Glassy carbon
GC	Gouy–Chapman (model)
GND	Ground (electrical circuit)
HMDE	Hanging mercury drop electrode
HOMO	Highest occupied molecule orbital
I^2C	Inter-Integrated Circuit (serial communication bus)
ICCP	Impressed current cathodic protection
IC	Integrated circuit
ICP-AES	Inductively coupled plasma atomic emission spectroscopy
ICP-MS	Inductively coupled plasma mass spectroscopy
IDA	Interdigitated electrode array
IE	Indicator electrode
IHP	Inner Helmholtz plane
IL	Ionic liquid
IPAD	Integrated pulsed amperometric detection

IPE   Ideal polarizable electrode
ISE   Ion-selective electrode
ISO   International Organization for Standardization
ITIES   Interface between two immiscible electrolyte solutions
ITO   Indium-tin-oxide
IUPAC   International Union of Pure and Applied Chemistry
LC   Liquid chromatography
LIB   Lithium-ion battery
LiBON   Lithium boron oxynitride
LIC   Lithium-ion capacitor
LiPo   Lithium-ion polymer battery
LiPON   Lithium phosphorous oxynitride
LOD   Limit of detection
LSV   Linear scan voltammetry
LTI   Linear, time-invariant (system)
LUMO   Lowest unoccupied molecule orbital
M-ENK   Methionine-enkephalin
MCFC   Molten carbonate fuel cell
MD   Molecular dynamics
MEA   Microelectrode array
MEMS   Microelectromechanical systems
MFC   Microbial fuel cell
MFE   Mercury film electrode
MIC   Microbiologically influenced corrosion
MIP   Molecularly imprinted polymer
ML   Monolayer
mPD   m-Phenylenediamine
mpy   Milli-inch per year
MWCNT   Multiwalled carbon nanotubes
NEMS   Nanoelectromechanical systems
NE   Norepinephrine (noradrenaline)
NHE   Normal hydrogen electrode
NIB   Sodium-ion battery
NPV   Normal pulse voltammetry
OCP   Open circuit potential
OCV   Open circuit voltage
OEM   Original equipment manufacturer
OHP   Outer Helmholtz plane
OOC   Organ-on-chip
opamp   Operational amplifier (integrated circuit)
ORR   Oxygen reduction reaction
P2G   Power-to-gas
PAD   Pulsed amperometric detection
PAFC   Phosphoric acid fuel cell
PANI   Polyaniline
PB   Prussian blue
PBR   Pilling-Bedworth ratio
PBS   Phosphate-buffered saline
PCB   Printed circuit board

PC	Propylene carbonate
PEDOT	Poly-3,4-ethylene dioxythiophene
PEMFC	Proton-exchange membrane fuel cell
PEM	Proton-exchange membrane
PF	Pitting factor
pHEMA	poly(2-hydroxyethyl methacrylate)
PPV	Polyphenylene vinylene
PPy	Polypyrrole
PU	Polyurethane
PVC	Polyvinyl chloride
PVD	Physical vapor deposition
QCM	Quartz crystal microbalance
QD	Quantum-nanodot
RDC	Reciprocal derivative chronopotentiometry
RDE	Rotating disk electrode
RE	Reference electrode
RFB	Redox flow battery
RHE	Reversible hydrogen electrode
RIE	Reactive ion etching
RPAD	Reverse pulsed amperometric detection
rpm	Revolutions per minute
RPV	Reverse pulse voltammetry
RRDE	Rotating ring disk electrode
RTIL	Room-temperature ionic liquids
SCCF	Sensing Cell Culture Flask
SCCW	Sensing Cell Culture Well
SHE	Standard hydrogen electrode
SIROF	Sputtered iridium oxide film
SI	International System of Units
SIB	Sodium-ion battery
SLRR	Surface-limited redox replacement
SMBG	Self-monitoring of blood glucose
SMDE	Static mercury drop electrode
SNG	Synthetic natural gas
SNR	Signal-to-noise ratio
SOEC	Solid oxide electrolysis cell
SOFC	Solid oxide fuel cell
SPI	Serial Peripheral Interface (serial communication bus)
STFT	Short-time Fourier transform
STM	Scanning tunneling microscope
SWCNT	Single-walled carbon nanotubes
SWV	Square-wave voltammetry
TBA	Tetrabutylammonium
TCNQ	Tetracyanoquinodimethane
THF	Tetrahydrofuran
TIA	Transimpedance amplifier
TIROF	Thermal iridium oxide film
TTF	Tetrathiafulvalene
UART	Universal asynchronous receiver-transmitter (serial communication)

UME	Ultramicroelectrode
UPD	Underpotential deposition
USB	Universal Serial Bus
UV/Vis	Ultraviolet-visible (spectroscopy)
VGND	Virtual ground (electrical circuit)
VRFB	Vanadium redox flow battery
WE	Working electrode
Wi-Fi	Wireless network protocols according to the IEEE 802.11 standard
ZRA	Zero-resistance amperemeter

Symbols (greek letters)

α	Bunsen absorption coefficient (1)
α	Symmetry factor, transfer coefficient (1)
δ	Diffusion layer thickness (m)
δ_h	Hydrodynamic boundary layer (m), context: RDE
ϵ_0	Permittivity of free space (8.854×10^{-12} F m^{-1})
ϵ_r	Relative permittivity (1)
ζ	Zeta potential (V)
η	Dynamic viscosity (Pa s)
γ	Activity coefficient (1)
ϑ	Temperature (°C)
ϑ_b	Boiling point (°C)
ϑ_m	Melting point (°C)
Λ	Reversibility (1), context: CV
λ	Air-fuel ratio (1), context: lambda sensor
μ	Chemical potential (kJ mol^{-1})
μ	Dynamic viscosity (Pa s)
μ	Electrical mobility (m^2 V^{-1} s^{-1})
μ	Shear modulus (kg m^{-1} s^{-2})
μ_{ep}	Electrophoretic mobility (m^2 V^{-1} s^{-1})
ν	Scanrate (V s^{-1})
ν_k	Kinematic viscosity (m s^{-2})
ρ	Density (kg m^{-3})
σ	Conductivity, specific conductance (S m^{-1} or Ω^{-1} m^{-1})
τ	Drop life time (s), context: DME
τ	Duration of one step or period (s), context: pulse voltammetry
ϕ	Electrostatic potential (V)
ϕ	Galvani potential, inner potential (V)
ϕ	Phase shift (rad or °)
Ψ	Power spectral density
ψ	Volta potential, outer potential (V)
χ	Surface potential (V)
ω	Angular frequency (rad s^{-1})
ω	Angular frequency of rotation (rad s^{-1}), context: RDE

Symbols (latin letters)

A	Area (m^2)
A	Frequency factor in the Arrhenius equation ($m\,s^{-1}$)
a	Electrode radius (m), context: UME
A_w	Warburg constant ($\Omega\,s^{-1}$)
C	Capacity (F)
C	Capacitor (circuit element)
C_{dl}	Double layer capacity (F)
C_f	Selectivity factor ($Hz\,kg^{-1}\,m^{-2}$), context: EQCM
c_∞	Bulk concentration ($mol\,l^{-1}$)
CR	Corrosion rate ($m\,s^{-1}$)
C_s	Solid content ($kg\,m^{-3}$), context: EPD
D	Diffusion constant ($m^2\,s^{-1}$)
$\hat{E}$	Voltage amplitude (V) of a sinusoidal waveform
$\mathcal{E}$	Electrical field strength ($V\,m^{-1}$)
E	Electrical potential (V)
E^0	Standard potential (V)
$E^{0\prime}$	Formal potential (V)
e^-	Electron
E_{corr}	Corrosion potential (V)
E_{diff}	Diffusion potential (V)
E_m	Donnan potential, membrane potential (V)
emf	Electromotive force (V)
ΔE_p	Pulse height (V), context: pulse voltammetry
E_p	Passivation potential (V), context: corrosion
E^{rev}	Reversible cell potential (V)
ΔE_s	Step height (V), context: pulse voltammetry
E_{str}	Streaming potential (V)
E_t	Transpassivation potential (V), context: corrosion
F	Faraday constant (96 485 $As\,mol^{-1}$)
f	Frequency (Hz)
f_0	Fundamental frequency (Hz)
ΔG	Change in Gibbs free energy ($J\,mol^{-1}$)
ΔG^0	Change in Gibbs free energy at standard conditions ($J\,mol^{-1}$)
$\Delta G^{\ddagger}$	Gibbs free energy of activation ($J\,mol^{-1}$)
G	Gibbs free energy ($J\,mol^{-1}$)
$\hat{I}$	Current amplitude (A) of a sinusoidal waveform
I	Current (A)
I	Ion strength ($mol\,l^{-1}$)
i	Current density ($A\,m^{-2}$)
i_0	Exchange current density ($A\,m^{-2}$)
i_{00}	Standard exchange current density ($A\,m^{-2}$), referring to [O] = [R] = 1 M
$\bar{I}$	Mean value of current over drop lifetime (A), context: DME
i_l	Limiting current density, context: mass transfer ($A\,m^{-2}$)
I_{rh}	Rheobase current (A)
I_{ss}	Steady-state current (A)
i_{ss}	Steady-state current density ($A\,m^{-2}$)
I_{str}	Streaming current (A)

J	Flux ($mol\,s^{-1}$)
j	Flux density ($mol\,s^{-1}\,m^{-2}$)
K	Dissociation constant (1)
k	Rate constant ($m\,s^{-1}$)
k_a	Charge transfer rate constant of the oxidation reaction ($m\,s^{-1}$)
k_{ads}	Rate constant for adsorption ($m\,s^{-1}$)
k_B	Boltzmann constant ($1.380 \times 10^{-23}\,J\,K^{-1}$)
k_c	Charge transfer rate constant of the reduction reaction ($m\,s^{-1}$)
k_H	Henry's law constant ($Pa\,m^3\,mol^{-1}$)
k_H^{-1}	Henry solubility ($mol\,Pa^{-1}\,m^{-3}$)
K_{ij}	Selectivity coefficient (1)
K_M	Michaelis–Menten constant ($mol\,l^{-1}$)
k_m	Mass transfer coefficient ($m\,s^{-1}$)
L	Characteristic length (m)
L	Inductance (H)
L	Inductor (circuit element)
Δm	Mass change ($kg\,m^{-2}$), context: EQCM
M	Molar mass ($g\,mol^{-1}$)
m	Mass (kg)
$\dot m$	Mass flow rate ($kg\,s^{-1}$)
Me	Fictive metal
N	Collection efficiency (1), context: RRDE
n	Number of electrons, $n = 1, 2, 3 \dots$ (1)
N_A	Avogadro constant ($6.022 \times 10^{23}\,mol^{-1}$)
N_h	Harmonic number (1)
n_{Me}	Number of metal atoms in the metal oxide (1)
O	Oxidized form of a fictive species
p	Pressure (Pa)
pH_{abs}	Unified pH scale (1)
p_N	Normal pressure (101.325 kPa)
Q	Charge (A s)
q	Charge density ($A\,s\,m^{-2}$)
q_e	Elementary charge ($1.602 \times 10^{-19}\,As$)
q_H	Charge required for one monolayer adsorbed hydrogen atoms ($A\,s\,m^{-2}$)
R	Gas constant ($8.314\,J\,K^{-1}\,mol^{-1}$)
R	Resistance (Ω)
R	Reduced form of a fictive species
R_{ct}	Charge-transfer resistance (Ω)
R_n	Noise resistance (Ω), context: ENA
R_p	Polarization resistance (Ω)
R_{sn}	Spectral noise impedance ($V\,A^{-1}$), context: ENA
Re	Reynolds number (1)
Rf	Roughness factor (1)
RG	Ratio of sheath to core diameter (m), context: UME
R	Resistor (circuit element)
r_T	Radius of the tip (m), context: UME
R_u	Uncompensated resistance (Ω)
t	Time (s)
t_c	Chronaxie (s)

t_p	Pulse duration (s), context: pulse voltammetry
$\mathcal{T}$	Dimensionless time (1)
v	Flow velocity (m s^{-1})
v	Reaction rate (m s^{-1})
v	Wave speed (m s^{-1}), context: QCM
V_0	Open-loop gain (1)
v_d	Drift velocity (m s^{-1}), context: EPD
v_i	Stoichiometric number of species i (1)
W	Energy, work (J)
Z	Atomic number, proton number (1)
Z	Impedance (V A^{-1})
Z	Z-factor (1), context: QCM
z	Valency, $z = \pm1, \pm2, \pm3 \ldots$ (1)

List of tasks

■■□ Task 2.1: Faraday's law
■□□ Task 2.2: Nernst equation
■□□ Task 2.3: Charge transfer close to equilibrium
■■■ Task 2.4: Cottrell equation
■■□ Task 2.5: Diffusion constant measurement

■□□ Task 4.1: Definitions of the normal hydrogen electrode (NHE)
■□□ Task 4.2: Diffusion potentials in simple electrolytes

■■■ Task 5.1: Cyclic voltammetry – simulation
■□□ Task 5.2: Influence of the double layer capacity on voltammetry

■□□ Task 6.1: Rotating ring disk electrode – collection efficiency
■■□ Task 6.2: Electrochemical quartz crystal microbalance

■□□ Task 7.1: Charge transfer resistance
■■□ Task 7.2: Randles circuit
■■■ Task 7.3: Electrochemical impedance spectroscopy – simulation

■□□ Task 9.1: Electrodeposition of metals

■■□ Task 10.1: Lactate biosensor

Please check www.electrochemical-methods.org for solutions of the tasks.

https://doi.org/10.1515/9783111488844-017

Acknowledgment

This book evolved from the first ideas to the final textbook a long way, initially without the thought of writing a book, with many people contributing directly or indirectly. On the professional side, many thanks go to Gerhard Jobst, who introduced me to electrochemistry first and taught me to fabricate and operate electrochemical microsensors. I owe my thanks equally to many others. Isabella Moser recruited me into science and expanded my engineering world view to include thinking at the molecular level. Peter Ebbesen and Erik Pettersen were the coordinators of two large European Union projects that enabled me to travel not only between the disciplines supporting cancer research by electrochemical sensors but also in the practical sense all over Europe, including research stays in Norway. Besides, Peter and Erik were great teachers on the real meaning of science. Alongside were Jan Villadsen and Joe Sandvik being my translators to the world of medicine and cell biology. Another vital prerequisite was the support and freedom that Professor Gerald A. Urban gave me to shape my research topics, develop my lecture on electrochemical methods, and finally write this book. We spent hours discussing the right way to teach sensors and electrochemistry; I learned a lot and sharpened my thoughts through the few points on which we disagreed. For the second edition, my new boss, Professor Stefan J. Rupitsch, supported the book's progress by granting me the freedom to continue my research and additionally deepened the attraction electrical instrumentation and embedded systems already had to me.

Over the years, I worked with many colleagues and students who shaped my way of seeing, applying, and teaching electrochemical methods: Kuppusamy Aravindalochanan, David Bill, Nils Bork, Patrick Campbell, Arne Dannenberg, Patrick Daubinger, Can Dincer, Moritz Döring, Johannes Dornhof, Barbara Enderle, Johannes Erhardt, Hubert Flamm, Dev Ganatra, Christopher Gelbke, Mohammad Halhouli, Steffen Hammer, Mark Jasper, Felix Kleiser, Katrin König, Julian Kübler, Camille Laurent, Fabian Liebisch, Andreas Marx, Julia Marzioch, Eleni Miethig, Angelina Müller, Shriya Nithyan, Marie Odenthal, Stefan Partel, Stefan Reinelt, Inka Schönfeld, Behrokh Shams, Kinga Slotwinski, Menghua Song, Florian Spies, Stephanie Sumrow, Yaara Tamari, Besnik Uka, Tobias Unmüssig, Sebastian Urban, Andreas Weltin, Ruchi Yadav, and Peter Zimmermann. Among them is Kupps, Can, Hubert, Julia, and Andreas W. to highlight. Especially with the latter, I was able to have the most intense and fundamental electrochemical discussions. It was him at the same time bringing me down to earth and critically questioning what of the theory is needed to advance microsensors and biomedical applications.

Thanks also go to the team of DeGruyter and VTeX, who enabled this book and supported all my wishes.

My wife, Gaby, is the most important person in my life – so also while writing this book. She accepted those endless hours that we would have spent together instead and freed me from many other tasks to find my time to write and still assured me she liked

https://doi.org/10.1515/9783111488844-018

the idea of writing a textbook. Our dog, Hira, will never read here; nevertheless, she contributed a lot, too. She slept many times next to my desk, giving me mental support to write, and hours later defined a strict timeout dragging me out for a walk or inviting me to play.

www.ingramcontent.com/pod-product-compliance
Lightning Source LLC
Chambersburg PA
CBHW080657220326
41598CB00033B/5235